MAGNETIC FIELDS IN ASTROPHYSICS

THE FLUID MECHANICS OF ASTROPHYSICS AND GEOPHYSICS

A series of books edited by PAUL H. ROBERTS,
University of Newcastle upon Tyne, U.K.

International Standard Serial Number (ISSN): 0260–4353

Volume 1   SOLAR FLARE MAGNETOHYDRODYNAMICS
             Edited by E. R. Priest

Volume 2   STELLAR AND PLANETARY MAGNETISM
             Edited by A. M. Soward

Volume 3   MAGNETIC FIELDS IN ASTROPHYSICS
             Ya. B. Zeldovich, A. A. Ruzmaikin, D. D. Sokoloff

Additional volumes in preparation

MAGNETIC FIELDS IN ASTROPHYSICS

by

Ya. B. ZELDOVICH

A. A. RUZMAIKIN

D. D. SOKOLOFF

Institute for Cosmic Studies, U.S.S.R.

GORDON AND BREACH SCIENCE PUBLISHERS

New York   London   Paris   Montreux   Tokyo

PHYSICS

Copyright © 1983 by Gordon and Breach, Science Publishers, Inc.

Gordon and Breach, Science Publishers

One Park Avenue
New York NY 10016
United States of America

42 William IV Street
London WC2N 4DE
England

58, rue Lhomond
75005 Paris
France

P.O. Box 161
1820 Montreux-2
Switzerland

48–2 Minamidama, Oami Shirasato-machi
Sambu-gun
Chiba-ken 299-32
Japan

Library of Congress Cataloging in Publication Data

Zeldovich, Ya. B. (Yacob Borisovich)
  Magnetic fields in astrophysics.

  (The fluid mechanics of astrophysics and geophysics, 0260–4353; v. 3)
  Translated from papers written in Russian.
  Includes bibliographical references and index.
  1. Magnetic fields (Cosmic physics)   2. Astrophysics.
I. Ruzmaikin, A. A. (Aleksandr Andreevich)
II. Sokoloff, D. D.   III. Series.   IV. Title.
QC809.M25Z39 1983   523.01'88   83–16294
ISBN 0–677–06380–6

Contents

Introduction to the Series

Anyone who surveys the shelves of an average science library will soon convince himself that fluid mechanics has taken more than its share of the rapid growth of scientific publication following World War II. And within fluid mechanics, publications of astrophysical or geophysical interest have enjoyed increasing prominence. There are a number of obvious reasons for this, ranging from the academic to the practical. Prominent is the long-term aim of understanding and partially controlling the Earth's atmosphere, with implications that range from the environmental one of minimizing the spread of harmful industrial wastes to the humanitarian one of maximizing agricultural yields. No less pressing in these days of developing ocean resources is the need to understand sea movements, how they might diffuse radioactive wastes placed on the ocean floor, how they can be harnessed to help quench man's thirst for energy, how they affect oceanic life and the harvest man takes from the sea. Seismology and vulcanology are of scarcely less practical significance to many populous areas, and they owe their existence to crustal stresses set up by the motion of an underlying mantle. Beneath the mantle lies the Earth's core in whose internal motions lies the cause of the Earth's magnetism.

Like geomagnetism, most topics in astrophysical fluid mechanics are hard to model in the laboratory. They are in any case principally of academic interest. Exceptions are the interaction of the solar wind with the Earth's external magnetic field, magnetic storms and their ionospheric consequences on radio reception. The number of fluid mechanical topics of academic interest in astrophysics are legion, and include the study of stellar winds in general, the motion of gas clouds and H II regions, the pulsation of stars, their rotation, their magnetism, and their formation by gravitational collapse, the propagation of density waves in galaxies, the maintenance of galactic magnetic fields, pulsar magnetospheres, supernovae explosions, the motion of planetary atmospheres, and so forth. Some of these subjects have an old and respected history, but many are young. Sometimes the impetus has come from discoveries made during space exploration or from orbiting observatories but often it can be understood as part of a natural progression; for example, as the classic problem of the structure and evolution of a static spherically symmetric star becomes increasingly well-solved, what more natural than to examine the 'exceptional' departures: the pulsation of a star, its distortion by rotation, tides, or magnetism?

A thread common to many of these topics is the fluid mechanics of rotating stratified (and perhaps electrically conducting) fluids. Not surprisingly, concepts useful in large-scale atmospheric dynamics and oceanography find parallels in solar, planetary and cosmical physics. A scientist in one of these disciplines may well find himself reading, comprehending, and perhaps being helped by a paper in another. It was with this idea in mind that the publishers encouraged me in 1977 to broaden the scope of the journal *Geophysical Fluid Dynamics* to include papers of astrophysical interest. The response has been so favourable that they urged me to start a series of books to complement it.

This is the third volume in that series on *The Fluid Mechanics of Astrophysics and Geophysics*, and follows books on *Solar Flare Magnetohydrodynamics* and on *Stellar and Planetary Magnetism*. The previous volumes were collections of reviews and articles that the editors (E. R. Priest and A. M. Soward, respectively) welded into unified state-of-the-art accounts by supplementary material of their own. The format of the present book is the traditional one. These authors, internationally respected for their own contributions to their topic, have joined forces in writing about some of the most exciting subjects in astrophysics, that is to say some of the most exciting subjects in the whole of physics. As well as transmitting this excitement to their readers, Academician Zeldovich and his two collaborators, Drs. Ruzmaikin and Sokoloff, have aimed to instruct the aspiring scientist who wishes to become knowledgeable in these areas. The book has, then, something to offer to expert and beginner alike. It will, I am sure, be judged a worthy successor to the two earlier volumes of the series.

P. H. ROBERTS
Editor

Preface

The present book deals with magnetic fields, how they are diagnosed, how they behave, and how they can be influential in astrophysical conditions. The magnetic fields of stars, pulsars and black holes are considered, as well as those that permeate the interstellar gas. Principal emphasis is placed on the problem of magnetic field generation, especially through the hydrodynamic motions of plasma, i.e. by the so-called "dynamo" process. This choice was dictated not only by the intrinsic importance of this subject in astrophysics but also by the recent rapid advances made in its theory and application, advances to which the authors have contributed.

Papers and books on cosmic electrodynamics published in the 1950s and '60s dealt mainly with the behavior of particles in a magnetic field, conditions of stellar equilibrium, and spiral arms in the Galaxy, etc., in the presence of magnetic fields. The magnetic field was assumed to be given and, as a rule, sufficiently strong to be comparable with gravity and pressure. Much was borrowed from laboratory experience gained in the attempt to confine plasmas in magnetic configurations, but the last decade has dramatically amended this approach.

Observations show that the magnetic field strength in many astrophysical objects is rather weak so that its contribution to their equilibria may be ignored. In contrast, it has been established that the magnetic field plays a vital rôle in nonequilibrium, dynamic, active processes. This is well illustrated by the example of solar, or more generally, stellar activity whose manifestations (spots, flares, heating of coronae and so on) are all attributable to magnetic fields.

The Sun's 22-year cycle is the best known example of a rapidly changing cosmic magnetic field. Its explanation provides the clue for a proper understanding of the origin of cosmic magnetism. Such rapid changes of the magnetic field in observed dynamic processes cannot easily be explained by ohmic diffusion. The characteristic time for such diffusion is proportional to the product of the conductivity and the square of the length scales of the fields, and so it is large on the astronomical scale. This fact has long been known. In astrophysics the fields are said to be "frozen into" a plasma. The rapid changes of the field can only be explained by plasma motions and, in this, the cosmic magnetic fields differ crucially from laboratory ones. Giant pulsar fields are clear evidence of the triumph of the frozen field concept. On the other hand, ohmic diffusion cannot be ignored. It appears on the scene whenever the motions have drastically reduced the field scales.

In the light of the above remarks the importance of hydromagnetic dynamo theory becomes apparent. The concept of dynamo action to explain the maintenance and amplification of magnetic fields by the motion of conducting fluids was put forward in 1919 by G. Larmor. The foundations of the theory were laid by Cowling, Parker, Herzenberg, Braginsky and other great researchers before the middle of the 1960s. The importance of dynamo theory was understood when it was used to explain the Earth's magnetic field and the solar cycle. Particularly worth noting are the pioneering works and enthusiasm of Elsasser, but on the whole the impact of dynamo papers was much less than that of papers that used the quasi-laboratory approach to cosmic magnetic fields.

The well-known series of papers by Steenbeck, Krause and Rädler on turbulent dynamo theory, published in the late sixties, seems to have been a decisive point. They developed a simple technique to apply (the so-called "mean-field electrodynamics") through which generation conditions and field structure could be easily calculated and which gave a clear explanation of the physics of generation in terms of the mean helicity of the turbulence. In the simplest case the mean helicity is determined by the correlation of the fluctuating velocity and its vorticity, which is typical of the motion of a collection of chaotically distributed and chaotically rotating screws with a preferred (for example, right-handed) direction of rotation. Such a correlation is absent in a mirror-symmetric motion—the right-handed screw under mirror reflection goes into a left-handed screw and vice versa. The convection produced by a radial temperature gradient in a rotating star does have, however, precisely such a correlation, *viz.* a helicity. This helicity has different signs in the northern and southern hemispheres of the star. The approach using helicity won quick popularity and has been employed ever since in a variety of astrophysical contexts. Dynamo theory has ceased to be an adopted child of geophysics and astrophysics.

It is not enough in astrophysics to know the mean fields. Observations show an abundance of interesting processes operating on a small scale. To describe the small-scale magnetic fields, one needs further developments of dynamo theory.

The linear theory tells us the conditions necessary to obtain exponential growth of the magnetic field. This theory gives a real exponent or—in some cases—a complex exponent with a positive real part. The imaginary part implies that the growth is accompanied by oscillations in which all fields change sign, but obviously the linear growth gives an answer depending on the initial primordial field. In many cases, due to the longevity of astrophysical objects, the exponential growth predicted by linear theory leads to absurdly large estimates of the field strengths. Clearly this means that the linear theory has greatly exceeded the limits of its applicability. One needs a nonlinear theory. In practice, one must consider the back-reaction of the magnetic field on the motion of the plasma. One may expect that the case of a real exponent will result in a static field distribution. On the other hand, if the growth rate is complex, then the real part will vanish, leaving only the imaginary part. This would mean that, finally, periodic non-linear oscillations with field reversal are predicted. This is the situation for the solar 22-year cycle. If the nonlinearity is included in a more exact, more subtle way by a new differential equation, one could explain the global stochastic behavior of the field that seems to appear in global minima of solar activity, such as the Maunder minimum.

Research into astrophysical magnetic fields also deals with other problems. One of these arose after the discovery of pulsars, powerful sources of pulsed radio waves, as well as X-rays, γ and other radiation. It is generally accepted that the energy of the pulses comes from the rotational energy of the neutron star. The magnetic field plays the rôle of a drive belt between the rotation and radiation. The field is not observed directly; it is assumed to have the configuration of a dipole inclined to the axis of rotation and the strength of about 10^{12} Gauss. The gyrolines predicted by Gnedin and Sunyaev (1974) and observed at an energy level of about 58 GeV by Trümper et al. (1978), give direct evidence that the field strength is $5 \cdot 10^{12}$ Gauss on the surface of neutron stars in the binary system, Her X-1. The field is assumed to be the remnant of an initial field, compressed during the transformation of the star into the pulsar. On the basis of such assumptions, refined pictures and detailed consequences for pulsar physics are constructed: the acceleration of electrons and positrons, their gamma radiation and the chain reaction when photons are transformed into pairs by the magnetic fields.

The problem of the origin of magnetic fields and their associated chemical anomalies in the peculiar A-stars has been a mystery for many years. Unlike the Sun, these stars have no extensive convective shells in which the magnetic field could easily be generated. This, and some other similar astrophysical problems, are hard nuts to crack with the dynamo theory. It is possible that a dynamo was working during the early evolutionary stages of these stars.

Pulsars are frequently surrounded by gaseous nebulae. A well-known example is the pulsar NP 05032 in the Crab Nebula. We recall that in the Crab Nebula the synchrotron radiation of relativistic electrons in the magnetic field was first predicted by Ginzburg, Shklovsky and Gordon in 1953. A major contribution to the study of the Crab's magnetic field was made by Kardashev in 1964 and 1970.

Magnetic fields have now been discovered in interstellar clouds, ranging from the very small, very dense clouds associated with maser sources to the elongated, diffuse ones. Their origin is associated with the Galactic magnetic field. This field has a random turbulent structure on a small scale (~ 100 pc) where it is in direct contact with the clouds. A large-scale (several kiloparsecs) component of the Galactic field is also of importance. At present, we have simple dynamo models of this mean field that explain its origin.

This is approximately the background from which we started to write this new book on magnetic fields in astrophysics. We should also like to make the reader aware of the realities of astrophysical magnetic fields. In fact, this has become possible due to the remarkable progress in ground-based and in space observation. In Chapter 2, the observational methods, the results available, and the predicted values of magnetic field strengths, are discussed.

General speculation about the origin of magnetic fields is given in Chapter 3. Then we hope that the reader will want to study dynamo theory. We do not discuss the fundamentals of the theory in a strictly mathematical manner. This has been done in, for instance, the review by Roberts (1971), and in the books by Moffatt (1978), Parker (1979), Krause and Rädler (1980), Vainstein, Zeldovich and Ruzmaikin (1980). Instead, much space is devoted to a discussion of qualitative aspects and theoretical results. The main purpose is to understand and learn how to apply dynamo theory to reality.

A scientific monograph is something between a review and a textbook. Compared with a review it allows one to avoid details, to mention all the results, and to give all the references. We give a rather small number of references, and, as a rule, they are not given in order to establish priorities. With their aid, however, a reader can gain more information about the subject. In contrast to a textbook a monograph does not compel one to expound only well-established and verified matters. By exercising this right we took the risk, possibly not always justified, of presenting novel, speculative ideas. A justification for this risk lies in the incomplete but developing nature of the science of cosmic magnetic fields. Along with numerically calculated models of magnetic field generation (the solar cycle and the Galactic dynamo), questions about magnetic fields of such hypothetical objects as, for instance, black holes are raised. Nor do we omit the urgent modern problems of vacuum polarization in strong magnetic fields and magnetic monopoles. On the other hand, not all astrophysical applications are presented. Neither the size of the book nor the authors' abilities allowed this.

This book differs from other books on similar subjects such as the monographs by Parker, "Cosmic magnetic fields" (1979) and by Vainstein et al., "The turbulent dynamo in astrophysics" (1980, in Russian), in that it approaches the fundamentals of dynamo theory somewhat differently, and contains numerous astrophysical examples as well as a discussion of the best method for making observations of the strength of the field. In the monographs previously mentioned the main astrophysical examples considered are the Sun and the Galaxy. We tried to make our book intelligible not only to those who are interested in dynamo theory but also to astrophysicists working in other areas, both theoreticians and observers. This partly explains the structure of the book. This was also the reason why we sometimes reversed the order of material, violating the logical rule of going from the known to the unknown, from arguments to results. Let the reader himself judge to what extent this method helps towards an understanding of the subject.

When writing this book we used the papers, advice and help of many friends and colleagues. We should especially like to express our gratitude to V. I. Arnold, S. I. Braginsky, U. Frisch, T. S. Ivanova,

H. K. Moffatt, S. A. Molchanov, A. Pouquet, M. R. E. Proctor, A. Shukurov, R. Sunyaev, V. V. Usov and N. O. Weiss. We also thank our translator N. A. Sokolova, as well as N. Gula and A. A. Guskova who typed the manuscript. Our sincere thanks are also due to the editor of the series, Paul Roberts Sc.D., F.R.S., whose deep understanding of our topic combined with his careful work in improving the scientific and pedagogical content of our book were decisive in the final presentation. Last, but not least, Professor Roberts moulded the (at times) broken English of the original manuscript into its present form. Nevertheless we must take responsibility for all mistakes that remain in the text. We will be glad to receive suggestions from readers for corrections and improvements to the text.

In conclusion we note that the part of the book devoted to the dynamo (Chapters 3 to 10) was discussed and written jointly by all three authors while the responsibility for the remaining chapters lies mainly with two of them (A.A.R. and D.D.S.). Chapter 14 was written by T. V. Ruzmaikina.

THE AUTHORS

CHAPTER 1

Introduction

Exploration and exploitation of magnetic phenomena by man have a long history, stretching back more than a thousand years. The compass, which was created in China (in the second century A.D.) and reached Europe in the thirteenth century, played an enormous rôle in navigation and in the development of geography. This relied, of course, on the natural magnetic field of the Earth.

In "Geschichte der Physik" (1950), von Laue described the fear of Christopher Columbus in his famous voyage of 1492 when he observed that the compass needle began to deviate to the West of North instead of to the East, as he had previously experienced in Southern Europe. William Gilbert made a great contribution to science by investigating, with the help of a magnetic needle, the pattern of field lines around magnetized iron spheres, and by discovering the similarity between this pattern and the Earth's magnetic field.

The effect of a magnet on a piece of iron had greatly impressed the ancient world. Thales of Miletus stated that "a magnet has a soul for it moves iron" (Arist. 405a). This gave rise to the concept of magnetism as something extrasensory and mystical. Mesmer magnetized his aristocratic patients, who often made wonderful recoveries. References to the powers of magnetism appeared in literature, for example, the Russian poet M. Yu. Lermontov (1814–1841) claimed that: "when a wife begins to magnetize him, happy will the husband be to dream". Einstein's reminiscences of his childish interest and astonishment, when at the age of four he saw a demonstration using a magnet, reflect the fascination also widely felt by lesser mortals. A remote echo of the mystical and extrasensory perception of magnetic fields is the aura still surrounding the so-called magnetic treatment of water, grain, cement, and anything else the believer can lay his hands on.

1

In the nineteenth century, the relationship between electricity and magnetism became clear. In 1820, H. C. Oersted found that a magnetic field is created by a current. In 1831 Michael Faraday discovered the induction phenomenon while studying the effect of a magnet on Faraday's induction law into a differential equation that fitted into the earlier consistent theory of J. C. Maxwell. In St. Petersburg, Professor B. S. Jacoby, a colleague of E. H. Lenz, himself an outstanding researcher into inductive processes, invented the first magneto-electrical machine which he immediately put into operation: in 1838 it drove a boat with 14 passengers upstream on the river Neva. The age of electricity was born! The march of progress revealed important uses for magnetic fields and electric currents in the laboratory, in engineering and in industry, and this overshadowed the problems of natural magnetism.

During the twentieth century the theory of relativity was developed mainly on the basis of electromagnetic field theory. Lorentz invariance was inherent in Maxwell's equations, as was established 50 years before Einstein, Lorentz and Poincaré. Electric and magnetic fields proved to be components of a common antisymmetric tensor of the second rank in four-dimensional Minkowski space and, as such, the two fields lost their individuality. A statement like: "there is a magnetic field here but not an electric one", or vice versa, cannot be absolute. If one of these statements is valid for one observer, it is invalid for another, in relative motion. This would seem to contradict Faraday's demonstrable concept of magnetic field lines. We shall see below however that the insight of Faraday's genius is regenerated in modern magnetohydrodynamics and astrophysics.

Combining the electromagnetic and quantum theories leads to the concept of photons i.e. quanta of light, quanta of electromagnetic radiation. The quantum theory of electronic motion in atoms, including quantum electrodynamics, is a complete theory. Excellent agreement with experiment has been achieved, to an accuracy of within 10^{-11}–10^{-12} for the atomic levels.

Quantum electrodynamics, including the theory of charged particles and antiparticles (electrons and positrons, etc.), predicted nonlinear effects which were beyond the competence of Maxwell's theory, for instance, the scattering of interacting photons and photon scattering by the Coulomb field of nuclei. These effects were however confirmed experimentally. The theory also predicted the creation of electron–

positron pairs by a strong static electric field, double refraction of photons, and the absorption of hard photons ($h\nu > 2mc^2$) with the creation of real pairs in a strong magnetic field. The effects of super-strong magnetic fields have still to be observed in the laboratory, but they are believed to be essential processes in pulsar magnetospheres. One can hardly doubt such forecasts of electrodynamics.†

In quantum electrodynamics there are no intrinsic contradictions or difficulties up to energies now attainable. The theory provides a well-defined procedure for calculating all the observed effects, and predicts the outcome of any experiment, hypothetical or real. It will not come as a surprise to anybody that, the answer to such questions is often statistical. Such is nature, and such therefore is quantum mechanics for it to be an adequate description of nature.

Today, the theory is developing in the following directions:

The consideration of successively more complex systems with greater numbers of electrons and nuclei, collective effects being taken into account. A brilliant example is the success of the theory of superconductivity. In this context, nearly the whole of macroscopic physics (as well as chemistry) is a part of electrodynamics.

The influence of other interactions (strong-nuclear and, particularly, weak) on atomic properties. This branch may be thought rather special and narrow, particularly as compared with the previous one. However, a weak interaction between electrons and a nucleus, through the neutral current mechanism, brings to atomic theory the idea of broken mirror-symmetry. Rotation of the plane of polarization by free atoms is predicted. Detection of this effect is of fundamental importance for particle physics. A slight difference in the energies of the right and left isomerides of organic molecules is also predicted. The mirror asymmetry of biochemistry is possibly associated with these effects.

Finally, quantum electrodynamics is the model used in the construction of modern theories of other interactions. It is the most prominent and successful member of a large family of theories, and

---

† It is worth noting that the Delbrück scattering of gamma-quanta by the electrostatic field of heavy nuclei has been observed experimentally. This may be considered as direct confirmation of nonlinear effects in electrodynamics. The categorical words "hardly doubt" are justified by the magnificent agreement between experiments and quantum electrodynamic calculations for the anomalous magnetic moment of electrons and the shift of atomic levels.

is therefore the model that is imitated. Such imitation is not always wise; one must allow for the individuality of each member. Nevertheless, the tendency to imitate is inevitable in view of the success of quantum electrodynamics.

In modern theory, weak interactions are accomplished through the so-called "intermediate vector bosons" which are similar to photons. The gluons responsible for the interaction between quarks, and more generally strong interactions, also resemble photons. Finally, the hypothetical particles, which give rise to the hypothetical process of spontaneous decay of the proton, are assumed to be similar to photons. Unlike photons, these intermediate particles have non-zero rest mass. Therefore, instead of Coulomb's law, they give rise to short-range forces. The gluons have no rest mass but they are "charged"; instead of Coulomb's law, they create an interaction whose strength does not diminish with distance; in fact, the potential energy is proportional to the distance!

It is impossible to extract a quark from a nucleon in the way that an electron can be pulled off an atom. Despite the dissimilarity of such external manifestations, a common feature of electrodynamics and the theories of the interactions listed above is the existence of vector potentials and the interaction of these potentials with their respective "currents" and "charges".

Electrodynamics has proved to be a pioneer in the creation of a unified field theory. The unification of various natural forces into a single theory is an old ambition of physicists. The resemblance between Coulomb's law of interaction for charged particles and Newton's law of gravitational interaction for masses early gave rise to the idea of combining electrodynamics and gravitation. Faraday set up some test experiments, but in vain. A deep difference between these two theories was noticed: on the one hand, gravitation always produces a mass attraction, on the other hand, in electrodynamics there are charges of both signs. Therefore, in gravitation it is impossible to connect two masses by field lines.

New hopes of unifying gravitation and electrodynamics were inspired by the successes of Einstein's general relativity that explained gravitation in terms of the curvature of space-time. But despite much hard work, these hopes were dashed. The advancement of science has followed a different course which was impossible to foresee before the study of radioactivity and cosmic rays, the development of accelerators, and the discoveries of new elementary particles. Only recently, after understanding the magnitude of the problem, has the possibility of combining gravitation with the theories of other fields and particles become feasible.

Against this background of great success in electromagnetism, progress in cosmic magnetism may appear rather modest. The magnetic field of the Sun was not discovered until 1908, i.e. a few hundred years later than the discovery of the Earth's magnetic field. The discovery was made by George Hale who died in 1938. It was Hale who persuaded the Chicago tram millionaire, Yerkes, to finance the construction of the record 40-inch refracting telescope. This same Hale was a founder of the Mount Wilson observatory and of Caltech, and was the first editor of the Astrophysical Journal. This man of great vision and outstanding organizational abilities was the first to realize the importance of investigating the Sun's magnetic field (Hale, 1908). Fields of other stars were not detected until half a century later by another brilliant astrophysicist, Harold Babcock, with the aid of the magnetograph he had invented. And it is only during the last few years, thanks to exciting developments in optical and especially X-ray astronomy, that it has become clear that magnetic fields are present around practically all stars.

Magnetic fields in the interstellar medium of the Galaxy became a subject for discussion only after the Second World War, due to an upsurge of interest in cosmic rays. The first studies were carried out by H. Alfvén, E. Fermi and other outstanding researchers, among them V. L. Ginzburg and S. I. Syrovatski. We should like especially to pay tribute to the fundamental contributions made to the understanding of interstellar magnetic fields by the Soviet astrophysicist S. B. Pikelner. Rapid progress in space exploration, which began in the 1960's, also led to the discovery of the magnetic fields of other planets.

Another young field is the theory of cosmic magnetism. The first attempt to understand the origin of the magnetic fields of the Earth and Sun was made by G. Larmor in 1919 at a time when special relativity was in its teens, and the foundations of general relativity were being laid. That year marks the birth of hydromagnetic dynamo theory. According to this theory magnetic fields are maintained and evolve through the hydrodynamic motion of a conducting fluid. The famous work by T. G. Cowling (1934) was a turning point of this theory. Though the title of his work was "The magnetic field of sunspots", its main result was a theorem on the impossibility of maintaining an axisymmetric magnetic field by fluid flows. Having read this work, theorists forgot the raison d'être of dynamo theory, and rushed into a mathematical labyrinth of theorems and formulae. And we should recognize that they achieved a measure of success. After a few extensions of the Cowling non-existence theorem, made, in particular, by one of the present authors (Zeldovich, 1956), the first mathematical dynamo models were constructed (Herzenberg, 1958; Backus, 1958). A number of interesting positive results were obtained through study of the non-existence theorems. As an example we may cite the discovery of turbulent diamagnetism (Zeldovich, 1956). Also a fine though intricate dynamo theory, for fields and flows deviating only slightly from axisymmetry, was developed by S. I. Braginsky (1964).

At that time (and even now) dynamo theory investigated the induction equation for a moving fluid. It is surprising that this was not done in the last century, in the Faraday–Maxwell–Hertz epoch. The reason was probably that the doors of a cosmic laboratory had yet to open. Even up to the 1950's, only a few investigators were fully engaged in the study of cosmic, astrophysical magnetic fields. They included W. M. Elsasser and E. N. Parker, the founder of the modern theory of the solar cycle.

A second turning point, which demonstrated to everybody the importance of applications of hydromagnetic dynamo theory, was a series of papers by M. Steenbeck, F. Krause and K.-H. Rädler in the late sixties. On the one hand, they created a practical and convenient mathematical approach—electrodynamics of mean fields—which considerably simplified the study of large-scale magnetic fields. On the other hand, they showed how the mechanism of large-scale field generation in a turbulent fluid could be understood. This proved to be associated with the helicity of the flow, a notion which played a crucial part in the earlier models of Parker (1955) and Braginsky (1964). Only after the work of Steenbeck and his colleagues, however, was it clearly understood and commonly accepted. The birth of the topological approach to the hydromagnetic dynamo problem (Moffatt, 1969) bore witness to the further progress of the theory.

Today these works, having attained their goals, have become classics. Now, along with the large-scale fields, elucidation of the rôle played by the small-scale, fluctuating fields has become important. In particular, a question asked long ago by G. K. Batchelor (1950) has come into prominence, concerning the generation of a magnetic field in turbulence lacking a mean helicity. Another important question concerns the effect of the back reaction of the field on the fluid motion (the nonlinear dynamo problem). Both questions are logical developments in the march of theory, as well as in answering the needs of cosmic reality. No longer are explanations of the mean large-scale behavior of stellar and galactic fields the only interesting and important goals. Calculations of the spectra of random fields and a comparison of the results with observation are sought. Also a nontrivial treatment of the back reaction of the magnetic field on turbulence or rotation is required to explain not only the mean amplitude of the magnetic field but also such wonderful phenomena as the prolonged stochastic variability of stellar activity like the Maunder minimum.

One of the greatest achievements of our century in physics has been the understanding of solid state magnetism. Theories of dia-, para- and ferromagnetism have gone far beyond the scope of pure electrodynamics. While magnetism and electricity are mostly reduced to the basic properties of elementary particles, to understand these new theories it is necessary to know, besides electrodynamics, quantum mechanics and statistical physics.

Compared with this, the nature of cosmic magnetism has proved to be very simple. There are almost no solid bodies in space. The giant planets (Jupiter, Saturn), stars, galaxies, the interstellar medium and other astrophysical objects consist of plasma. Temperatures are high, and thermal energies are larger than the magnetic energies of the fluid particles. (Pulsar surfaces provide an exception.) Therefore only (collective) conductivity and collective motion play a role. Magnetic fields are born, live and die in a plasma. Their fate depends on the properties of the plasma. For example, the conductivity of the plasma determines, as we know, a characteristic time for ohmic decay of fields. Violation of barotropy, e.g. the noncoincidence of surfaces of constant pressure with those of constant density, or the difference between electron and ion interactions with radiation, may explain the presence of magnetic fields in a plasma. Hydrodynamic motions of the plasma, however, play the major rôle. It is these motions that lead to rapid changes in cosmic magnetic fields. Therefore dynamo theory has become a natural and necessary tool for learning about cosmic magnetic fields.

When we speak here of dynamo theory we mean, perhaps, something more than what is usually called "hydromagnetic dynamo theory". In the narrow sense this term means the kinematic dynamo problem: to construct a velocity field capable of the indefinite maintenance and amplification of an initially given magnetic field. Today, when this question has basically been answered (there are many examples of working kinematic dynamos), the connotations of the term "dynamo theory" have widened. On the one hand, it is still interesting to attempt a general solution of the kinematic problem, in other words, to formulate all the conditions for the velocity field which, when satisfied, would provide for the maintenance of a magnetic field indefinitely. On the other hand, great efforts are now being made to take into account the back reaction of the magnetic field on the motion, that is to solve a self-consistent nonlinear problem. Also there is a rapid growth of astrophysical applications where ideal simple theoretical models are mixed with observational evidence, hypotheses, and plausibility arguments.

R. Z. Sagdeev once observed, half in jest, that early astrophysicists used only $H^2/8\pi$ to explain magnetic phenomena and amazingly achieved reasonable results. To-day this naïve and happy period of cosmic science is over. For further advancement it is now necessary to understand, and be able to calculate, detailed plasma effects. The same may be said about dynamo theory. How simple the initial kinematic problem seems now in comparison with the nonlinear self-consistent one! And how trivial the generation of the azimuthal magnetic field by non-uniform rotation looks when compared with the nonlinear (and so far unknown) process producing global minima in stellar activity!

The main subject of this book is astrophysical magnetic fields. Much can be said about them with the help of dynamo theory; much, but not all. Observational reality is more diverse than theory. A major challenge is to understand strong magnetic fields whose energies greatly exceed that of hydrodynamic motions. This problem is discussed in the final chapter. Who knows, what in the future may be the basic topics in the science of cosmic magnetism? So far much is hypothetical. On the other hand, the attractiveness of problems of stellar magnetism, cosmic magnetohydrodynamic turbulence, and many other "simple" old problems is unlikely to fade.

CHAPTER 2

Evidence of Cosmic Magnetism

I Introduction

Unlike the fields that arise in laboratory conditions, cosmic magnetic fields are very remote. Their measurement is nontrivial and has become possible only recently. It is the purpose of this chapter to describe the principles of measurement, technical details are omitted and no attempt is made to review all the available techniques. We prefer to show an inveterate theorist, by a few simple examples, how the strength and configuration of a field can be determined, and to give an observer not only some theoretical insight into measurement techniques but also indicate to him what sorts of observations are most profitable for a better understanding of the nature and origin of magnetic fields.

Within the solar system, in situ measurements of magnetic fields are possible using spacecraft. In 1973, during the Pioneer-10 mission, Jupiter's field was measured from distances between 2.84 and 6 planetary radii, where it has a dipolar form with a strength of 4.2 Gauss at the planetary equator (Smith *et al.*, 1974). Magnetic fields as large as 0.2 Gauss were discovered near Saturn's equator by Pioneer-11 (Acuna *et al.*, 1980; Smith *et al.*, 1980; Ness *et al.*, 1981). Although sophisticated magnetometers were used, a field of such strength would have been detected even with a common compass! To do this, however, one must first fly close to Jupiter or Saturn. In fact, very sensitive magnetometers are needed to measure magnetic fields in inter-planetary space, where the field is of the order 10^{-5} Gauss. Besides, some planets have very weak magnetic fields. The field near Mars, for example, is only 7 to 10 times stronger than that it is in inter-

10

planetary space, according to the Mars-2 and Mars-3 measurements (Dolginov *et al.*, 1973). Mercury's field is also weak (Mariner-10), $3.3 \cdot 10^{-3}$ Gauss at the equator (Ness *et al.*, 1975), and the Venusian field has not been detected at all.

Information about the solar magnetic field or the fields outside the solar system can be gained only through radiation (or cosmic rays). One then measures the radiation generated during the interaction of charged particles, atoms, molecules or a whole plasma with the magnetic field. The simplest and best-known example of such an interaction is the motion of an electron (or some other charge) in the magnetic field.

Let an electron move freely along the field lines. The radius of its orbit and associated frequency of its motion in the transverse plane are determined by the field strength and the electron energy (Landau and Lifshitz, 1962)

$$r = pc/eH, \qquad \omega_H = (eH/mc)(\mathscr{E}/mc^2)^{-1},$$

where m and e are respectively the electron mass and charge, c is the velocity of light, $p = mc[(\mathscr{E}/mc^2)^2 - 1]^{1/2}$ is the particle momentum. Its centripetal acceleration causes the electron to radiate with total intensity proportional to the square of the field strength:

$$\mathscr{I} = (2e^2/3c^2)(eH/mc)^2(p/m)^2.$$

This radiation, called 'magnetic bremsstrahlung' (or, in the relativistic case, synchrotron radiation), is typical of many astrophysical objects. For example, it is thought that relativistic electrons in the Crab Nebula and in the Galactic disc lose their energy in this way. Unfortunately, one cannot determine the magnetic field strength directly from the bremsstrahlung intensity, because the electron energy is usually unknown. The field strength can be found by studying more refined properties of the radiation—its spectrum and polarization. (The spectrum of an electron radiating in a magnetic field consists of separate lines, with frequencies ω_H where $n = 1, 2, 3; \ldots$, that depend on the field strength.) Despite the simplicity of the idea, its utility in field determination has only recently been proved due to advances in measurement techniques for polarizations and spectra over a wide range of wavelengths. The strong fields of white dwarfs and neutron stars were, however, much earlier measured successfully in this way.

We postpone discussion of the method and the results it has yielded to the end of the chapter, and begin with more traditional methods.

One of these, the Zeeman effect, is based on the study of the lines which arise through the interaction of an atom or a molecule (more precisely, the atomic or molecular electrons) with a magnetic field. By measuring the value of the Zeeman splitting, one can determine directly the field strength. More refined measurements of the relative intensity and polarization of the Zeeman components allow one to obtain information about the orientation of the magnetic field.

More subtle effects discussed in connection with laboratory spectroscopy (Aleksandrov, 1972) are associated with the Zeeman effect and may, in principle, be used in astrophysics (see Sections III and IV). The integrated effect of a plasma on a magnetic field is discussed in Section V for the example of the Faraday rotation of the plane of polarization of cosmic radio radiation. This method can be applied successfully to determine very weak, large-scale magnetic fields.

Investigators with a systematic turn of mind can distinguish four ways of deriving information about the magnetic field from the observed radiation. First, there is the determination of the typical frequencies (or wavelengths) of the lines formed in the magnetic field. This involves the measurement of Zeeman splitting and cyclotron radiation frequencies. Second is the measurement of the linear and circular polarizations generated in the magnetic field and transmitted arross a magnetized medium. For linearly polarized radiation, the degree of polarization and the orientation of the polarization plane depend on the strength and orientation of the field relative to the direction of propagation of the radiation. This involves (a) an investigation of the Faraday rotation of the polarization plane of the radio waves; (b) the study of the partial polarization of starlight that has passed through dust orientated by a magnetic field, though this study is so far morphological in character; (c) the determination of the degree of polarization of magneto-bremsstrahlung, and in particular, synchrotron radiation; and (d) the polarization of lines. A third technique is to determine the magnetic field from the intensity of synchrotron radiation which by its nature is essentially linked to the magnetic field. Unfortunately, poor knowledge of the density of the emitting relativistic particles introduces a great uncertainty into this method. The trained reader will note that the second and third techniques involve the determination of the so-called "Stokes para-meters" describing the intensity and polarization of the radiation. And, finally, the fourth method, skilfully applied by observational astronomers of the older generation, may be described as morphologi-cal. Conclusions about the existence of a magnetic field are drawn, based on the patterns (i.e. the configuration and structure) of the objects observed. For example, Shajn (1955) deduced the existence of a large-scale Galactic field from the elongated configuration and orientation of gaseous dust nebulae. Some active solar formations (prominence arches, coronal streamers, etc.) are usually associated with magnetic fields. It is clear that such morphological deductions about the magnetic fields are qualitative and rather crude. As a rule, they are based on the assumption that the magnetic field plays an essential part in the equilibrium or dynamics of the object concerned. The case of the magnetic field in Galactic spiral arms is a good illustration of how such factors have been exaggerated. Starting from the well known work by Chandrasekhar and Fermi (1953a), it was for a long time thought that the Galactic spiral arms were maintained by a large-scale magnetic field of about 10^{-5} Gauss, orientated along the arms. After the nature of the spiral arms (as density waves) was understood, and the strength and configuration of the magnetic field was determined by more accurate methods, their ideas were relegated to the historical archives. But the work of these high-level scientists has left a legacy. A large-scale component of the magnetic field, directed along the spiral arms, has actually been proved to exist, though it appears to be weak ($2 \cdot 10^{-6}$ Gauss, see Chapter 12) com-pared with the chaotic one.

II The Zeeman Effect

It is natural to start with the simplest and most accurate method of determining magnetic fields, that based on the Zeeman effect. Though this effect has been known for a long time and is expounded in standard textbooks, we take the liberty of recalling briefly salient points which are too often forgotten or overlooked by practitioners.

In the absence of external fields, atomic energy levels do not depend on the direction of the total angular momentum, i.e. they are degenerate. In a magnetic field $\mathbf{H}$, that defines a preferred direction in space, an atom acquires an additional energy $-\mu(\mathbf{L}+2\mathbf{S}) \cdot \mathbf{H}$ (where $\mu = e\hbar/2mc = 7 \cdot 10^{-21}$ erg/Gauss is the Bohr magneton and $\mathbf{L}$ and $\mathbf{S}$ are, respectively, the electronic orbital and the spin angular momenta), which removes the degeneracy. The energy levels split, each split level being defined by the conserved projection of total angular momentum $\mathbf{M}$ on the field direction (Landau and Lifshitz, 1958). In astrophysics, magnetic fields are usually weak (however, see Chapter 17), and the additional energy,

$$\Delta \mathscr{E} = \mu g \mathbf{M} \cdot \mathbf{H},$$

which they impart to an atom is small, even compared with the distance between the hyperfine levels. The multiplier g (the Landé splitting factor) allows especially for the spin. For an atom with zero total spin $g = 1$, and in the magnetic field the atom behaves exactly like a classical rotator.

The splitting of energy levels leads to a corresponding splitting of the spectral lines of an emitting or absorbing atom. In the simplest case, where the Landé factor is the same for the upper and lower energy levels, the line is split into a triplet:

$$\nu_\pi = \nu_0, \qquad \nu_\sigma = \nu_0 \pm g(e/4\pi mc)H, \tag{1}$$

(normal Zeeman triplet). The unshifted π-component corresponds to a transition that leaves the projection of the atomic angular momentum unchanged. It is linearly polarized, while the symmetric σ-components correspond to transitions in which M changes by ± 1 and have, respectively, right-handed and left-handed circular polarizations. For illustrative purposes, they may be represented by two rotators with angular velocity (divided by 2π) equal to $\nu_0 \pm g(e/4\pi mc)H$. The total intensities of the Zeeman components, when integrated over all directions, are equal to each other. But astrophysical observations are only possible in fact along a certain direction relative to the magnetic field. If the magnetic field is directed towards the observer only the circularly polarized σ-components can be seen. If the field is perpendicular to the line of sight, however, all three components are visible, the intensities of the σ-lines appearing to be half those of the first case. Moreover they are now linearly polarized in a direction perpendicular to the field. Hence, the relation between intensities and polarization patterns of Zeeman components depends on the magnetic field orientation relative to the line of sight. But the absolute magnitude of the Zeeman splitting is defined by the magnetic field strength alone. It is in order to emphasize these conclusions that we have described here the Zeeman effect in more detail than is usual in the astrophysical literature. Sometimes one may hear or read the erroneous statement that, in principle, only the value of the longitudinal magnetic field component can be measured by the Zeeman effect. In fact, the total magnetic strength can be found by measuring the separation of the Zeeman components; knowledge of the relative intensities and polarizations of these components yields information about the orientation of the magnetic field.

Actual implementation of this method in astrophysics is a rather complicated matter because of the smallness of Zeeman shifts and the largeness of the Doppler broadening of lines. In the optical range, where the terminology of wavelengths is conveniently adopted, the distance between the σ-components is (in Ångströms)

$$\Delta\lambda = \frac{eg}{2\pi mc^2}\lambda^2 H \simeq 2.10^{-5}\, g \left(\frac{\lambda}{4500\,\text{Å}}\right)^2 \left(\frac{H}{\text{Gauss}}\right),$$

while the line width usually exceeds 0.1 Å, corresponding to an emission temperature of 10^4 K. Therefore, direct measurements of Zeeman splitting are only possible for magnetic stars and sunspots, where $H \gtrsim 10^3$ Gauss. The pioneering measurements were made by G. Hale and H. W. Babcock. In practice, the broadened lines are photographed, and then the line profiles are carefully measured. For weaker magnetic fields, direct photographic measurements of Zeeman splitting cannot be made. Instead a photoelectric method is used. The narrow slit of an electrophotometer is adjusted to the line edge and, by using a polarizing filter, the circularly polarized components are detected one by one. The brightness of the light in the slit will vary because the σ-component intensities differ at the edge of the line. An amplified signal can then be detected even for magnetic field strengths of the order of 1 Gauss. By this means one is able to get information about the longitudinal magnetic component, since this is what determines the intensities of the circularly polarized σ-components. For more complete information about the magnetic field one must know the entire line profiles (for example, see Beckers and Schröter, 1968).

The Zeeman effect allows one to measure much weaker magnetic fields from lines in the radio-frequency range. In 1957 Bolton and Wild predicted that fields of the order of 10^{-5} Gauss, typical for gas clouds in the interstellar medium, could be detected by measuring the Zeeman splitting of the 21 cm hydrogen line. The distance between the σ-components is (in Hertz, the accepted terminology for the radio-frequency range)

$$\Delta\nu \simeq 2.8 \cdot 10^6 \, H \text{ Hz,}$$

i.e. only 2.8 Hz per μGauss, while the Doppler width of the line is usually about 20 k Hz (Kaplan and Pikelner, 1970). Here again the differential method, based on the measurement of the difference between the right- and left-handed circular components at the line edge, is of great utility. The line profile has a Gaussian form so the strongest signal will come from the narrowest and most intense line. As a rule, this is an absorption line. In practice, a radio telescope is used to measure the temperature difference

$$\Delta T / T_a = \Delta\nu / \nu_G,$$

where T_a is the antenna temperature at the maximum of the absorption line and ν_G is the Gaussian width of the line profile. For example, for the radio source Cassiopeia A, $T_a = 10^4$ K, $\nu_G = 10^4$ Hz, and $\Delta T = 3.10^6$ H. In practice, the Zeeman signal can be detected for absorption lines with an antenna temperature of hundreds of degrees. The measurements take a long time and the radio telescope must meanwhile continually track the source. The results of such measurements will be given in Chapter 12.

So far we have discussed atoms. The Zeeman measurements of molecular lines can also be used to determine the magnetic field (Heiles, 1976). These appear to be invaluable in the case of cosmic maser sources, the most powerful of which are associated with H_2O and OH. Very intensive maser radio lines of OH appear to be circularly polarized, the explanation being that only one Zeeman component of the molecule OH is subject to maser amplification. For example, if maser pumping is brought about by an infrared line then, due to its Doppler shift, one of the Zeeman components of OH falls onto a wing of the pumping line thus becoming weaker (Varshalovich and Burdjuzha, 1975). Such a polarization mechanism is efficient in a source with density between 10^5 and 10^8 cm^{-3} and magnetic field strength between 10^{-2} and 10^{-3} Gauss.

Since the Doppler broadening significantly exceeds Zeeman splitting, it presents the main obstacle to the use of the Zeeman effect for determining magnetic fields in astrophysics. It appears that the magnetic field produces other, even finer, effects in quantum systems e.g. for the simplest triplet, effects which are negligibly affected by Doppler broadening (Aleksandrov, 1972). The idea is as follows.

Let ν_1 and ν_2 be the frequencies of transitions from the first and second levels to the ground state as a consequence of Zeeman splitting (Figure 2.1(a)). For astrophysical magnetic fields the resonance frequency

$$\nu_{12} = \nu_1 - \nu_2 = (e/mc)H \qquad (2)$$

is far less than either of the frequencies ν_1 and ν_2, the Doppler broadening being very small

$$\Delta\nu_{12} = \nu_{12}v/c \ll \Delta\nu_1.$$

Hence, when measuring the frequency difference ν_{12}, one may ignore Doppler broadening.

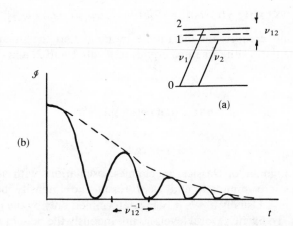

(a)

(b)

$$\leftarrow \nu_{12}^{-1} \rightarrow$$

Figure 2.1 When the superposition of the upper levels of a three-level system (a) is excited, the intensity (b) of radiation with polarization **e** beats with frequency $\nu_{12} = eH/2\pi mc$. Levels 1 and 2 are the result of the Zeeman splitting of the level shown by the dashed line in (a).

We consider two such effects below for the simplest three-level system (see Figure 2.1(a)). These can be used, in principle, as a source of information about magnetic fields in astrophysics.

III   Beats in Atomic Transitions

It is possible to determine the frequency difference ν_{12}, and hence to obtain in a very direct manner the strength of a magnetic field, by observing the time dependence of the line intensity. Since the line is the superposition of two lines with adjacent frequencies, the intensity beating observed must oscillate with the frequency difference. Indeed, the wave function for an atom excited at an initial time t_0 into the superposition of adjacent states 1 and 2,

$$\psi(t - t_0) = C_1|1\rangle \exp 2\pi i\nu_1(t - t_0) + C_2|2\rangle \exp 2\pi i\nu_2(t - t_0),$$

will (in the simplest case of dipole radiation) emit radiation at intensity (Figure 2.1(b))

$$\mathscr{I} \propto |\langle 0|\mathbf{d} \cdot \mathbf{e}|\psi\rangle|^2$$

$$\propto \exp\left[-\gamma(t - t_0)\right]\{a + 2|b| \cos\left[2\pi\nu_{12}(t - t_0) + \phi_{12}\right]\}, \qquad (3)$$

where $\mathbf{d}$ and $\mathbf{e}$ represent the dipole moment and the polarization vector, γ is the width (assumed equal) of levels 1 and 2, and

$$a = \sum_{1,2} |C_i|^2 \langle 0|\mathbf{d} \cdot \mathbf{e}|1\rangle|^2,$$

$$b = C_1 C_2^* \langle 0|\mathbf{d} \cdot \mathbf{e}|1\rangle\langle 2|\mathbf{d} \cdot \mathbf{e}|0\rangle,$$

where

$$\phi_{12} = \arg b.$$

In the language of classical mechanics, oscillations with adjacent frequencies are summed, and as a result the intensity beats at frequency ν_{12}. Usually $a = b$, a being determined only by the population levels. To excite several levels simultaneously the field must have a sufficiently wide spectrum, and the harmonics of the exciting field must be phase correlated and have the proper polarization. It is evident that, for the σ-components of the Zeeman triplet considered earlier, such a correlation arises when the exciting radiation is linearly polarized. Let us recall that a plane-polarized wave may be represented by a phased sum of right- and left-handed polarized waves.

Note that the intensity spectrum of the beats consists of the main line $\nu = 0$ and two satellite lines located at the frequencies $+\nu_{12}$ and $-\nu_{12}$, with amplitudes four times smaller

$$\Phi_\nu = \left| \int_{-\infty}^{\infty} \mathscr{I}(t) \exp 2\pi i \nu t \, dt \right|^2$$
$$\simeq (\gamma^2 + \nu^2)^{-1} + \tfrac{1}{4}[\gamma^2 + (\nu - \nu_{12})^2]^{-1} + \tfrac{1}{4}[\gamma^2 + (\nu + \nu_{12})^2]^{-1}.$$

Thus, in order to determine the field strength, it is sufficient to measure the beat frequency ν_{12} of the line intensity. Obviously, it is necessary that several beat periods (ν_{12}^{-1}) occur during the characteristic time of the intensity attenuation γ^{-1}, that is

$$\gamma < \nu_{12},$$

where γ is the natural line width.

In the case of elementary beating (single atom) just considered, the situation is extremely simple. In practice, however, there are an enormous number of atoms whose beats in intensity may overlap, thus smearing the picture. The important practical question is: Can a great number of atoms be "cooked", i.e. be in about the same phase?

The beating phase (3) depends on the three parameters: ν_{12}, ϕ_{12} and t_0. It may be assumed that the magnetic field changes little within the volume considered, i.e. ν_{12} is equal for all atoms. The initial phase ϕ_{12} is determined by atomic parameters and the polarization of the emitted quanta. Since levels 1 and 2 are states with different angular momentum projections for given moment, then as mentioned above the exciting source must be polarized, and a certain polarization should be selected for observation.

It should be noted that, in the case where the flux of emitting radiation is directed, polarization occurs due to an alignment of the atomic angular momenta, even when the exciting light is unpolarized. The reason for this is that the magnetic sublevel populations, and hence the coefficients of attenuation of radiative flux, along the field and perpendicular to it, depend on the angle between the direction of the field and the flux. This fact may itself be used for determining the direction of the magnetic field (Varshalovich, 1970).

Finally, the initial moment of excitation for all atoms can be made about the same by using a short pulse of duration $\Delta t \ll \nu_{12}^{-1}$. Differences between the beating of individual atoms will be approximately Δt, i.e. they will be phased.

So far the method considered here has found no practical application in astrophysics, even though it is not new. Ruzmaikin (1975) proposed that the beating of forbidden line intensities in the filaments of a nebula around a pulsar should be observable. The radiation is excited by the short, sharp pulses of the pulsar. To avoid the picture being smeared by interference from different parts of the nebula, the telescope would need to have a high angular resolution (about 0.01″). But, if one selected a very fine filament, the angle would be sharply defined by the radiating region itself. Appropriate observations could be conducted, in fact, from observatories enjoying a good astroclimate or by a space telescope.

IV   The Hanle Effect

Another effect that is not smeared by Doppler broadening is based on the sensitivity of atomic fluorescence to weak magnetic fields. As early as 1922 Rayleigh found that the resonant radiation, observed along a line perpendicular to the direction of a beam of natural light that provided the excitation, is polarized. In 1924 Hanle showed experimentally using the 2537 Å mercury line (transition $^3P_1 \rightarrow {}^1S_0$) that the polarization of the resonant radiation is very sensitive to the magnetic field. Specifically, when the magnetic field is perpendicular to the beam and directed towards the observer, the polarization plane rotates, the degree of polarization decreasing simultaneously, provided the magnitude of the field lies within certain bounds. Breit (1933) determined from quantum theory the polarization of the scattered radiation. A simple semi-classical representation, and a discussion of possible applications in astrophysics, can be found in a paper by Ruzmaikin (1976).

Let a beam of radiation (Figure 2.2) be directed along the axis towards an oscillator (atom), supposed situated at the origin and having a natural frequency ω_0 and damping constant γ. For simplicity,

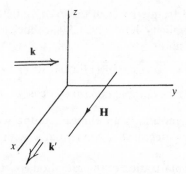

Figure 2.2 Incident (**k**) and scattered (**k′**) light beams, when *H* is the magnetic field component along the line of sight.

assume that the magnetic field is orientated along the *x*-axis, along which the observations are also made. The incident radiation will excite oscillations along the *x*- and *z*-axes. An oscillator with motions along the *x*-axis does not emit in the *x*-direction, and neither does the field act for this direction of motion. Therefore we shall assume, for simplicity, that the electric vector of the incident radiation is directed along the *z*-axis. Then, with due allowance for radiative damping of the oscillator, its equation of motion is

$$\ddot{\mathbf{r}} + \omega_0^2 \mathbf{r} = em^{-1}[\mathbf{E} + c^{-1}\dot{\mathbf{r}} \times \mathbf{H} + (2e/3c^2)\dddot{\mathbf{r}}].$$

Since the Lorentz force, the radiation damping force and eE/m are all small we can make the approximation $\ddot{\mathbf{r}} = -\omega_0^2 \mathbf{r}$ and introduce

$$\gamma = e^2\omega_0^2/3mc^3, \qquad \tfrac{1}{2}\omega_H = eH/2mc,$$

for the damping constant and the Larmor frequency respectively. It is convenient to describe the two-dimensional oscillator by using the complex variable $\xi = z + iy$, in terms of which we now have

$$\ddot{\xi} + (2\gamma - i\omega_H)\dot{\xi} + \omega_0^2\xi = eE/m.$$

In the absence of an external electric field the oscillator will have three natural frequencies ω_0, $\omega_0 \pm \tfrac{1}{2}\omega_H$.

When excited by monochromatic radiation with frequency $\omega = \omega_0$ (resonant excitation), the frequency of the oscillator will be that of the external forcing, its natural frequencies being therefore irrelevant. In this case the magnetic field will not influence the degree of polarization of the radiation from the oscillator. The situation is quite different when the frequency ω of the exciting field is unequal to ω_0. For the sake of simplicity, let it have all frequencies, $eE/m = a_0\omega_0\delta(t)$. Then, for $\gamma \ll \omega_0$, and $t > 0$,

$$y = a_0 \exp(-\gamma t) \sin \omega_0 t \sin \tfrac{1}{2}\omega_H t,$$

$$z = a_0 \exp(-\gamma t) \sin \omega_0 t \cos \tfrac{1}{2}\omega_H t.$$

The trajectory of motion is an ellipse rotating with frequency $\tfrac{1}{2}\omega_H$, whose dimensions decrease exponentially with the characteristic time γ^{-1}.

The degree of polarization of the radiation emitted by the oscillator depends essentially on the magnetic field, according to

$$p \equiv (\mathscr{I}_z - \mathscr{I}_y)/(\mathscr{I}_z + \mathscr{I}_y) = p_0/(1 + \tfrac{1}{4}\omega_H^2\gamma^{-2})$$

$$= p_0/[1 + (geH/\gamma mc)^2], \qquad (4)$$

where $\mathscr{I}_z \propto \int z^2\,dt$, $\mathscr{I}_y \propto \int y^2\,dt$, and p_0 is the polarization in the absence of a magnetic field, and is equal to unity in our case. The final expression for p above has precisely the form it should have according to quantum theory (g is the Landé factor and γ^{-1} is the damping constant).

There exists a ϕ-direction in the zy-plane, for which the polarization is maximal. To find this direction it is sufficient to calculate the value of $\mathscr{I}_z - \mathscr{I}_y$ and to find its maximum in the reference frame rotated through an angle ϕ. The result we obtain is

$$\tan 2\phi = \omega_H/2\gamma = geH/\gamma mc. \qquad (5)$$

If the magnetic field is very weak, $\omega_H \ll \gamma$, it does not affect the polarization. In a strong field, $\omega_H \gg \gamma$, the polarization plane manages to perform multiple rotations during the characteristic time of energy loss by the oscillator (γ^{-1} sec), i.e. the radiation diagram becomes spherically symmetric and the polarizaton, measured by an actual device, approaches zero. But of most interest is the case when

$$geH/\gamma mc = O(1). \qquad (6)$$

To reduce the degree of polarization (see Eq. (4)) the magnetic field rotates the polarization plane through a finite angle ϕ during the time γ^{-1}. By measuring this angle and p one can determine (using the known g, γ and p_0) the strength of the field.

Let us explain in terms of quantum theory the physical significance of the occurrence of linearly polarized radiation in a rotated plane of polarization. Again we turn to the simple triplet (Figure 2.3). Transitions from states 1 and 2 to the ground state produce the circularly polarized σ-components that may interact between themselves when the level separation, is comparable with the splitting eH/mc created by the field [see condition (6)]. The interacting σ-components, like two rotators having angular velocities $\omega_0 \pm \frac{1}{2}\omega_H$, give together linearly polarized radiation with a rotated polarization plane. Note that for $eH/mc \gg \gamma$, when the levels are not degenerate, intensity beating may be effective. In this limit, one can consider the population of each magnetic sublevel separately. The intensity and polarization of unphased atoms depend on the field direction only, not on its strength (Charvin, 1965; Varshalovich, 1970). In particular, when the magnetic field is directed towards an observer, $p = 0$.

Let us formulate the conditions under which this method of determining fields might be applicable:

1. The radiation providing the excitation must be anisotropic (as in a beam) and must have sufficient intensity at the resonant or higher frequencies (resonance scattering or fluorescence) for the probability of excitation to exceed the probabilities of excitation and damping by other mechanisms, e.g. by collisions. Note that anisotropic beams

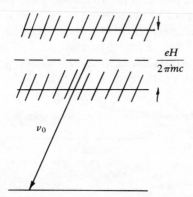

$$\frac{eH}{2\pi mc}$$

ν_0

Figure 2.3 Magnetic sublevels $m = \pm 1$ of the simplest quantum system. Dashes show the width of the sublevels, compared with the magnitude, $eH/2\pi mc$, of the splitting caused by the magnetic field.

of particles can cause similar effects. The maximum polarization occurs in the direction perpendicular to that of the incident flux.

2. Condition (6) must be satisfied. Since we have no *a priori* knowledge of the magnitude of the field, a few lines with different g and γ must be observed. Obviously, the lines chosen should be polarized in the absence of a field ($p_0 \neq 0$).

3. The rotation of the plane of polarization is affected by the field component parallel to the line of sight.

The transition probabilities corresponding to condition (6) are estimated in Table 2.1. It is clear that the allowed dipole transitions can be used only for measuring a field of about 1 Gauss (planetary or stellar fields). The probabilities typical for forbidden transitions at optical frequencies correspond to the other limiting case—a hypothetical intergalactic field of order 10^{-9} Gauss or a galactic field of order 10^{-6} Gauss. At intermediate strengths, the necessary probabilities are specific for molecular transitions.

So far, actual measurements of line polarizations have been performed only for the Sun. Hyder (1964) measured the degree of polarization and the angle through which the polarization plane had been rotated for the H_α line from a prominence loop. He obtained $p \approx 0.8\%$ and 20–25°, the latter being measured from the tangential direction of the solar limb. The magnetic field, H, of the prominence was estimated to be 45–60 Gauss. During the occultation on March 7, 1970, Mogilevsky *et al.* (1973) detected the polarization of the solar corona in the $\lambda 5303$ green line of Fe XIV and concluded that the direction of the magnetic field was nearly radial at an altitude of about 0.1 $R_\odot$ from the surface (the region investigated). Note that, at field strengths of the solar corona, $eH/mc \gg \gamma$ so that the method can only be used for determining the field configuration (Charvin, 1965).

Table 2.1   Line radiation probability estimates for various astrophysical magnetic fields.

	Planets, stars	Comets, solar corona, galactic nuclei, molecular clouds	Galaxies	Intergalactic field
H, Gauss	1	$10^{-3} - 10^{-5}$	10^{-6}	10^{-9}
γ, sec^{-1}	10^7	$10^4 - 10^2$	10	10^{-2}

V Faraday Rotation

If the magnetic field is so weak that it is difficult to detect by local methods, use can be made of the fact that magnetic fields in astrophysics are large-scale, and hence capable of exerting a significant effect on radiation passing through them. The Faraday rotation of the polarization plane due to a magnetic field is the most important of these effects. A similar method is used in the food industry to determine the concentration of an optically active ingredient, e.g. sugar, in solutions. There are two kinds of mirror-asymmetric molecules in sugar, and these cause right- and left-handed rotations of the polarization plane of light passing through the solution. A plasma, of course, has no such molecules. The non-equivalence of the left- and right-handed rotations arises in a plasma through the ambient magnetic field $\mathbf{H}$ (a pseudovector). The important quantity is the pseudoscalar $\mathbf{H} \cdot \mathbf{k}$, where $\mathbf{k}$ is the direction of wave propagation. Such pseudoscalars reverse sign under the reflection $\mathbf{r} \to -\mathbf{r}$, and this explains the different behaviors of the two polarizations. We now give a simple derivation of this effect. (The reader interested in a more detailed and exact justification may refer, for example, to the book by Ginzburg, 1970.)

We first consider circularly polarized waves of frequency, ω, propagating along the x-axis. In the absence of any magnetic field the electrons in the plasma, which is assumed cold, will under the action of the wave fields, circulate in the yz-plane; the coordinates of the electron encircling the origin are given by

$$z + iy \equiv \xi = |e| m^{-1} \omega^{-2} E \exp i\omega t.$$

The displacement of the ions can be neglected. As is generally known, the waves re-radiated by the electrons have the same effect as a change in the wave propagation speed. The effective phase speed is thus not the velocity of light but $c(1 - \omega_p^2/\omega^2)^{-1/2}$, where $\omega_p^2 = (4\pi n_e e^2/m)$ is the plasma frequency.

When a magnetic field is present, the plasma electrons have yet another frequency $\omega_H = |e|H/mc$, corresponding to the circulation of electrons round the magnetic field lines. In the presence of an electromagnetic wave, their motion becomes still more complicated. For simplicity we shall assume that the magnetic field is also directed along the x-axis. Then the equation of motion for the electrons is

$$m\ddot{\mathbf{r}} = e\mathbf{E} \exp i\omega t + ec^{-1}\dot{\mathbf{r}} \times \mathbf{H},$$

or

$$\dot{\xi} + i\omega_H \xi = (i|e|/m\omega)E \exp i\omega t.$$

From this one immediately sees that, in addition to the Larmor rotation with frequency ω_H, the electrons will perform a circulation due to the fields of the wave, this being in the same sense as that of the wave fields:

$$\xi = |e|[m\omega(\omega + \omega_H)]^{-1}E \exp i\omega t.$$

Compared with the unmagnetized plasma, a new qualitative effect has arisen. Now the amplitude of the electron's motion depends on the sign of the wave frequency. It is greater for the right-handed polarized wave ($\omega < 0$) than for the left-handed one and when $\omega = -\omega_H$ the right-handed wave comes into resonance with the Larmor rotation. The velocities of the left- and right-handed polarized waves in a plasma are evidently different. It is easy to see (compare the expressions for ξ) that, in the case under consideration, ω^2 can be replaced by $\omega(\omega + \omega_H)$, and so the phase velocities are

$$u_\pm = c[1 - \omega_p^2/\omega(\omega \pm \omega_H)]^{-1/2},$$

where the signs "$\pm$" stand, respectively, for the left- and right-handed polarized waves. It is now easy to derive the formulae of interest for the Faraday effect.

Consider a linearly polarized wave propagating along the magnetic field (Figure 2.4). Linear oscillations of the electric field of the wave may be written as the sum of two circular oscillations with opposite rotations: $E = E_+ + E_-$. Due to the non-equivalence of the left- and right-handed polarizations, these waves will propagate along the magnetic field with different velocities, and

$$(E_y, E_z)_\pm = \tfrac{1}{2}E_c[\cos \omega(t - xu_\pm^{-1}), \pm\sin \omega(t - xu_\pm^{-1})].$$

Let the vector $\mathbf{E}$ at the origin $x = 0$ be orientated along the y-axis at $t = 0$, i.e. $E_z(0) = 0$. After the waves have travelled a distance dx we sum them once again. Since the phase of the waves has changed, their sum will now have a z-component in addition to the y-component, i.e.

$$E_z = E_y \tan \phi.$$

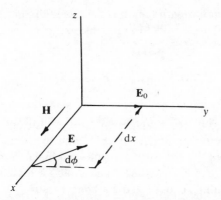

Figure 2.4 A linearly polarized wave propagating along the magnetic field (the x-axis). At $t = 0$, when the wave is at the origin $x = 0$, the plane of polarization is parallel to $z = 0$.

By substituting into the sum $\dot{E}_+ + E_-$ we obtain

$$\sin\left[\phi + \tfrac{1}{2}\omega(u_+^{-1} - u_-^{-1})x\right] = 0,$$

which gives, for the rotation angle for the full wave,

$$\mathrm{d}\phi = \tfrac{1}{2}\omega(u_+^{-1} - u_-^{-1})\,\mathrm{d}x.$$

We shall take $\mathrm{d}\phi$ as positive when the magnetic field is directed towards an observer. In astrophysics, magnetic fields and electron densities are usually small, so that $\omega \gg \omega_p$ and ω_H. The formula for the rotation angle of the polarization plane then takes the very simple form

$$\mathrm{d}\phi = (\omega_H/2c)(\omega_p/\omega)^2\,\mathrm{d}x.$$

Thus, the rotation angle is proportional to the magnetic field, the electron density and the distance from the source of the wave, and inversely proportional to the square of the wave frequency. It is customary to express the rotation angle in terms of the squared wavelength, as

$$\phi = (RM)\lambda^2 + \phi_0. \tag{7}$$

The factor RM is termed "the rotation measure". In astrophysical applications, it can be represented in an especially clear and simple

$$RM = .81 \int n_e\,B \cdot d\vec{r}$$

way; if the radiation path length is measured in parsecs, the electron density in cm^{-3}, and the field in μGauss,

$$RM = 0.81 \int n_e \, \mathbf{B} \cdot \mathbf{dr} \, [\text{rad m}^{-2}]. \qquad (8)$$

The coefficient 0.81 (that replaces $e^3/\pi m^2 c^3$) is close to unity, so the formula is easy to memorize. It will come in useful in Chapter 12 when we discuss the measurement of weak galactic magnetic fields.

We will now explain how information about a magnetic field may be extracted from an observed Faraday rotation. Imagine linearly polarized radiation propagating through a medium that lies between an observer and the source. By measuring the angle between the polarization plane and the (fixed) direction of incident radiation for several (at least two) wavelengths, one can determine the rotation measure, i.e. the tangent of the angle (7), and then by extrapolating to $\lambda = 0$ find the angle ϕ_0 at the source. Given the distance to the source and the electron density of the medium, the field projection onto the line of sight can be found. It is necessary to choose the correct range of wavelengths in which to carry out the measurements. For the weak fields typical of intergalactic and interstellar space, the rotation measures do not exceed a few hundred radians per square meter. From this it follows that the radio frequency range is relevant (here $\Delta\phi \sim 1$ to 2π). More precisely, measurements of the Faraday rotation are carried out in the approximate range 0.9 cm to 74 cm for extragalactic radio sources (radio galaxies, quasars) and pulsars. For longer wavelengths, ϕ may rotate through many radians and unpolarized radiation will be observed. For shorter wavelengths (e.g. the optical range) the Faraday rotation is evidently small.

Various factors that lead to loss of polarization should be taken into account when making the actual measurements, especially for small angles $\Delta\phi$. It must be recognized that detectors record an entire band $\Delta\lambda$ (the band pass) and not a specific wavelength. This obviously results in a signal dispersion $\Delta\phi/(\phi - \phi_0) = 2\Delta\lambda/\lambda$ and it is necessary that this dispersion is only a small part of the rotation at the given wavelength. Detectors with bandwidths of a few kHz are usually used to observe solar radiobursts. For Galactic and extragalactic radio sources the bands from 10 kHz to a few MHz are sufficient. The discussion of this and other causes of de-polarization can be found in a review by Verschuur (1974).

Despite the apparent simplicity of Formula (7), its practical implementation encounters some serious difficulties. First, the angle ϕ cannot be determined uniquely since, for any given ϕ, $\phi \pm k\pi$, $k = 1, 2, \ldots$, is also a valid solution. The measurements are usually carried out at separate discrete wavelengths λ_i, and the angles ϕ_i are obviously subject to errors $\delta\phi_i$ (in radio astronomy, $\delta\phi_i \sim 10^0$). Under these circumstances the determination of RM and ϕ_0 generally has a non-unique answer.

Figure 2.5 shows that three different straight lines (see Eq. (7)) can be drawn if, at three wavelengths, the measurements of ϕ_i are made with errors $\delta\phi_i$. From the physical point of view, however, the RM and ϕ_0 values must be unique. The choice of the minimum RM [the

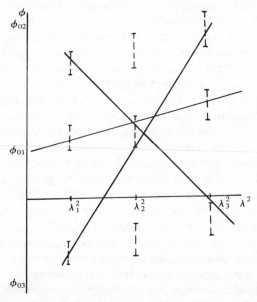

Figure 2.5   An illustration of the ambiguity of the rotation measure and the determination of the intrinsic position angle. If the position angles $\phi_i \pm \delta\phi_i$ at the three wavelengths $\lambda_1, \lambda_2, \lambda_3$ are determined with accuracy $\pm\pi$ then one can draw three different straight lines: $\phi = RM\lambda^2 + \phi_0$ through the data. Note that the presence of the measurement errors $\delta\phi_i$ is essential. If $\delta\phi_i = 0$, ambiguity can arise only when the wavelengths λ_1, λ_2, λ_3 are comparable.

tan ϕ of Eq. (7)] does not help since other straight lines are often inclined at nearby angles. An obvious way of evading this difficulty is to take measurements at sufficiently close wavelengths. How close these wavelengths should be, and what should be done when measurements exist at a number of isolated wavelengths, are questions discussed in a paper by Ruzmaikin and Sokoloff (1979), where references can also be found to other work on the determination of Faraday rotation measures in the radio range.

The other difficulty is due to the fact that Faraday rotation also takes place within the source itself where the radiation is generated. This, as was first pointed out by V. A. Razin, results in the violation of the simple law (7) relating the type of radiation to the source structure (see, for example, Ruzmaikin and Sokoloff, 1979; Vallée, 1980). Moreover, the polarized radiation of cosmic sources is often variable, and this requires nearly simultaneous measurements of ϕ to be made at several different wavelengths. For example, half the quasars and radio galaxies exhibit variable or nonlinear $\phi(\lambda^2)$. Measurement of the Faraday rotation of these sources will yield invaluable information about magnetic fields at distances of thousands of megaparsecs from the Earth. Such investigations seem to require no new instruments, and could be performed today.

The method of magnetic field determination by Faraday rotation has been applied to the Galaxy and has yielded excellent results (see Chapter 12). Here, hundreds of radio sources are used to scan interstellar space in various directions. It is possible to obtain reliable data on the magnitude and direction of the large-scale magnetic field and to draw some conclusions about the fluctuating components of the field. This method of "radioscopy" may also be applied to determine the magnetic field of the solar corona (Stelzried et al., 1970; Ruzmaikin and Sokoloff, 1978; Berlin et al., 1978).

Information about the rotation of the polarization plane, complemented by data on the degree of radiation polarization, may tell us much about the magnetic fields in the sources themselves, particularly in those parts of the Galaxy in which continuous radio emission occurs. The polarization of the continuum of the Galaxy was discovered by Dutch scientists for short radio waves and by the Soviet radio-physicist V. A. Razin for long waves. Unfortunately, the calculations of the degree of polarization of radiation associated with Faraday rotation have so far not reached a level that makes them satisfactory for

practical use. The well-known work of Burn (1966) is rather preliminary. It is worth noting (Razin, thesis, 1975 unpublished) that the
degree of polarization of a source having a purely chaotic magnetic
field is non-zero, since the degree of polarization is an essentially
positive quantity and therefore does not vanish on averaging over
many turbulent cells. (Although it is proportional to $N^{-1/2}$, where N
is the number of cells, it is not very small in the Galactic case because
N is not very large.)

VI Gyrolines

In some astrophysical objects, e.g. X-ray sources formed in the vicinity
of neutron stars, the magnetic fields can be very strong, perhaps
10^{12} Gauss. In these regions the temperatures are as a rule so high
that the atoms are fully ionized and the Zeeman effect cannot be
used to determine the magnetic field. The field strength can be
estimated however by taking into account the specific character of
the spectral distribution and the polarization of magneto-bremsstrahlung from the heated nonrelativistic plasma.

It is known [see, for example, Landau and Lifshitz (1962)] that the
radiation spectrum from a particle (electron), moving with energy $\mathscr{E}$
in a magnetic field H, consists of discrete lines located at frequencies

$$\omega_n = \omega_H n, \qquad n = 1, 2, 3, \ldots .$$

These lines are called gyrolines and their detection leads immediately
to the magnetic field strength. The line frequencies are actually large
in comparison with the separation ω_H between lines, i.e. the spectrum
is nearly continuous (since the lines are closely packed) with a
maximum at the frequency $\omega_{\max} \simeq (eH/2mc)(\mathscr{E}/mc^2)^2$. In general, the
magneto-bremsstrahlung of a particle is elliptically polarized. The
polarization is circular when observed approximately along the field
and it is linear when observed transversely (if motion along the field
is neglected).

The spectral properties and polarization of a magnetically active
plasma (a set of particles) in astrophysical conditions are considered
in detail in Chapter 6 of the monograph by Dolginov, Gnedin and

Silantev (1979), where many diagrams and useful formulae are also presented. Here we shall consider briefly some of the main applications.

The circular polarization of the continuum detected for some white dwarfs means that they have rather strong magnetic fields, $H \simeq 10^6 - 10^8$ Gauss, (Kemp et al., 1970; Angel, 1978). It is interesting that the sign of the degree of circular polarization depends on the relation between ω_H and the measurement frequencies: $p(\omega) > 0$ if $\omega < \omega_H$, and $p(\omega) < 0$ if $\omega > \omega_H$ (Dolginov et al., 1979). Optical observations of the white dwarf G195-19 established that the polarization maximum in the blue-green part of spectrum matches a corresponding minimum in the red part. This at once gave the transition frequency ω_H and hence the projected magnetic field strength along the line of sight, $H_\parallel \simeq 10^8$ Gauss.

The bremsstrahlung of X-ray sources (for example, Her X-I, Cen X-3) also provides information about their magnetic fields. These sources represent a white dwarf or a neutron star that is accreting matter with concomitant release of gravitational energy. For example, in the case of a neutron star which reveals itself as an X-ray pulsar in a binary system, the ionized gas falls from the visible star. The strong magnetic field of the neutron star decelerates the gas near the so-called "Alfvén surface", on which the gas and magnetic field pressures are comparable. It then channels the gas onto a polar area of the magnetized neutron star, in which the gas gives up its energy ($10^{36}-10^{37}$ erg/sec). The polarization and direction of the emitted radiation depend on the magnetic field, the gas flux and the parameters of the system (Gnedin and Sunyaev, 1973). By investigating the directivity diagram, one can draw certain conclusions about the strength of the field and its geometry. Observations suggest strong fields ($10^{10}-10^{13}$ Gauss) in dipolar forms. An important property of X-ray pulsar models, which are diagnostics for these strong fields, are gyrolines (Gnedin and Sunyaev, 1974) with strong linear and circular polarizations, depending on the direction of the field. A gyroline has been detected in the radiation from the pulsar Her X-I (Trümper et al., 1976). Mazets et al. (1980) discovered an absorption line at 30–60 kev in the γ-spectrum of a neutron star with a gravitational potential of about $0.1c^2$. By identifying this with a gyroline, they obtained an estimate of the stellar magnetic field strength ($2.7 \cdot 10^{12}$ G).

The determination of a magnetic field through the gyrolines is, in a sense, one of the simplest of all methods (even compared with the Zeeman method) provided only the characteristic frequency of an individual particle in the magnetic field is involved. Complications arise because of necessity of considering a large set of particles (i.e. a magnetically active plasma).

In order to determine weak magnetic fields, gyroline frequencies are measured in the radio range. In this way, for example, the fields of the Earth, Jupiter and Saturn were estimated. Their peak frequencies were found to be, respectively, 0.3, 8.8 and 1.1 MHz. The resulting field estimates for Jupiter and Saturn were obtained before in situ measurements of the planetary magnetic fields were made from spacecraft, which later confirmed these estimates. In 1981 the Gorki radioastronomers Kaverin et al. (1980) measured gyrolines in order to detect the magnetic field of the solar corona.

VII   Other Methods

The simplest way to determine magnetic fields in astrophysics is to measure the intensity of magneto-bremsstrahlung, called the synchrotronic radiation at X-ray energies. This well-known method has been used to evaluate the field strength of supernova remnants, both in the Galaxy and in other galaxies (Ginzburg and Syrovatsky, 1963; Pikelner, 1966; Shklovsky, 1976; Chibisov and Ptuskin, 1981). Scant knowledge about the density of the emitting relativistic electrons, n_{re}, is responsible for the main uncertainty in this method. The intensity is proportional to the product of this density and a power of the field strength, such as

$$\mathscr{I} \sim n_{re} H^{1/2(\tilde{\gamma}+1)},$$

where $2 < \tilde{\gamma} < 3$ is the index of the energy distribution of electrons. Perhaps this relation would be more profitably used to evaluate n_{re} from the field determined by other methods. For example, in the Galactic disk it yields $n_{re}(>1 \text{ Gev}) \simeq 5 \cdot 10^{-13} \text{ cm}^{-3}$ (Kaplan and Pikelner, 1970). It is essential that the distribution of relativistic

Table 2.2   Magnetic fields in astrophysics.

Astrophysical object	Field value, Gauss	Characteristic scale of the field and its orientation	Method
Intergalactic field	$<10^{-9}$	?	Faraday rotation (RM) of extragalactic radio sources
Galaxy			Faraday rotation of extragalactic radio sources and pulsars; optical polarization of dust; polarization of magnetobremsstrahlung of interstellar medium
regular field	2.10^{-6}	Several kiloparsecs $l = 100$ pc, orientated along spiral arms	
random component	$\delta H/H \sim 1$		
Interstellar clouds	10^{-5}	10 pc	Zeeman effect in the 21-cm line
Maser sources, dense cool clouds, galactic nuclei	$10^{-2} - 10^{-3}$	$<10^{16}$ cm	Zeeman effect on OH-molecules
Quasars and radio galaxies	100	~ 1 pc	Intrinsic Faraday rotation and polarization
Sun: general poloidal field	1	$0.1-1\,R_\odot$, Dipole+weak quadrupole component	Zeeman effect in optical lines; Hanle effect; Faraday rotation

sub-photospherical azimuthal field corona	$>10^3$ 10^{-5}	Dominant odd field relative to reflection in equatorial plane	In situ measurements by spacecraft; intensity and polarization of radio radiation
Planets: Jupiter Saturn Earth Mercury Mars	4 0.2 1 $3 \cdot 10^{-3}$ $6 \cdot 10^{-4}$	Dipole slightly inclined to the rotation axis	
Ap-stars	10^4	Rotating dipole or spot structure	Zeeman effect in optical lines
Some white dwarfs	10^6–10^8	Dipole	Circular polarization of magneto-bremsstrahlung
X-ray sources near black holes in binary stellar system	$\sim 10^9$	3–$100 r_g$ (r_g is the gravity radius of the black hole)	Energy considerations and mechanism of angular momentum transport
Pulsars	10^{12}	Dipole-type field	Estimate by energy considerations and direction of the radiation
X-ray sources near neutron stars	10^{10}–10^{13}	Dipole-type field distorted near poles in "columnar" form	Gyrolines; polarization of magneto-bremsstrahlung

electrons be highly inhomogeneous (Heiles, 1976), while the field should vary only on a large scale, i.e. be smooth.

Some methods (e.g. the study of the orientation of interstellar clouds or the configuration of the solar corona, etc.) are morphological in nature, and are of little help in gaining quantitative information about the field. Usually only the field direction can be reliably obtained. The most popular of these methods determines the optical polarization of starlight through scattering by interstellar dust. It is assumed that the anisotropically shaped dust particles are orientated by the magnetic field, the particle flux and the radiation (Greenberg, 1968; Dolginov *et al.*, 1979). The degree of linear polarization which arises is given by

$$p = p_0 d \sin^2 \theta,$$

where p_0 is a constant depending on the properties of the grain, θ is the angle between the orientation axis of the grain (which appears to coincide with the field direction) and the line of sight, and d is the distance to the star.

Though the mechanism that orientates dust particles is incompletely understood, the observational results from more than 7000 stars (e.g. Mathewson and Ford, 1970) are impressive. The dust particles have a regular orientation on a rather large scale, like iron filings in the field of a school magnet. This allows one to deduce the direction of the large-scale Galactic magnetic field (Spoelstra, 1977).

VIII   Strength and Scale of Magnetic Fields in Astrophysics

In conclusion, without claiming a particularly high degree of accuracy or completeness, we would like to display the state of knowledge about magnetic fields in astrophysics (Table 2.2). It is amazing that we know more about the weak field of the Galaxy than the strongest fields of pulsars.

CHAPTER 3

Origin of Magnetic Fields

I Specifics of the Cosmic Medium: Large Reynolds Numbers

By a cosmic medium we mean, in the broad sense, the matter comprising planets, stars, galaxies and the spaces between them. If one disregards relatively small regions, e.g. the solid parts of planets, the cosmic medium can be regarded as a plasma, most often as a gaseous plasma. The range of temperatures in this plasma is wide: from about 10 K for dense molecular clouds in galaxies, through the tens of millions degrees of the thermonuclear stellar reactors, to the gigantic 10^{10} K and more, typical of the early stages of evolution of the hot Universe. Table 3.1 presents estimates of these and other relevant parameters for a number of astrophysical objects. It may be seen that the conductivity of the cosmic plasma is comparable to that of poor metals.[†] In one sense there is nothing remarkable in this. Even cold clouds have some appreciable conductivity caused by the nonthermal ionization of gas by the UV background and—within dense objects—by the penetration of cosmic and γ radiation. As a rule, the hydrodynamical velocities of plasmas are not very great; for example, the rotational velocities of young stars do not exceed a few hundred km/s, and the random velocities of the interstellar gas are of the order 10 km/s. The exceptions are relativistic objects (neutron stars and black holes); the gas velocities in their gravitational fields may approach that of light. The main difference between the cosmic

---

[†] The conductivity of copper (a good conductor), at room temperature is $\sigma = 5.10^{17}$ C.G.S.E., corresponding to $\nu_m = c^2/4\pi\sigma \simeq 10^2$ cm^2/s; for iron it is 6.10^2 cm^2/s and for mercury 6.10^3 cm^2/s. For a fully ionized plasma heated to a million degrees, $\nu_m = 10^{13} T^{-3/2} = 10^4$ cm^2/s.

37

$\sigma_{IRON} = 1 \times 10^{17} \ s^{-1}$

$\sigma_{MERCURY} = 1 \times 10^{16} \ s^{-1}$

Table 3.1 Typical Parameters of the Cosmic Medium.

	ρ g cm^{-3}	T K	Ω s^{-1}	L cm	v cm s^{-1}	ν cm^2 s^{-1}	ν_m cm^2 s^{-1}	Re	R_m
The Earth's liquid core	10	$4 \cdot 10^3$	$7.3 \cdot 10^{-5}$	$3.5 \cdot 10^8$	$4 \cdot 10^{-2}$	$\sim 10^{-2}$	$3 \cdot 10^4$	10^9	$5 \cdot 10^2$
Jupiter's core	1	10^4	$1.8 \cdot 10^{-4}$	$5 \cdot 10^9$	$1-10$	$3 \cdot 10^{-2}$	10^4	10^{12}	10^6
Solar convection zone	10^{-3}	10^5	$3 \cdot 10^{-6}$	$2 \cdot 10^{10}$	10^5	0.3	10^7	$5 \cdot 10^{15}$	$2 \cdot 10^8$
Accretion disk around black hole (Cyg X-1):									
(a) inner region, r ~ 30r$_g$	10^{-5}	$2 \cdot 10^6$	40	10^7	$5 \cdot 10^7$	$5 \cdot 10^6$	15	10^6	$3 \cdot 10^{13}$
(b) intermediate region, r ~ 300r$_g$	10^{-6}	$3 \cdot 10^5$	$4 \cdot 10^{-2}$	10^8	$5 \cdot 10^5$	$3 \cdot 10^2$	$4 \cdot 10^3$	10^{11}	10^{10}
Interstellar medium:									
(a) HI region	$2 \cdot 10^{-23}$	10^2		$3 \cdot 10^{19}$	$3 \cdot 10^5$	$2 \cdot 10^{19}$	$5 \cdot 10^{21}$	$5 \cdot 10^5$	$2 \cdot 10^3$
(b) HII region	$2 \cdot 10^{-25}$	10^4		$3 \cdot 10^{20}$	10^6	$5 \cdot 10^{18}$	$3 \cdot 10^{20}$	$5 \cdot 10^7$	10^6
(c) tunnel	$5 \cdot 10^{-27}$	10^6		$3 \cdot 10^{20}$	10^6	$3 \cdot 10^{26}$	$5 \cdot 10^{17}$	1	$6 \cdot 10^8$
Galactic gaseous disk (on average)	$2 \cdot 10^{-24}$	10^4	10^{-15}	10^{21}	10^6	$5 \cdot 10^{17}$	10^{21}	$5 \cdot 10^9$	10^6
Expanding Universe at z = 1400	$3 \cdot 10^{-20}$	$4 \cdot 10^3$		$4 \cdot 10^{22}$	10^6	10^{17}	$5 \cdot 10^7$	$5 \cdot 10^{11}$	10^{21}

The radius of a spherical body and the half-thickness of a disk-like object define the typical length scale, L. For a completely ionized medium the kinematic viscosity and magnetic diffusivity are evaluated using the formulae $\nu = 1.2 \cdot 10^{-16} T^{5/2} \rho^{-1}$ cm^2 s^{-1} and $\nu_m = 10^{13} T^{-3/2}$ cm^2 s^{-1} respectively. For the accretion disk we have adopted the value Ma = 0.1 for the Mach number ($\alpha \simeq \frac{1}{3}$ for the viscosity parameter); ν in the inner region is the radiation viscosity. The magnetic diffusivity of the partially ionized interstellar gas is due to ambipolar diffusion. In the "tunnels" a resonant interaction of thermal ions with magnetic fluctuations is the dominant dissipative mechanism for interstellar turbulence and here we cite the corresponding value for ν. The parameters for the cosmological plasma (the last line) are calculated for a flat Universe on the scale of a protocluster of galaxies (10^{15} M$_\odot$); the amplitude of the adiabatic density perturbations is assumed to be $\delta \rho / \rho = 10^{-3}$ at z = 1400 and ν is calculated for the molecular viscosity of neutral hydrogen.

medium and ground-based laboratory conditions lies in the "astronomical" scales, which in the former case are enormous. This leads to a sharp increase in the characteristic time $t \simeq L^2/\nu_m$ and to large values of the dimensionless Reynolds numbers which decides the nature of hydrodynamic and magnetohydrodynamic phenomena in astrophysics. In the theory that is the subject of this book, a special rôle is played by the magnetic Reynolds number. It is defined by

$$R_m = Lv/\nu_m,$$

where L is a characteristic dimension of the region occupied by the conducting plasma, v is a characteristic velocity of its hydrodynamic motion, and ν_m is the magnetic viscosity or, in other words, the magnetic diffusivity of the plasma. As often happens in similarity theory, the determination of the magnetic Reynolds number is somewhat arbitrary. Indeed, what should one choose for L? In the case of a star (say), is it the radius, the scale-height of the density gradient[†], the size of the convection zone or some other dimension? The same questions can arise concerning the velocity amplitude and even the magnetic viscosity ν_m, because the latter can be affected by several different mechanisms, e.g. by ohmic diffusion or plasma effects.[‡] Actually, uncertainties may be eliminated if R_m is determined in every specific problem by the exact assignment of the dimension, the velocity and the most efficient mechanism of conduction. A common feature of the problems considered below is that, under astrophysical conditions, the dimensionless magnetic Reynolds number always proves to be large. And, since this determines the relative importance of inductive and dissipative effects, one can draw a most important conclusion about the dominant rôle of motions in the cosmic medium.

---

† This is the height difference over which the density changes by a factor of e, i.e. $L = (d \ln \rho/dr)^{-1}$

‡ We also note that the conductivity of the plasma depends on the magnetic field, which introduces anisotropy. The magnitude of this effect depends strongly on the plasma density: it is small in the depths of the Sun but can be great in the interstellar gas (Spitzer, 1956).

II   Frozen-in Magnetic Fields

As a first step it is natural to ignore the ohmic dissipation of the magnetic field, i.e. to consider the medium to be a perfect conductor, $R_m = \infty$. In this situation the behavior of the field is controlled completely by the motion. One usually says that the magnetic field is "frozen into the medium." The concept of frozen-in fields very naturally extends Faraday's ideas about magnetic field lines. It was introduced into modern theory by Kiepenheuer and Alfvén, and has proved to be very useful for the qualitative understanding of the behavior of the magnetic field in the cosmic medium. Let us clarify the meaning of this concept.

The absence of electrical resistance means that the number of field lines encircled by each contour (thin ring) in imaginary isolation in a moving fluid is conserved. The contour transports this value of the magnetic flux $\int \mathbf{H} \cdot d\mathbf{S}$ in its motion (see Figure 3.1).

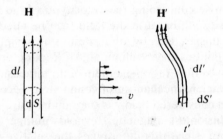

Figure 3.1 Transformation of an artificially extended tube of magnetic field lines in a perfectly conducting fluid moving with the velocity profile shown. In an incompressible fluid the reduction of the tube cross-section is followed by a proportional increase in the number of magnetic field lines per unit area so that the magnetic flux remains constant.

There is another interpretation of the frozen-in concept, proposed by Walén (Cowling, 1957), that has proved to be more useful. Let us imagine a thin pencil of magnetic field lines with cross-section $d\mathbf{S}$, the so-called magnetic flux tube. Let dl be the distance along the flux tube between two adjacent cross-sections. During the motion the following two quantities are preserved: the mass $\rho\, dl\, dS$, where ρ is the fluid density, and the magnetic flux $\mathbf{H} \cdot d\mathbf{S}$. Hence, $H/\rho \sim dl$, i.e.

any change dl in the distance between adjacent particles on the same field line induces a corresponding change in the value of H/ρ. It is evident that, during the motion of an incompressible fluid, $\mathbf{H} \sim d l$. This idea may be formulated in a more general way. Take two adjacent fluid particles in an incompressible flow such that

$$\delta \mathbf{r} \equiv \mathbf{r}_2 - \mathbf{r}_1 = \varepsilon \mathbf{H},$$

where ε is a pseudoscalar. In other words, we travel a small distance $\varepsilon |\mathbf{H}|$ from the first particle in the direction of $\mathbf{H}$ along which the field strength does not noticeably change. It turns out that the above relation also holds over larger distances in an infinitely conducting medium, the value ε being small but constant for each such pair of particles. In particular, this means that a pair of particles on the same field line will always remain on this line, and a pair of particles which are not on the same line never find themselves on the same line. Similar remarks hold for a compressible fluid with H/ρ replacing $\mathbf{H}$. We may apply this theory also to unsteady motion provided we bear in mind that the separation, $\delta \mathbf{r}$, between the fluid particles is infinitesimal: application to finite $\mathbf{r}_2 - \mathbf{r}_1$ may lead to erroneous results.

The frozen-in condition means that, in a moving medium, currents develop whose fields add to the initial field, changing the latter so that the field lines move in concert with the fluid particles. In this sense the frozen-in picture may be understood as a manifestation of the collective interaction between charged particles of the plasma. Of course, it is a description of the same phenomenon but in another language. An interesting discussion of the "field" and "current" approaches to the description of MHD-phenomena can be found in a review by Alfvén (1977).

The frozen-in condition is applied to continuous media of high conductivity in which displacement currents are ignored, i.e. when the notion of magnetic field lines makes sense. It is absolutely clear that it cannot be used in any form when the field is in a vacuum, in a poorly conducting medium, or (more precisely) whenever the magnetic Reynolds number is small.

It is easy to picture the behavior of the magnetic field using the frozen-in concept. For example, it is clear that in the compression of a gas layer, the magnetic field along the layer must be strengthened in direct proportion to the density, and in the isotropic compression of a spherical cloud $H \sim \rho^{2/3}$, the magnetic energy density $H^2/8\pi$

increasing in proportion to $\rho^{4/3}$. This law also holds for isotropic adiabatic compression of electromagnetic fields of other types: the energy of the Coulomb field of charges and the energy of the radiation field increase in the same manner. In Chapter 5 we shall demonstrate another, less trivial use of the frozen-in field concept for non-isotropic motions when proving antidynamo theorems and formulating the notion of the so-called "slow dynamo".

III   Relict Field Hypothesis

The conservation of the magnetic flux due to the frozen-in condition was the basis for an explanation of the origin of observed cosmic magnetic fields. According to this, the magnetic fields are a direct result of the compression of an initial ancient "relict" field (Hoyle, 1958; Piddington, 1972) whose origin is postulated by, or referred back to, the cosmological problem. Let us explain why this approach to the origin of magnetic fields has proved to be unsatisfactory.

Observations of the background 3 K radiation show that on large spatial scales the Universe is homogeneous and isotropic (Zeldovich and Novikov, 1975). A global magnetic field can be embedded in it without violating this homogeneity. According to the Hubble expansion law $\mathbf{v} = \mathscr{H}\mathbf{r}$, where $\mathscr{H}^{-1} \simeq 10^{10}$ years is the Hubble constant, the homogeneous magnetic field generates an electric field $c^{-1}\mathbf{v} \times \mathbf{H} = (c\mathscr{H})^{-1}\mathbf{r} \times \mathbf{H}$ at each point of space. An observer following the expanding matter does not see this field and the Universe appears to him to be homogeneous. The magnetic field thus does not violate the homogeneity of the Universe but violates its isotropy. In the modern epoch a homogeneous field contradicts the observed isotropy, but when travelling into the past the magnetic field grows because, due to the conservation of the magnetic field, all dimensions are reduced (see Chapter 15).

A very small relict field is quite enough to explain the large-scale magnetic fields now observed in galaxies. Indeed, the large-scale component of the Galactic field is about 2.10^{-6} Gauss. To obtain such a field, by condensing the Galaxy to a density of 1 cm^{-3} from inter-galactic matter with a density of 10^{-5} cm^{-3}, a field of

$2.10^{-6}(10^{-5}/1)^{2/3} \simeq 10^{-9}$ Gauss is sufficient. It is interesting that even this weak magnetic field could be detected by observations. Due to the Faraday effect, the polarization plane of the linearly polarized radio radiation received from remote radiogalaxies and quasars manages to turn through a finite angle. By measuring the polarization plane positions for several wavelengths one can determine $\int n_e \mathbf{H} \cdot \mathbf{dr}$ (see Chapter 2). Then, by knowing the distances to the sources (these can be found from the red-shifts of the optical lines in the spectra) and the independently determined electron density of the intergalactic medium, an estimate can be made of the intergalactic magnetic field. Analysis of the observational data obtained from a few hundred remote radio sources does not reveal the existence of any noticeable relict field. An upper limit for the strength of the hypothetical homogeneous field is about (or even less than) 10^{-9} Gauss. The main uncertainty arises from a poor knowledge of the density of the intergalactic medium. Nevertheless, it should be admitted that observations do not confirm the existence of a hypothetical homogeneous intergalactic magnetic field (for details see Chapter 15).

The main argument in favor of the relict nature of the galactic magnetic fields was the fact that large-scale fields attenuate very slowly due to the ohmic dissipation. For example, for a field on a 400pc scale, based on the semi-thickness of the gaseous Galactic disk, the characteristic time of ohmic dissipation (assuming that the mean gas temperature is about 10^4 K) is about 10^{26} years, i.e. much greater than the age of the Galaxy. But it must be borne in mind that magnetic field can be destroyed very much more efficiently by the turbulent diffusion associated with small-scale chaotic motions. In the Galaxy the characteristic mean square velocity of such motions is about 10 km/s on the 100-pc scale (Chapter 12), which corresponds to a turbulent diffusion coefficient of 10^{26} cm^2/s (the coefficient of ohmic diffusion is only 10^7 cm^2/s). In a medium with this turbulent diffusion coefficient, the large-scale fields decay in a time of about 10^9 years which is less than or comparable with the age of the Galaxy (about 10^{10} years).

On the one hand, turbulent motions are fatal for large-scale fields. And what about the frozen-in concept and the corresponding conservation of magnetic flux? It is because of the frozen-in condition, the attachment of the field lines to the moving fluid particles, that the characteristic field scale becomes smaller, with the concomitant decrease in effective magnetic Reynolds number. Of course, this is not the end of the matter, and the last word on the dissipation of the smaller scale magnetic fields is left to ohmic diffusion or some other kind of dissipative process at the kinetic (non-hydrodynamic) level.

On the other hand the motion of the medium may also play a creative role for the large-scale magnetic fields, and in this the generosity of Nature reveals itself. It forms the basis of the other concept of field origin developed in this book, the notion of hydromagnetic dynamo action (section IV). This hypothesis must replace the suggestion of a primary origin for astrophysical magnetic fields as relics.

But may we leave the hypothesis of relict origin totally for the archives? It certainly proved useful, by raising the question of the existence of the maximal-scale magnetic fields in the Universe regardless of the problem of Galactic field generation. An answer should be sought in the analysis of observational data obtained during the investigation of remote cosmic sources as well as in the study of physical processes in the strongly anisotropic solutions to early-stage cosmological situations. The hypothesis of relict origin is still alive and still discussed in the scientific literature, not in its initial form, but at another level in its application to stellar magnetic fields. The problem addressed is the origin of the strong magnetic fields of some peculiar stars (Ap-stars, see Chapter 11). Turbulent flows observed at the surfaces of these stars prove to be insufficient to influence the large-scale magnetic field effectively. But we prefer the other point of view that asserts that the magnetic field of these stars is not a remnant of the Galactic field, but the result of generation at certain stages of evolution when the motions were more efficient. Even a more striking example is provided by pulsars whose magnetic fields are about 10^{12} Gauss. It is assumed that these fields are produced by the compression of the magnetic fields of the stars from which the pulsars emerged. This hypothesis will be discussed in Chapter 17.

IV The Necessity for Dynamo Action

The question of the origin of cosmic magnetic fields can be divided into several parts. One should explain how these fields are maintained or amplified over long periods of time and why they can suffer drastic changes over short periods. Also one should theoretically find the field configuration and current distribution. In addition to these, one should also provide quantitative estimates for the characteristic times of amplification and variation of field, as well as its amplitude and configuration. As a rule one should take into account the influence of the field on the motion.

When considering such extended astronomical structures as galaxies, of most interest is the problem of the amplification and maintenance of the large-scale magnetic field against turbulent tangling. Small-scale seed fields arise naturally in a plasma from the action of the thermal e.m.f. or various plasma instabilities. The occurrence of large-scale magnetic fields, even those small in magnitude, is less easily explained. Their amplification may be considered in a manner similar to that used for treating stability problems. Namely, it is necessary to find field configurations with a given velocity distribution which grow exponentially with time. If the product γt in the exponent is large enough, the problem of seed fields loses its importance but for completeness we shall now mention some mechanisms of seed field excitation. One of these was proposed by Harrison (1970) and developed by Mishustin and Ruzmaikin (1971), who applied it to galaxies in formation. It is based on the difference between the Compton electron and ion interactions with the background radiation filling space. Light electrons in a rotating gas cloud (the forming galaxy) are swept along by photons and a vortex motion of the electrons relative to the ions (i.e. an electric current) starts, resulting in magnetic field excitation. The electric field induced by the magnetic field tends to equalize the velocities of the electrons and ions, and a quasistationary process is thus set up. The amplitudes of the magnetic fields excited in this way over cosmological times are small (about 10^{-18} to 10^{-20} Gauss); however, the field scales are comparable to the galactic dimensions. For details see Chapter 15.

The problem of seed fields is not as urgent as that concerning the origin of the magnetic fields of stars, planets and other bodies of relatively small-structure. A galaxy can provide the initial fields for its stars, and a star can provide the initial fields for its planets. Nevertheless, even in these cases, the process of magnetic field amplification must be very rapid (exponential) for a steady state to be established. A characteristic time for the growth of a stellar field should be compared not to the total lifetime of the star but to typical periods in its evolution.

The occurrence of processes which rapidly change the magnetic field can be clearly illustrated for planets. For example, it is known that the electric currents maintaining the Earth's magnetic field flow in its core where the magnetic viscosity is rather large (about $3.10^4 \, cm^2/s$). A characteristic time for the ohmic decay of the field in the outer liquid-metal core of the Earth, which forms the spherical layer $(0.19-0.55) \, R_E$, is thus only about 10^4 years. On the other hand, a study of the magnetism of rocks shows that the Earth's magnetic field has existed for not less than 10^9 years. Moreover, paleomagnetic analysis (Cox, 1969) reveals random reversals of the direction of the geomagnetic dipole (with a mean characteristic time of about 10^5 years), followed by quicker (10^3-10^4 years) deflections of the dipole relative to the axis of rotation. At present the dipole axis deviates from the rotation axis by the small angle of 11^0, and on a time scale in excess of 10^4 years it is orientated, on average, along the rotation axis.

Jupiter provides another example. The angular velocity of its surface is $1.8 \cdot 10^{-4} \, s^{-1}$; its radius is $7.10^9 \, cm$. According to modern knowledge (Gehrels, 1976; Hide, 1980) Jupiter has a liquid-metal core of radius about $R_c = 5 \cdot 10^9 \, cm$. The conductivity of the core, consisting mainly of hydrogen with the addition of a small percentage of helium, may be considered to be close to that of liquid metals (for example, mercury), say $\nu_m \simeq 10^4 \, cm^2 \, s$. Thus, the characteristic time of ohmic dissipation of the large-scale field in the Jovian core is $R_c^2/\nu_m \simeq 10^7$ years, which is considerably less than the planet's age, i.e. the magnetic field must be self-maintained. The magnetic axis of Jupiter also deviates from the rotation axis by a small angle. In contrast, the magnetic dipole of the similar planet, Saturn, is orientated nearly along its rotation axis (Acuna et al, 1980). Possibly we observe to-day two different phases of the oscillations of the magnetic axes of these planets.

Motion in the conducting parts of planets offers a natural explanation of magnetic field variations. In the case of the Earth one can estimate the velocities required for generation from the observed westward drift rate (0.2 degree per year) of the non-dipole component of the magnetic field ($H_{nd} \simeq 0.02$ Gauss). The resulting value, $v \simeq 4 \cdot 10^{-2}$ cm/s, is a million times greater than the characteristic mantle velocities near the Earth's surface, derived from geological data. Thus, the westward drift of magnetic disturbances is an indication of strong motion, possibly convection or nonuniform rotation, in the core. The velocities in Jupiter's core may be estimated by assuming that in the stationary state the magnetic stresses are balanced by the Coriolis force

$$H_\phi H_p \simeq 4\pi\bar{\rho}v\omega R_c,$$

where $\bar{\rho} = 1.9$ g/cm^3 is the mean density, and $\omega = 1.8 \cdot 10^{-4}$ s^{-1} is the angular velocity of the planet. Supposing a quasi-dipole character for the poloidal component H_p in the core, $H_p \simeq H(R_j)(R_j/R_c)^3$ and, postulating the origin of the azimuthal component as due to the twisting of H_p by nonuniform rotation, we obtain the estimate $v \simeq 0.05$ cm/s. A similar estimate for Saturn ($R_c \simeq 0.5R_s$, $\bar{\rho} \simeq 1.5$ g/cm^3, $\omega_s \sim \omega_j$, $H_{po} \simeq 0.2$ G; Acuna et al, 1980) yields the smaller velocity 5.10^{-3} cm/s. One should bear in mind that these planets are rotating rather fast (they make approximately one revolution every 10 hours, which is a hundred times greater than the angular velocity of the Sun), so the rotation must have an influence on convection as happens in the Earth's atmosphere (in particular, resulting in cyclone formation). In a rapidly rotating fluid a symmetry arises with respect to translation along the rotation axis. Convection cells take the form of rolls, elongated in the meridians. The corresponding flow proves to be favorable for magnetic field amplification (Busse, 1976).

Rapid changes are typical of stellar magnetic fields. Moreover, such field changes are responsible for all manifestations of stellar activity. After the general rotation of a star its most interesting features are periodic alterations of field.

The ability of a conducting medium to convert the energy of its hydrodynamic motion into the energy of its magnetic field is usually called 'the hydromagnetic dynamo' or simply 'dynamo.' It operates naturally in the cores of planets, stars and other astrophysical objects, i.e. without the application of external electromotive forces, the use of wires, coils, etc. So far such a dynamo machine has not been put into operation in laboratory conditions, the main reason being the rather modest dimensions of laboratory installations. In other words, the magnetic Reynolds number is small (≤ 30, Gailitis et al., 1977), i.e. the action of the motion is weaker than the action of ohmic diffusion. To increase the magnetic Reynolds number for fixed dimensions one needs a very vigorous motion. These difficulties, however, are of a purely economical, non-fundamental, nature.

On the other hand, under cosmic conditions (see Table 3.1) the magnetic Reynolds numbers are great, so even motions with small velocities can be efficient. The basic problem, first highlighted by Cowling (1934), is to know what kind of motion is capable of amplifying or maintaining a magnetic field. It is worth remembering that motion need not necessarily amplify the field; it may instead accelerate its diffusion. For example, a large-scale magnetic field quickly disappears in a strong, homogeneous, isotropic, mirror-symmetric turbulence. A considerable part of this book is concerned with formulating the conditions under which the dynamo is possible and constructing specific dynamo models.

Before we formulate these conditions and consider the problem strictly, we would like to display a few examples of dynamos (Chapter 4). By looking at their qualitative aspects, unobscured by mathematical formulae, one can, we believe, gain an immediate and sufficiently deep penetration into the essence of dynamo theory. A reader interested in the mathematical basis of these examples can find detailed and strictly formal presentations in the papers quoted in the references.

Dynamos

Before speaking of dynamo theory and its applications to astrophysics, we will present a few simple examples of dynamos. This will allow us to understand the main principles (including rather intricate aspects) without going through the torture of analyzing mathematical formulae and complex equations. From this point of view, the present chapter is in the nature of a tentative discussion. Exact solutions and detailed analyses can be found in the relevant references.

The word "dynamo" originates from the term "dynamo-machine", now forgotten (or, more precisely, replaced by the term "generator"). From our school days it evokes memories of a rotating frame placed between magnetic poles to produce an electric current. Thus mechanical energy is transformed into electrical energy. It may also be recalled that the electromotive force (e.m.f.) of the dynamo-machine is proportional to the product of the magnetic field and the rotational velocity. These two ideas are valid when considering magnetic field generation in a moving conducting medium (the hydromagnetic dynamo). The main difference is that in the case of the hydromagnetic dynamo there are no external magnets, frame, or wires. We will begin by considering a "laboratory" dynamo-machine which, while it can only be constructed with the help of wires and a disk, demonstrates some essential features of the hydromagnetic dynamo.

I The Homopolar Dynamo

One can construct a simply acting dynamo (Bullard, 1955) by exploiting the phenomenon of unipolar (or homopolar) induction, discovered

49

by Faraday. The phenomenon consists of the appearance of an e.m.f. in a closed circuit which includes both a moving and a stationary conductor in a magnetic field. (We shall close the circuit for these conductors with the aid of sliding contacts so that the electric current can flow through the circuit.) The explanation stems from the fact that the electric fields in the moving and fixed reference frames are unequal. The electric field is absent in the rotating conductor and is equal to $-\mathbf{v} \times \mathbf{H}/c$, where $\mathbf{v}$ is the rotation velocity, in the stationary conductor. Thus an electromotive force develops due to the difference in angular velocity when passing from one part of the circuit to the other. Elsasser (1946) was the first to note what this effect might mean for magnetic field generation in the Earth's core and in differentially rotating stars. In these cases, we are dealing with a rotating plasma instead of a solid conductor. To confirm the fact that the homopolar electromotive force arises during the rotation of a conducting fluid in a magnetic field, Lehnert (1958) carried out a laboratory experiment, using a cylinder with a diameter of 40 cm filled with 58 liters of liquid sodium. The latter was forced into motion by a rotating copper disk with vanes.

For the rotating conductor it is convenient to use a conducting disk set in rotation by an applied couple, $\mathcal{K}$ (Figure 4.1). The stationary conductor is represented by a wire with resistance R and inductance L, twisted into a helix and connected to the edge of the disk and to its axis by means of sliding contacts. The classical experiments on unipolar induction use a rotating magnet instead of the conducting disk, but the unipolar machine shown in Figure 4.1 does not require a permanent magnet. It is sufficient that a magnetic field is provided at the initial instant. For example, we may temporarily place the system in a constant magnetic field or lead an initial current I_0 into the wire. What will happen next?

For the sake of simplicity we shall ignore the inductance of the disk. It is rather large for a solid disk, but may be greatly reduced by replacing the disk with alternating conducting and non-conducting radial spokes. Then the voltage balance in the circuit is simply

$$c^{-2}L \, dI/dt + RI = \mathcal{E}, \tag{1}$$

where L, R are the inductance and resistance of the electric circuit and

$$\mathcal{E} = c^{-1} \int \mathbf{v} \times \mathbf{H} \cdot d\mathbf{l} = c^{-1} \boldsymbol{\omega} \cdot \int_0^a \mathbf{H} r \, dr \tag{2}$$

$F = q v \times B$

See Feynman

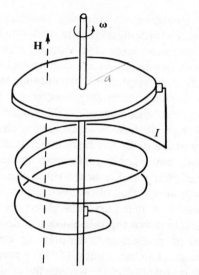

Figure 4.1 The homopolar dynamo, composed of two conductors: one moveable (a disk rotating with an angular velocity ω) and one fixed (a wire coiled as a helix and connected to a disk by sliding contacts).

is the homopolar e.m.f. supporting the current, I, in the circuit; a is the disk radius. The magnetic field determining the homopolar e.m.f. is generated by a current in the twisted part of the wire. According to the Biot-Savart law

$$\boldsymbol{\omega} \cdot \mathbf{H} = Ic^{-1} \int \boldsymbol{\omega} \cdot (\mathbf{dl} \times \mathbf{r}) r^{-3} = Ic^{-1} \int r^{-3} \mathbf{v} \cdot \mathbf{dl},$$

i.e. the sign of the e.m.f. is determined by the projection of the direction of the twist in the wire onto the rotation velocity. It is convenient to express the e.m.f. through the mutual inductance of two contours, the twisted one and the rim of the disk: $\mathcal{E} = \pm M\omega I/2\pi c^2$. Now, equation (1) may be solved to give

$$I = I_0 \exp \gamma t,$$

$$\gamma = \omega M L^{-1} (\pm \tfrac{1}{2}\pi^{-1} - c^2 R/M\omega).$$

Thus, when the wire is twisted in the same direction as the rotation and $M\omega/c^2 R > 2\pi$, the current, and consequently the magnetic field and flux through the plane of the disk, will grow exponentially. Clearly the result does not depend on the direction of the initial magnetic field. We have a self-exciting magnetic dynamo. It should be noted that the growth rate γ increases as the value of $M\omega/c^2 R$ (the magnetic Reynolds number) increases. We call the dynamo fast if $\gamma \not\to 0$ when the magnetic Reynolds number becomes infinite (see Chapter 7).

It is clear from energy considerations, say, that the exponential growth of the current and magnetic field cannot continue for ever. The amplified magnetic field will produce a back-reaction on the angular velocity of the disk. The case in which the velocity (or ω) is given is called the "kinematic dynamo problem" (kinematic dynamo), and the self-consistent problem when the action of the field on the motion (or ω) is also included, is called the "dynamical dynamo problem". In our case it is easy to account for the back-reaction. Each current element flowing out along the disk radius across the magnetic field H perpendicular to the plane of the disk exerts a force $I \, dr \, H$ on the disk. The net torque of these forces over the entire radius is $I \int_0^a Hr \, dr = MI^2/2\pi$. Consequently

$$C \, d\omega/dt = \mathcal{K} - (2\pi)^{-1}MI^2 - \nu_f\omega, \qquad (3)$$

where C is the moment of inertia of the disk, ν_f is a friction, and $\mathcal{K}$ is the applied couple. Equations (1) and (3) have the stationary solution

$$I_s^2 = 2\pi\mathcal{K}M^{-1} - \nu_f 4\pi^2 RM^{-2}, \qquad \omega = 2\pi RM^{-1}.$$

Bullard (1955) investigated the system (1)–(3) in detail for $\nu_f = 0$, i.e. he studied the stability of the stationary state. Now it is important to exhibit some general properties of the dynamo.

As Moffatt (1978) first noted, the homopolar dynamo has two features typical of the hydrodynamic dynamo. The first is the different angular velocity of disk and wire which produces the unipolar e.m.f. Inhomogeneous, or differential, rotation proves to be a widespread property of astrophysical bodies be they planetary nuclei, stars or galaxies. The second property is the direction of the twist in the wire. We obtain a positive rate of field growth γ only for the correct choice of direction of wire winding. For the reverse winding $\mathcal{E} = -MI$, and γ is always negative! In this case, exponential growth can be achieved only by reversing the direction of rotation. The twist of the wire is evidently a reflectionally non-invariant feature. In hydrodynamics there are no wires. The presence of a reflectionally non-invariant part of the hydromagnetic dynamo proves, however, also to be important. It can be created by a helical motion of the fluid (see Chapter 8).

II Double Differential Rotation

Differential rotation is only one of the key elements of the homopolar dynamo. (The twist is the other key element.) It is tempting to try to construct a dynamo by employing only the differential rotation. Having a cylinder, disk or a sphere without helicity is evidently insufficient. The differential rotation would create an azimuthal (or, what is the same, a toroidal) magnetic field from the poloidal one (a term originating from the word "pole"). In other words, it would generate an e.m.f., and hence a current with which an azimuthal field is associated. There would be no feedback, however, i.e. there would be nothing to maintain the poloidal field and to implement self-excitation. Let us consider two rotating spheres immersed in a conducting medium. We shall call these "stars", though for real stars the effects considered appear to be problematical (see the end of this section). The two-disk dynamo (with wire helices added) has been studied as a model for geomagnetism (Rikitake, 1958; Allan, 1962; Cook and Roberts, 1970). A system of two rotating cylinders was investigated in the laboratory (Lowes and Wilkinson, 1963, 1968) also as a model for geomagnetism. As is well known, two spheres were used in the original model due to Herzenberg (1958).

The idea is simple. When the rotation axes of stars are parallel it is evident that no qualitatively new field is generated as compared with the case of a single sphere. If the axes are inclined to each other, the azimuthal field generated by the differential rotation of one of the stars is poloidal with respect to the second star (at least partially) and vice versa (Figure 4.2). So there is feedback in a system of two rotating stars. It remains to find the conditions for self-excitation. For simplicity, each star may be assumed to be in solid body rotation. Then, as was the case with the unipolar machine, there is a jump in the angular velocity on the star's surface.

The picture is so transparent that it may easily be understood in the language of lines of force without using mathematics. Let us have

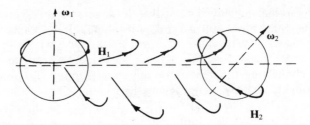

Figure 4.2 Two rotating stars imbedded in a conducting medium. The azimuthal field H_1, generated by the angular velocity jump near the surface of the first star, serves as the poloidal field of the second star. The interaction of this field with the rotation of the second star generates the azimuthal field H_2 which is poloidal for the first star.

at the first star a poloidal field $H_p^{(1)}$ directed, say, along its axis of rotation. From this we can construct the azimuthal field $H_\phi^{(1)}$, which we can "extend" to the second star with the help of diffusion; there $H_\phi^{(1)}$ has (with respect to the second star) a poloidal part $H_p^{(2)}$ which, in like fashion generates an $H_\phi^{(2)}$ that returns to the first star to give $H_p^{(1)}$. A nontrivial matter, however, arises here. The trouble is that the initial field generates an azimuthal field with a certain symmetry. This field changes sign on reflection through the equatorial plane of the first star. Near the second star this field $H_\phi^{(1)}$ has, in addition, a component that does not change sign under reflection. If the initial poloidal field at the first star has an even symmetry, the second star also acquires a field of odd symmetry. It is worthy of note that an even azimuthal field is wound from a poloidal field of even symmetry. Thus the field harmonics of the two symmetries must both be recognized in the argument.

The odd component of the azimuthal field corresponds to the symmetry of a dipole-type poloidal field, and the even to a quadrupole-type field. In the vicinity of each star the odd component is certainly greater than the even one. If the stars were surrounded by vacuum the azimuthal field would vanish outside the stars. In a medium of finite conductivity, the diffusing even harmonic of the first star, whose variation with polar angle is twice as slow as that of the first odd harmonic, decreases with distance more slowly (as r^{-2}) than does the odd harmonic (r^{-3}). Therefore, near the second star these components, being now poloidal, reverse their rôles. The even component that was weaker at the first star generates the stronger dipole poloidal component at the second star.

As a qualitative example we consider the case when the stellar rotation axes are inclined to each other but are perpendicular to the vector connecting the centers of the stars [Moffatt (1978, Section 6.9); for a discussion of the general case, see Gibson (1968) and Roberts (1971)]. The conductivity of the stars and the medium is assumed to be the same and equal to $\sigma = c^2/4\pi\nu_m$. It is convenient to use the dimensionless magnetic Reynolds number $R_m = \omega R^2/\nu_m$, where ω and R are the angular velocity and radius of the stars, assumed to be the same for both. One can rather easily obtain a crude estimate of the critical magnetic Reynolds number at which generation is possible. Let the poloidal field of the first star at some epoch be H_0. Due to the differential rotation (a unipolar effect) this creates the azimuthal field $H_0 R_m$. While diffusing to the second star this field will be reduced to $H_0 R_m (R/d)^3$, then be strengthened by the differential rotation of the second star to $H_0 R_m^2 (R/d)^3$ which on returning to the first star is decreased by a further factor of $(R/d)^3$. The returned field must be equal to, or greater by an order of magnitude than, the initial field H_0. Therefore, $R_{m_{crit}} \sim (d/R)^3$. These arguments, however, cannot guarantee that the constant of proportionality in this relation differs from zero. To clarify the sense and the meaning of the multiplier one must turn to symmetry considerations.

The system under study has, at first sight, reflectional symmetry through any plane. Indeed, let us choose any axes x, y, z and perform the reflections $x \to -x$, $y \to -y$, $z \to -z$ successively. At each stage, the angle between the rotation axes of the stars is the same. For field generation, however, not only the mutual orientation of the angular velocities of the stars is important but also the orientation of the vector **d** that connects them. From the three vectors $\boldsymbol{\omega}_1$, $\boldsymbol{\omega}_2$ and **d**, the pseudoscalar $\mathbf{d} \cdot (\boldsymbol{\omega}_1 \times \boldsymbol{\omega}_2)$ is uniquely defined, and changes its sign on the mirror reflection $\mathbf{r} \to -\mathbf{r}$. Hence the system is mirror asymmetrical! This asymmetry plays a crucial role in magnetic field generation. Indeed, the azimuthal field of the first star is proportional to $\boldsymbol{\omega}_1 \times \mathbf{d}$. At the second star, this makes a contribution to the poloidal field which is proportional to $\boldsymbol{\omega}_2 \cdot (\boldsymbol{\omega}_1 \times \mathbf{d}) = -\mathbf{d} \cdot (\boldsymbol{\omega}_1 \times \boldsymbol{\omega}_2)$. From this it is immediately clear that generation is not possible when the axes of stellar rotation are parallel or when one of these axes is parallel to the line connecting them.

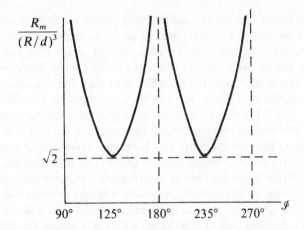

Figure 4.3 The dependence of the critical magnetic Reynolds number on the angle between the angular velocities of the stars. The magnetic field is generated neither in the trivial case $\phi = 180°$ when the axes of the stars are parallel nor in the less obvious case $\phi = 90°$ when the dipole components are ineffective. The system is easiest to excite (with minimal magnetic Reynolds numbers) when $\phi = 125°$ and $235°$. The symmetry $\phi \to 2\pi - \phi$ is evident and corresponds to a mirror reflection along the axis of the first star. It is interesting that due to this symmetry, ω_1 and ω_2 must form an acute angle, i.e. one cannot change the direction of rotation of one of the stars (otherwise the generation would become an accelerated dissipation).

It is natural to assume that there is no essential difference between the two stars. Then for each star the field is the sum of the first two (or maybe higher) harmonics.† The critical magnetic Reynolds number at which the stationary dynamo is possible is given in Figure 4.3. It is assumed that the conductivities of the stars and the surrounding medium are equal and that the stellar radii, R, are small compared with the distance, d, between them. First of all we see that the critical magnetic Reynolds number is large $R_{mc} \sim (R/d)^3$, i.e. the effect of diffusion is weak ($\omega R^2 \gg \nu_m$). It cannot, however, be set to zero! It is

---

† If we reject the condition that the angular velocities are perpendicular to the connecting vector, then an orientation of ω_1 and ω_2 can be found such that, for one of the stars, only the odd component of the azimuthal field is present, while at the other there are both even and odd components.

due to diffusion that the azimuthal field created by one of the stars can reach the other. The smallness of R_m^{-1} has arisen because, in the stationary state, diffusion and dissipation of the stellar field must balance a source proportional to the product of the differential rotation and the weak "tail" field of the other star. On the other hand, as was stressed by Herzenberg (1958), it is only the remoteness of the stars that allows us to ignore the higher harmonics, preserving only the first two. Because of the large magnetic Reynolds number, the magnetic field components which are asymmetric with respect to the axes of rotation cannot penetrate the star, due to the skin-effect. Therefore, the magnetic field is axisymmetric within each star. It is quite clear that, if the magnetic Reynolds number exceeds the critical one, the dynamo will become nonstationary. The growth rate γ, of the field will essentially be determined by the magnetic diffusion: $\gamma \sim \omega R_{cr}^{1/2} R_m^{-3/2}$ (Gailitis, 1973) and will vanish when $R_m \to \infty$. Such a dynamo is called "slow". It may be noted that the flow lines are unknotted, while the magnetic field lines are linked to each other and to the current lines.

It is interesting that the problem can be generalized to the case of three or four or an ensemble of rotating spheres (Roberts, 1971). One may consider this to be a step towards turbulence, a set of rotating vortices. Turbulent vortices are however, moving in space instead of being pinned as in our case. Therefore, there is no direct analogy. It is natural to assume, however, that the pseudo-scalar $\omega_2 \cdot (\mathbf{d} \times \omega_1)$ arising earlier will, when replaced by $\mathbf{v} \cdot (\nabla \times \mathbf{v})$ for the ensemble, play the principal role. If the number of vortices (rotors) with clockwise rotation is unequal to the number of vortices with counterclockwise rotation then the mean over the ensemble will differ from zero and is called "the mean helicity". It is known that this quantity plays a definite part in the generation of mean magnetic field. But, even if $\langle \mathbf{v} \cdot \nabla \times \mathbf{v} \rangle = 0$, it may be possible, under the conditions of Chapter 8, to generate a small-scale magnetic field in a reflectionally-symmetric medium.

Can we apply the dynamo mechanism considered here directly to astrophysics? It is natural to consider two areas of application; (1) double stars and (2) stellar clusters.

For a solar-type star ($R_\odot = 7 \cdot 10^{10}$ cm, $\omega = 3 \cdot 10^{-6}$ s^{-1}, $\nu_m \simeq 10^7$ cm^2/s), the magnetic Reynolds number is $R_m \simeq 10^9$. In a double system of such stars, where the distance between components $d \lesssim 10^3 R_\odot$, we have $R_m > R_{mcr}$, and generation is possible in principle, although the growth rate will be very small, $\gamma^{-1} \approx \omega^{-1} R_m (R_m / R_{cr})^{1/2} \gtrsim 10^7$ years. The transition to the steady state $\gamma = 0$ requires a time of the order d^2 / ν_m, which is greater than the star's age. The situation may be improved by considering a large-scale magnetic field diffusing under the action of turbulence. One model of field generation in close double systems with dwarfs, observed for example in the formation of new stars, is based on this idea (Dolginov and Urpin, 1979). Here it may be estimated that $R \simeq 10^{10}$ cm, $d \simeq 5R$, $\nu_T \simeq 10^{12}$ cm^2/s, $\omega \sim 10^{-3}$ s^{-1}, i.e., $R_m \sim 10^5$, $R_{cr} \sim 10^2$. And then $\gamma^{-1} \sim 1$ year. The actual situation is more complicated: we should take into account the orbital motion, the difference between the diffusion coefficients in the stars and the ambient space, and other factors. From a qualitative point of view, however, the result will be the same.

As an example of a stellar cluster let us consider the nucleus of the Galaxy. In other stellar clusters (open or spherical) the density is small, i.e. the stars are too far from each other to interact magnetically. The mass of the Galactic nucleus is approximately 10^{10} M$_\odot$ and the density near its center is about 10^7 M$_\odot$/(pc)^3 (Oort, 1977), i.e. the mean distance between the stars is of order 10^{-3} pc. The main part of the stellar mass is in the form of dwarfs with radii $R \simeq (1-3)10^{10}$ cm $\sim 10^{-8}$ pc. Thus the critical magnetic Reynolds number is extremely large here: $(d/R)^3 \sim 10^{15}$ and the condition $R_m > R_{m\,cr}$ does not hold. We could try to decrease $R_{m\,cr}$ and increase R_m by replacing the dwarfs with young stars whose radii are significantly greater (up to 10^{14} cm) and which rotate faster than dwarfs (their velocity of rotation may approach 400 km/s). It is known, however, that there are at most a few dozen O-type stars in the Galactic nucleus. Besides, in this case one should take into account the outflow of matter from the surfaces of these stars, and this might change the picture entirely. It would be interesting to see whether generation is possible in the nuclei of quasars where stellar densities are much higher than in the Galaxy. It should be borne in mind that, associated with the increase in the density of stars in a cluster, their relative velocity will grow in compliance with gravitational laws, and the stars may escape from each other before any mechanisms of magnetic field amplification can be brought to bear.

III   The Helical Dynamo

Let us now consider a dynamo whose main element is the helical motion of fluid. A "pancake" helix in which the flow is two-dimensional is totally ineffective as a dynamo (see Chapter 6). The simplest relevant spatial helix is that given by helical flow along a cylindrical surface. The next step is to bend that cylinder into a torus. A more intricate configuration may be obtained by connecting the ends of the cylinder after tying one or two knots. In any case, locally one may imagine the helix as a motion along a line of double curvature (Figure 4.4). We consider in detail the case of a cylinder, in which one of the radii of curvature is infinite.

Flow in a helix has a remarkable property. It is mirror-asymmetric. Let us consider a flow wound onto a cylinder orientated along the z-axis. When reflected in the xy-plane ($z \to -z$) for example, the direction of rotation remains unchanged, but the motion along the axis reverses its direction. This property of oddness may be characterized by the pseudo-scalar $\mathbf{v} \cdot (\mathbf{\nabla} \times \mathbf{v})$.

In cylindrical coordinates (r, φ, z), helical flow has components

$$\mathbf{v} = (0, \omega r, v_z); \tag{4}$$

it cannot, when ω and v_z are constant, generate magnetic field in the nonrelativistic case. It is easy to see that in this case the magnetic field does not change under transformation to a frame of reference moving in the helical manner

$$\phi \to \phi - \omega t, \qquad z \to z - v_z t.$$

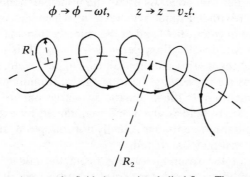

Figure 4.4   The trajectory of a fluid element in a helical flow. The path has a double curvature, R_1^{-1} and R_2^{-1}.

It is impossible, however, to postulate that such a flow exists in all space. If the volume is infinite, the velocity must diminish at infinity. In the case of a finite volume, a change in velocity will occur at the boundary. Due to the resulting velocity gradient, electromotive forces can arise that generate a magnetic field.

The first example of a stationary dynamo with a helical-type velocity field was given by Lortz (1968b). It was a significant theoretical achievement but lacked illustrative interpretation. Ponomarenko (1973) proposed a very simple example of helical flow (4) that acted as a nonstationary dynamo. Here ω and v_z are non-zero constants in $r \leq r_0$, and are zero in $r > r_0$. An attractive simplicity of this example is the fact that, in each region, the field is subject to diffusion only (in $r \leq r_0$ one must first transfer to the helical frame of reference). The whole generation process is concentrated at the tangential discontinuity of the flow at $r = r_0$. The fields in the two regions are related by the usual electrodynamic jump conditions for discontinuity from one medium to another. Suppose, for simplicity, the conductivity is constant everywhere. Then the magnetic field is continuous at the boundary $r = r_0$ but its radial derivative is not. A discontinuity arises because the electric field changes in the transformation from the rotating fluid frame of reference (where $\mathbf{E} = \mathbf{j}/\sigma$) to the stationary frame of reference, where $\mathbf{E} = \sigma^{-1}\mathbf{j} - \mathbf{v} \times \mathbf{H}$. Hence, at the boundary $r = r_0$, the electric current suffers a jump. Accordingly, the magnetic field derivatives suffer like jumps since $\mathbf{j} = (c/4\pi)\nabla \times \mathbf{H}$.

The magnetic field naturally decomposes into cylindrical harmonics $\exp(\gamma t + im\phi + ikz)\mathbf{H}(r)$, where the radial part $\mathbf{H}(r)$ is made up of Bessel (cylinder) functions. As usual, the principal mode corresponds to a solution with maximal symmetry in which the pitches of the helices of the magnetic field and of the velocity coincide in magnitude but are opposite in direction, i.e. $m/k = -v_z/\omega$. This last statement is justifiable because this situation appears to be the most dynamically advantageous for preserving the total (hydrodynamic plus magnetic) angular momentum. In the kinematic dynamo problem, however, i.e. when the velocity field is given and maintained by external forces, the total angular momentum is actually not conserved, and harmonics of opposite twist may be excited.

Diffusion of the (pseudo-) vector magnetic field in a stationary conducting medium is different from the transport of a scalar, for example, temperature or a smoke cloud in air. In axisymmetric geometry, unlike planar geometry, the individual field components cannot in general diffuse independently. In our case ($\nu_m = c^2/4\pi\sigma$)

$$\partial H_z/\partial t = \nu_m \nabla^2 H_z,$$

$$\partial H_r/\partial t = \nu_m[(\nabla^2 - r^{-2})H_r - 2r^{-2}\,\partial H_\phi/\partial\phi], \tag{5}$$

$$\partial H_\phi/\partial t = \nu_m[(\nabla^2 - r^{-2})H_\phi + 2r^{-2}\,\partial H_r/\partial\phi],$$

and only the H_z-component diffuses independently. Nevertheless the helical symmetry of the problem allows a choice of other combinations of components that do diffuse independently. Indeed, let us turn from the cylindrical wave (H_r, H_ϕ) to the left- and right-polarized waves

$$H_\pm = H_r \pm iH_\phi. \tag{6}$$

The components $H_\pm$ will, like H_z, diffuse independently

$$\left[\frac{1}{r}\frac{\partial}{\partial r}\left(r\frac{\partial}{\partial r}\right) - \frac{(m\pm 1)^2}{r^2} - q^2\right]H_\pm = 0, \tag{7}$$

where $q^2 = \nu_m^{-1}\gamma + k^2$. For the sake of simplicity we shall assume that ν_m is constant throughout all space. It should be pointed out that the separation of the field components is possible only in the absence of a velocity gradient. Therefore, at the boundary $r = r_0$, these components prove to be in a tangle.†

We now turn to the discontinuity conditions. As has been noted above, the field (H_z, H_+, H_-) is continuous, and the continuity of the tangential (ϕ, z)-components of the electric field means that the quantities $\nu_m\,\partial H_\phi/\partial r + r_0\omega H_r$ and $\nu_m\,\partial H_z/\partial r - v_z H_r$ are continuous. From the condition $\nabla \cdot \mathbf{H} = 0$ it follows that dH_r/dr is continuous. Therefore, at the interface,

$$(dH_\pm/dr)_{r_0+0} - (dH_\pm/dr)_{r_0-0} = \mp(i\omega/2\nu_m)(H_+ + H_-). \tag{8}$$

Under astrophysical conditions the magnetic Reynolds number is usually very large,

$$R_m \equiv r_0 v_0/\nu_m \gg 1, \qquad v_0^2 = \omega^2 r_0^2 + v_z^2.$$

---

† If the angular velocity and v_z depend on the radius they may be approximated by step functions, i.e. we may consider a nest of mutually enclosing cylinders with helical motions and apply the matching conditions at each boundary. In a similar way one may take into account radially-varying conductivity. These generalizations do not lead, however, to any qualitative changes in the results.

So the solutions of most interest are those with $qr_0 \gg 1$. By solving Eqs. (7) by successive approximations in powers of the small parameter $(qr)^{-1}$, we obtain

$$H_{\pm} = \exp\{im(\phi - \omega z/v_2)\}$$

$$\times \begin{cases} A_{\pm}r^{-1/2}\,e^{qr}(1-[4(m\pm 1)^2-1]/8qr+\cdots), & r \le r_0, \\ B_{\pm}r^{-1/2}\,e^{-qr}(1+[4(m\pm 1)^2-1]/8qr+\cdots), & r \ge r_0. \end{cases}$$

The solution is concentrated near the boundary $r = r_0$ within the characteristic width, $\sim q^{-1}$, of the skin-effect.

Substituting this solution into the boundary conditions (8) we obtain for the maximal growth rate of the solution

$$\gamma = (1+i3^{1/2})v_m r_0^{-2}[\tfrac{1}{2}(m\omega r_0^2/2v_m)^{2/3} - v_m m^2(\omega/v_z)^2 + \cdots]$$

$$= \omega(1+i3^{1/2})R_m^{-1/3}[\tfrac{1}{2}(m^2 v_z/4r_0\omega)^{1/3} - \cdots].$$

The dynamo obtained is slow, $\gamma \to 0$ (like $R_m^{-1/3}$) as $R_m \to \infty$. The helical dynamo is also of interest at small magnetic Reynolds numbers because, of all dynamos known, this one seems to have the lowest threshold for excitation, viz, $(R_{m_c})_{\min} \approx 17$. This may therefore be the easiest to implement under laboratory conditions (Gailitis et al., 1977). Development of the relevant dynamo-experiment is under way at the Institute of Physics of the Latvian Academy of Sciences. As a working medium, they suggest the use of liquid sodium moving in a helical channel. With a total length of apparatus of about 3 m, they expect $R_m \simeq 50$ will be attainable. The design value of the excitation threshold for this apparatus is $R_{m_c} \simeq 24$.

IV   The Rope Dynamo

A very simple example of a dynamo was proposed by Ya. B. Zeldovich (Vainshtein and Zeldovich, 1972). Anybody who wishes can construct it with the usual "string or shoe-lace". The dynamo algorithm consists of making a figure "eight" out of a closed loop of rope, imitating a bundle of magnetic field lines, and then folding it. We then have two

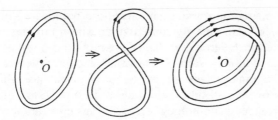

Figure 4.5 The loop transformation procedure. Note that the two resulting loops have similar directions.

loops (Figure 4.5). Let us repeat this act, each time stretching the loops to twice their circumference (which is hardly possible with the shoe-lace but quite allowable for a magnetic field line), at the same time decreasing by half the cross-section of the resulting torus. At each step, then, the magnetic flux doubles in value. By repeating the algorithm n times, amplification by a factor of 2^n is achieved. In other words, we have an exponential growth of the field, $\exp(n \ln 2)$, where the number n of operations is the analogue of the time.

But the simplicity of beautiful things can be deceptive. Look carefully at the details of the process. First of all, the topology of our closed line changes at every doubling. It returns to its initial state first with one turn, then with two, and finally with n turns. Speaking mathematically, the action of the rope dynamo results in the growth of the linkage coefficient between the rope and the axis perpendicular to the plane of the paper passing through the point O. To return the bundle of field lines to its initial topology it is necessary to introduce a reconnection of the lines in a small neighborhood of r, say, with the help of ohmic diffusion. So diffusion proves to be absolutely necessary. Notice the order of the actions: *first* we carry out the doubling procedure and *then* we introduce an arbitrarily small (but finite) diffusivity, but not *vice versa*. This circumstance plays an extremely important rôle in dynamo theory at small diffusivities (large magnetic Reynolds numbers): the limits $t \to \infty$ and $R_m \to \infty$ must be performed in the order indicated.

We also note two other details. At first sight it may seem that only one (toroidal) component of field is intensified by the rope dynamo. During doubling, however, the bundle actually comes out of the plane of the paper, i.e. a poloidal component of the magnetic field appears (and becomes stronger at each successive reappearance). Another peculiarity of the rope dynamo is that a nonstationary motion is responsible for the growth of field. The motion is nonstationary only within one doubling step. On averaging over repeated cycles, it becomes stationary. This is reminiscent of turbulence which is non-stationary over small times and scales, but on average may be kept statistically steady and homogeneous. Let us illustrate the pattern of the motion in the rope dynamo by a simplified one-dimensional description.

Let a closed bundle of field lines initially have the form of a circle (or more exactly, a ring). We consider a particle whose crosswise dimension is equal to that of the bundle.† We shall mark the particle's position by the angle ϕ normalized by dividing by 2π and reckoned positive in (say) the counterclockwise direction $(0 < \phi < 1)$. The operation of turning the ring into a figure "eight" requires that the angles in the third quadrant go to the fourth quadrant and vice versa. Superposition of one half of the figure "eight" onto the other implies a transformation of double extension $\phi_1 = \{2\phi_0\}$, where the symbol $\{\cdot\}$ stands for the fractional part of the number inside the brackets. Therefore, the angular coordinate evolves according to the law

$$\phi_{n+1} = \{2\phi_n\}.$$

The variable $n = 0, 1, 2, \ldots$, marks the steps of doubling and parallels the time. After the completion of each step a re-identification of angles, i.e. a re-assignment of the angles to the interval $(0, 1)$, takes place. The above law means that any two, initially adjacent, points quickly separate with increasing n (check it!). It may be shown that, with the growth of n, the angular coordinates, ϕ_i and ϕ_j, of any two points become statistically independent. Evidently, there are some points which, after a number of steps, return to their original positions, i.e. describe closed periodic center-linked trajectories. But such points correspond only to the rational ϕ and are therefore countable. The remaining nonperiodic open trajectories have the density of a continuum, and the sequence of corresponding mappings is random. It is easy to see that open trajectories are knotted. The stochastic

---

† Due to the smallness of this dimension, as compared with the ring's radius, it is difficult to prevent the ohmic diffusion in the transverse direction and keep the lines of the bundle separate.

behavior of the trajectories ϕ_n is related to the instability. Indeed, for two initially adjacent points separated by the arc $\delta\phi_0 \ll \frac{1}{2}$,

$$\delta\phi_n = 2\delta\phi_{n-1} = \cdots = 2^n\delta\phi_0,$$

i.e. $\delta\phi$ becomes $O(1)$ in the characteristic "time" $\ln |\delta\phi_0|/\ln 2$. During this time the points of the arc $\delta\phi_0$ will fill the whole circumference uniformly. Due to the frozen nature of the magnetic field, the exponential growth

$$H_n = H_0 \exp{(n \ln 2)}$$

occurs with the characteristic time $\tau \approx 0.7 \times$ (the time required to complete one doubling step); $0.7 \approx \ln 2$. If the process operates in a highly conducting fluid (with small ν_m) on the characteristic scale l and velocity v independent of the diffusion coefficient, then we are dealing with a fast dynamo with a characteristic time of $\tau \sim l/v$.

Finally, we should remark that the rope dynamo discussed in this section and illustrated by Fig. 4.5 above has sometimes† been confused with a mechanism proposed by Alfvén and illustrated in Fig. 1 of his paper.‡ Briefly, in the first step of the Alfvén process, a closed flux loop is expanded by fluid motions to twice its original length. Unlike the rope dynamo, the second step does not twist that loop out of its plane: instead, further motions carry opposite ends of a diameter of the loop into close proximity, so that its shape resembles an ∞ sign, though without the crossing at the center of the ∞. The high field gradients at the center and enhanced ohmic diffusion there are, however, favorable for field line reconnection. The ∞ therefore separates into two loops, each resembling the starting loop. As for the rope dynamo, the flux has been doubled. There is, however, an important distinction: as the central rôle of diffusion in Alfvén's mechanism correctly suggests, his is a slow dynamo, in sharp contrast to the rope dynamo, which is fast. Confusion between the two models is unfortunate, therefore, not merely for its historical inaccuracy!

---

† e.g. see Moffatt's article in Volume 2 of this series.
‡ H. Alfvén, *Tellus* **2**, 74–82 (1950).

CHAPTER 5

Conditions for Magnetic Field Generation

I Introduction

In the previous chapter we became acquainted with examples of magnetic field generation by the motion of a conducting medium. The purpose of this chapter is to provide a consistent description of this phenomenon. This description will not require new mathematical techniques—the basic equations of dynamo theory could have been written down in the days of Faraday.

The dynamo process is possible because of the transport of magnetic fields by hydrodynamic motions. The field derives its energy from these motions. Here we shall talk about amplification and maintenance, and not about the creation, of field. The point here is that magnetohydrodynamics is symmetric with respect to the replacement of $\mathbf{H}$ by $-\mathbf{H}$, the velocity and force fields being held fixed. One solution is $\mathbf{H} \equiv \mathbf{0}$, and to generate a magnetic field one must introduce an initial field (or current) or an interaction between charged particles of the plasma. A number of such interactions are known that generate weak seed fields. The amplification (ignoring the back-reaction of the magnetic field on the velocity) or the maintenance of the seed fields is a quite specific process; it must carry on for an infinite time and satisfy proper boundary conditions corresponding to the absence of electromagnetic forces of external (i.e. of non-hydrodynamic) origin.

Dynamo theory was born when the desire arose to understand the nature of the magnetism of celestial bodies (Larmor, 1919). Even now, cosmic applicaẗions are the only realistic ones for the theory. The relatively modest dimensions of laboratory installations do not allow one to attain large magnetic Reynolds numbers. In contrast, in cosmic conditions this number is very large (Chapter 3). It is due to their large spatial scales that the motion of a cosmic medium becomes an essential factor in transforming its magnetic field. It was for the same reason that the concept of a field frozen into the moving plasma, proposed by Kiepenheuer and Alfvén, arose.

The concept of magnetic field lines has proved to be an extremely fruitful one for dynamo theory in particular, and for magnetohydrodynamics in general. The field lines of electrodynamics are of a less general nature because they are not Lorentz-invariant. For example, if two points in space are connected by a field line for one observer, they are not in general so connected for another observer in another frame of reference. But in a plasma where relativistic effects (in particular, the displacement current) can be ignored, the magnetic field does not depend on the reference frame and the concept of magnetic field lines acquires an invariant physical meaning. Here a natural reference frame, the one defined by the plasma, appears. Faraday's ideas about magnetic field lines become real and alive. Applying the field line concept one can not only correctly interpret magnetohydrodynamic solutions, but also make progress in understanding problems which are difficult from a mathematical point of view.

The topological properties of magnetic field lines play a significant part in dynamo theory. It is known that (in the absence of magnetic poles) magnetic field lines unlike those of the electrostatic field must either be closed, go to infinity, or fill everywhere two- or three-dimensional manifolds. Unexpectedly, the presence of linked field lines proves to be important in dynamo theory (Chapter 6). A field line with a simple knot is equivalent to two linked field lines (imagine two linked loops of a chain). The appearance of linked field lines is a typical feature of all known dynamo solutions. The number of such lines per unit volume is proportional to the pseudoscalar $\mathbf{A} \cdot \mathbf{H}$, where $\mathbf{A}$ is a vector-potential of the magnetic field (Woltjer, 1958; Moffatt, 1978). No less important in modern dynamo theory are the topological properties of the velocity field. Formally the magnetic field is analogous to a vortex, therefore in hydrodynamics we can speak of linked vortex ropes or knotted vortex lines. Quantitatively, this property is described by the pseudoscalar $\mathbf{v} \cdot (\nabla \times \mathbf{v})$, i.e. the density of the helicity of the velocity field. Here the term "helicity" is used in a way similar to particle physics (where it is the product of the spin of a particle and its velocity). The total helicity of a fluid volume determines the number of linkages of vortex lines. In the absence of viscosity and other forces (in particular, the magnetic one) and assuming the normal component of velocity at the boundary of the volume is zero, the total helicity does not change in time; it is a topological invariant of the velocity field (Moffatt, 1978). The commonly occurring dynamo motions are of non-zero helicity, and one may consider the helicity to be a fundamental quantity on which the possibility (to be or not to be?!) of a dynamo process hinges. There are, however, motions of conducting fluid over planar (z = constant) or spherical (r = constant) surfaces for which the helicity is not zero although dynamo action is impossible (see Sections II and V). These examples also show that the topology of the trajectories followed by the fluid particles is an essential factor for dynamo regeneration. In other words, in a stationary flow the topological properties of the flow lines seem to be important.

It is quite understandable that theoreticians should attempt to reap a reward by reasonably simplifying a problem. Considerable successes have been achieved in dynamo theory by appealing to one symmetry or another, or by considering one- or two-dimensional approximations, and so on. But even the first researchers into the dynamo problem ran into a striking circumstance—simplification of the problem could decisively change the result. Strict proofs exist of non-existence theorems, e.g., statements that dynamo action is impossible when the magnetic field and the velocity field do not depend on one of the spatial coordinates; for the case of axial symmetry, see Cowling (1934), Backus and Chandrasekar (1956) and Braginsky (1964), and for the case of planar symmetry, see Zeldovich (1956) and Lortz (1968a). These results are true even when none of the three velocity components are zero. Also, the validity of an antidynamo theorem is obvious for spherical geometry for the case when $v_r = 0$ and $\partial/\partial r = 0$.

The situation is more complicated when the velocity field is two-dimensional, i.e. when $v_n = 0$, where n is the normal to a given family of surfaces, but depends on all three coordinates, i.e. when we completely exclude translational symmetry of the type $\partial/\partial z = 0$. It has been proved that dynano action is impossible in the spherical case when $v_r = 0$ (Bullard and Gellman, 1954; Backus, 1958) and in the planar case when $v_z = 0$ (Moffatt, 1978; Zeldovich and Ruzmaikin,

1980); in more general geometries when a conducting fluid moves over a system of stationary surfaces, a dynamo (if at all possible) must be slow (Zeldovich and Ruzmaikin, 1980; Ruzmaikin and Sokoloff, 1980, see Section VII). It is important that a dynamo generated magnetic field cannot be axisymmetric. Hide and Palmer (1982), by a straightforward extension of Cowling's neutral point argument, showed that no axisymmetrical magnetic field can be maintained even by motions of a compressible fluid.

The consideration of such non-existence theorems is useful not only in itself; it also allows us to draw some important positive conclusions. One of these is evident from the above discussion: dynamo processes must be nonsymmetric, in a broad sense. This immediately suggests a possible method for constructing dynamos. Let us assign a state symmetrically and consider small deviations from that symmetry. The dynamo of Braginsky (1964) was constructed in this manner with the inclination of the Earth's magnetic dipole to its rotational axis as a small parameter.

A dynamo process depends essentially on the vectorial (or more exactly, the pseudovectorial) character of the field. A scalar, e.g. temperature or smoke density, passively transported by a fluid will be asymptotically homogeneous or decay as $t \to \infty$ under the action of thermal conductivity or diffusion. It is interesting that the motions can create a temporary (and sometimes very strong) growth of the scalar gradients. As an illustrative example one may imagine fluid particles with different temperatures approaching one another. The point of the non-existence theorems for magnetic field generation is that the field is transported like a scalar in some fluid flows. It is therefore expedient to begin our study with the scalar case, with temperature as an example.

II Transport of a Scalar

We consider the temperature field, T, in a fluid with thermal conductivity $\kappa \geq 0$. For $\kappa \neq 0$ the temperature will be equalized asymptotically $(t \to \infty)$ at all points of the fluid. The temperature will tend to $\bar{T}$ in a bounded isolated volume, or to zero in infinite space if $T \to 0$ at infinity. Indeed, let us consider a point ($r = 0$, say) where the temperature is a maximum. In its vicinity $\nabla^2 \bar{T} \sim [\bar{T} - T(0)] r_\varepsilon^{-2} < 0$, where r_ε is the radius of a small sphere and $\bar{T}$ is the mean temperature over this sphere. Consequently, according to the heat conduction equation,

$$\partial T / \partial t = \kappa \nabla^2 T, \tag{1}$$

the maximum temperature decreases, and similarly the minimal temperature increases, with time.

Now let the fluid move with a velocity $\mathbf{v}(\mathbf{r})$ and be incompressible ($\nabla \cdot \mathbf{v} = 0$). Then on the left-hand side of equation (1) we should write the Lagrangian derivative $d/dt = \partial/\partial t + \boldsymbol{v} \cdot \nabla$, so when $\kappa = 0$ we just have

$$T(\mathbf{r}, t) = T(\mathbf{r}_0, 0), \tag{2}$$

where $\mathbf{r}_0$ is the position of the particle at the instant $t = 0$.

The temperature of each fluid particle is conserved; in other words, the temperature is "frozen" to the fluid. When $\kappa \neq 0$ the Lagrangian approach becomes inconvenient since the diffusion term in (1) is changed in a complex way during the transformation to Lagrangian coordinates. Nonetheless the conclusion about the behavior of the maximum and minimum temperatures remains valid. The heat conductivity, as in a stationary fluid, smoothes the temperature: it lowers the temperature of the hot matter and raises that of the cold. The smoothing effect of the heat conductivity may be described quantitatively with the aid of integral consequences of the equation

$$\partial T / \partial t + \mathbf{v} \cdot \nabla T = \kappa \nabla^2 T. \tag{3}$$

It is evident that in an unbounded medium or in a medium with thermally insulated walls $\partial_n T|_S = 0$, the mean temperature $\bar{T} = v^{-1} \int T d^3 \mathbf{r}$, which is proportional to the total heat energy, is conserved. The temperature inhomogeneity can conveniently be assessed by the value of the integral

$$\int T^2 \, d^3 \mathbf{r} = \bar{T}^2 V + \int (T - \bar{T})^2 \, d^3 \mathbf{r}.$$

One readily sees that this integral is a minimum when the temperature is uniform everywhere, i.e. $T - \bar{T} = 0$. Equation (3) allows us to follow the evolution of this integral. Multiply Eq. (3) by T and integrate over the volume occupied by the fluid taking into account the boundary conditions $v_n = 0$, $\nabla_n T = 0$. We obtain (Zeldovich, 1937)

$$(d/dt) \int T^2 d^3\mathbf{r} = -2\kappa \int (\nabla T)^2 d^3\mathbf{r}. \tag{4}$$

This relation is true regardless of the motion of the fluid.

Thus the temperature inhomogeneities must be smoothed away with time, and the system tend to the state, $T = \bar{T}$. We call this statement a *dissipation theorem*. The integral $\int T^2 d^3\mathbf{r}$ is, in a sense, an entropy but with the opposite sign. Indeed, let $T = \bar{T} + T'$, $\bar{T}' = 0$, $\bar{T} \ll T'$ (a linear theory of heat conduction). Then the entropy is

$$\int \ln T \, d^3\mathbf{r} = \int \ln \bar{T} \, d^3\mathbf{r} + \int \ln (1 + T'/\bar{T}) \, d^3\mathbf{r}$$

$$\approx \text{constant} - \bar{T}^{-2} \int T'^2 \, d^3\mathbf{r}.$$

Hence relation (4) is equivalent to the statement that, in a system free from the action of external forces, the entropy grows due to dissipation ($\kappa \neq 0$).

The dissipation theorem (4) can be extended to the case of open systems. Let the temperature on one part, S_1 of the surface be T_1 and $T_2(<T_1)$ on another part, S_2, where $S_1 + S_2 = S$; $\nabla_n T$ is not given on S_1 and S_2, but elsewhere the surface is thermally insulated ($\nabla_n T = 0$). Then instead of (4) we have

$$(d/dt) \int T^2 d^3\mathbf{r} = -2\kappa \int (\nabla T)^2 d^3\mathbf{r} + 2Q(T_1 - T_2), \tag{5}$$

where $Q = \kappa \int \nabla T \cdot d\mathbf{S}$ is the flux of heat across the surface S_1. Due to energy conservation, the flux across the surface S_2 is $-Q$. The above formula may be interpreted as an expression for the dissipation of external heat sources. Following from it one may show (Zeldovich, 1937) that, in a stationary state ($\partial/\partial t = 0$) and with the geometry, κ, T_1 and T_2 specified, the flux Q is a minimum when the fluid is at rest. The dependence of this flux on the velocity has the form $Q = Q_0 (1 + \text{Pe}^2)$, where $\text{Pe} = lv/\kappa$ is the Péclet number, and l and v are the characteristic scales of length and velocity.

Note that the velocity does not enter into the dissipation theorem, i.e. the fluid motion does not directly influence dissipation. The temperature distribution throughout space, however, and hence $\int (\nabla T)^2 \, dV$ depend on the motion. By transporting the temperature from one place to another, fluid particles can intensify local temperature gradients and thus accelerate the dissipation. This happens most efficiently during intensive turbulent mixing. Considering such a motion as turbulent heat conduction, one can easily obtain an estimate for the mean square temperature gradient in the stationary state. Indeed, by setting $d/dt = 0$, $Q = \kappa_T S_1 \, \Delta T$, $\kappa_T \sim lv$, and $\Delta T = T_1 - T_2$ in (5), we obtain

$$\overline{(\nabla T)^2} = (\kappa_T/\kappa) \, \Delta T = \text{Pe} \, \Delta T. \tag{6}$$

The Péclet number is here similar to the Reynolds number $\text{Re} = lv/\nu$. It is interesting to read this formula from right to left, i.e. to assume that the mean temperature gradient is not maintained but is being determined. Then we find that, in a region of intensive turbulent mixing ($\text{Pe} \gg 1$), the mean temperature gradient is small and $\sim \text{Pe}^{-1/2}$. Turbulent motion causes particles of one temperature T and particles from elsewhere with an essentially different temperature to be present in the same neighborhood. The temperature gradient grows. Only molecular heat conduction can smooth out the temperature inhomogeneities on a small scale and halt this growth. We wish to estimate the temperature contrast $T' = T - \bar{T}$ on small scales. Let $T' = \text{Pe}^\alpha (l \, \Delta T)$ on the scale $\lambda = l \, \text{Pe}^\beta$. According to Kolmogorov $\beta = -\frac{3}{4}$ in a uniform isotropic turbulence ($\text{Pe} \sim \text{Re}$). Then, according to (6), $\alpha = \frac{1}{2} + \beta = -\frac{1}{4}$.

Thus, temperature smoothing in a moving medium is determined not only by the thermal conductivity but depends essentially on the flow. In the turbulent case, an effective turbulent coefficient appears on the scene. In laminar flows the temperature smoothing process can also be very fast. We illustrate this by an example where the velocity field depends linearly on the coordinates

$$v_i = c_{ij} x_j, \qquad (i, j = 1, 2, 3). \tag{7}$$

In general such an expression can be regarded as the first term in the expansion of the velocity field in the neighborhood of $\mathbf{x} = \mathbf{0}$. Now, for the sake of simplicity, we shall assume it to have this form in all space. The constant component in (7) is removed by transforming to a moving frame of reference. We note that, since it is possible to eliminate the constant component, the motion (7) does not violate the assumed homogeneity, i.e. the point $\mathbf{x} = \mathbf{0}$ is no way distinguished. (It is interesting that such a velocity field describes the cosmological recession of galaxies in the Universe first discovered by Hubble.) We assume, in addition, that the fluid is incompressible, i.e. $\nabla \cdot \mathbf{v} = 0$ and, consequently, the trace of the tensor c_{ij} vanishes,

$$c_{ii} \equiv c_{11} + c_{22} + c_{33} = 0. \tag{8}$$

Let us illustrate the role of the velocity field (7) by the example of a potential flow, i.e. $c_{ij} = c_{ji}$. In this case c_{ij} may be reduced to diagonal form, with three eigenvalues connected by the relation $c_1 + c_2 + c_3 = 0$. We shall assume for definiteness that $c_1 > 0$; $c_2, c_3 < 0$. Let us imagine that at the initial time there is a temperature excess in the fluid localized in a spherical region of radius L. Due to the stretching along the x-axis $(v_x = c_1 x)$ and the squeezing along the other two axes $(v_y = -|c_2|y, \ v_z = -|c_3|z)$, the sphere will elongate with time into an ellipsoid. The stretching-squeezing processes are exponential since $dx/dt = c_1 x$, etc. Evidently the squeezing will take place until the transverse dimensions of the ellipsoid reach the value $L\,\text{Pe}^{-1/2}$, i.e. the characteristic scale determined by the heat conductivity, but the stretching along the x-axis will continue. The frozen-in condition is violated and the volume of the ellipsoid will grow exponentially. Since the mean temperature (which is proportional to the total heat energy) is conserved, the temperature in the ellipsoid must be rapidly $[\sim\exp(-c_1 t)]$ equalized to the ambient temperature of the surrounding medium. For a velocity of the form $v_y = cx$ (plane Couette flow), the recession of neighboring particles in the flow will be linear rather than exponential. Therefore the temperature of the distorted ellipsoid will decrease as an algebraic power of time. This case is, however, degenerate; in general, even if there is vorticity $(c_{ij} \neq c_{ji})$, exponential stretching will take place.

The above qualitative picture allows us to choose the correct mathematical form for the solution of the heat conduction equation (3), for the velocity field (7) and (8). It is clear that each harmonic of the temperature, with wave vector $\mathbf{k} = (k_x, k_y, k_z)$ say, will be transformed in time into a harmonic with larger k_y and k_z but smaller k_x. This means a decreasing scale along the y and z axes. Therefore it is natural to seek a solution of the form

$$T(x, y, z, t) = \int T_{\mathbf{k}}(t) \exp[i\boldsymbol{\lambda}(\mathbf{k}, t) \cdot \mathbf{r}] \, d^3 k, \tag{9}$$

where $\boldsymbol{\lambda}(\mathbf{k}, 0) = \mathbf{k}$, and $T_\mathbf{k}(0)$ determines the initial temperature distribution. By substituting (9) and (7) into Eq. (3) and equating the terms involving the zeroth and first power of x_i, we obtain

$$\dot{\lambda}_j + c_{ij}\lambda_i = 0, \qquad \dot{T}_\mathbf{k} - \kappa\lambda^2 T_\mathbf{k} = 0, \tag{10}$$

where a superposed dot stands for differentiation with respect to time. To solve these equations one should find the eigenvalues and eigenvectors of the matrix c_{ik}. In general this matrix has three different eigenvalues, but in special cases two or three eigenvalues may be equal.

Leaving aside the general solution, we consider two important cases: (1) potential flow in which the matrix c_{ik} can be reduced to diagonal form with eigenvalues c_1, c_2, c_3; (2) Couette flow in which the matrix has a single non-zero element, say c_{21}, and hence only two eigenvectors.

In the first case the solution has the form

$$\lambda_i(t, \mathbf{k}) = k_i \exp\,(-c_i t),$$

$$T_\mathbf{k}(t) = T_\mathbf{k}(0) \exp\left\{ -\tfrac{1}{2}\kappa \sum_{i=1}^{3} k_i^2 c_i^{-1}[1 - \exp\,(-2c_i t)] \right\}. \tag{11}$$

For definiteness let $c_1 > 0$; c_2, $c_3 < 0$. Then for $t > (2|c_i|)^{-1}$, each harmonic whose wave vector is not parallel to the x-axis decreases sharply:

$$T_\mathbf{k}(t) \simeq T_\mathbf{k}(0) \exp\left(-\frac{\kappa k_x^2}{2c_1} \right)$$

$$\cdot \exp\left\{ -\frac{\kappa}{2}\left[\frac{k_y^2}{|c_2|} \exp\,(2|c_2|t) + \frac{k_z^2}{|c_3|} \exp\,(2|c_3|t) \right] \right\}. \tag{12}$$

It is easy to see that, for a wave packet localized at the initial instant, the solution (9) will describe an ellipsoidal distribution of temperature with a Gaussian profile over y, z. The length of the ellipsoid will grow exponentially $(\exp c_1 t)$ along the x-axis, and the temperature will decrease, also exponentially, by the same factor.

For Couette flow $(c_{21} \equiv c)$

$$\lambda_x = k_x, \qquad \lambda_y = k_y - ctk_x, \qquad \lambda_z = k_z,$$

$$T_\mathbf{k}(t) = T_\mathbf{k}(0) \exp\left[-\kappa(\tfrac{1}{3}c^2 t^3 k_x^2 - \tfrac{1}{2}ct^2 k_x k_y + t k_z^2) \right]. \tag{13}$$

Evidently, for localized $T_k(0)$, the maximum temperature $T(r, t)$ will decline like $t^{-5/2}$, whereas in the absence of motion it falls as $t^{-3/2}$, as $t \to \infty$.

III Magnetic Field Transport

The consideration of heat conduction in a flow was only a preparatory step to the analysis of magnetic field behavior in a moving, conducting medium. For a given conductivity, the larger the dimensions of the region occupied by the field, the smaller the rôle played by the ohmic resistance. In astrophysics, therefore, a special significance attaches to the approximation of infinite conductivity, when the magnetic field is frozen to the medium (see Chapter 3). This concept allows one to draw a number of conclusions about the behavior of the magnetic field without the use of mathematics.

The frozen-in condition means that in an incompressible fluid the field **H** is transported like the infinitesimal vector separation δx between two neighboring particles. At the initial time ($t = 0$) let the particles on a particular field line be separated by the vector distance δx_0. According to the equations of motion, $x^i = x^i(x_0, t)$, this vector is transformed during time t into the vector $\delta x^i = (\partial x^i / \partial x_0^k) \delta x_0^k$. The magnetic field undergoes a corresponding transformation

$$H^i(\mathbf{x}, t) = H^k(\mathbf{x}_0, 0) \, \partial x^i / \partial x_0^k, \tag{14}$$

where H^i is the contravariant form of the field vector. In this form the law (14) is valid for any curvilinear coordinates. It should be noted, however, that the physical components of the field are $(H_1 H^1)^{1/2}, (H_2 H^2)^{1/2}$ and $(H_3 H^3)^{1/2}$. In orthonormal coordinates this means that the contravariant components are multiplied by the respective Lamé coefficients. Also, for a compressible fluid H^i must be replaced by H^i/ρ, where ρ is the density.

Comparing (14) with the corresponding law (2) for the transport of a scalar, we see that relation (14) is identical to relation (2) for all three components only when the fluid is at rest or in uniform motion. If the velocity is directed along some coordinate axis, x say, and depends on a transverse coordinate [say y], it follows from (14) that the transverse components of the field are conserved while the longitudinal one is proportional to the time $[H_x = H_{ox} + H_{oy}(dv_x/dy)t]$. This result is revealing especially when the frozen-in condition is applied. Since the velocity is directed along the x-axis for any pair of adjacent particles (to which the field line is attached) δy and δz are conserved, and δx changes in proportion to t. Note that the slow growth ($\sim t$) is due to the stationary nature of the velocity. It is also significant that the growth of H_x depends on H_y, but that there is no back reaction of H_x on H_y. So even in this simple example magnetic field amplification can be observed. It is not, however, a dynamo process because when dissipation is taken into account the transverse component of the field ultimately vanishes after which field growth will cease.

We now assume that the fluid flows along the planes $z = $ constant with a velocity that changes from plane to plane, i.e. which depends on all the coordinates. It is clear that the projection, of the infinitesimal vector separating any two adjacent fluid particles onto the normal to these planes, is conserved: $(\delta x^1)_n \equiv \delta z = $ constant. Hence the normal field component H_z is also conserved (Figure 5.1). This conclusion may be generalized to flow over arbitrary stationary surfaces. Certainly, in the general case the projection onto the normal $(\delta x^i)_n$ is not constant. For example, imagine motion over the ellipses $(x^2/4a^2) + (y^2/a^2) = 1$. For two neighboring elements that lie on surfaces a and

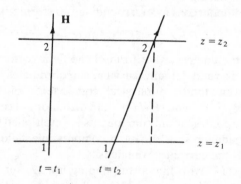

Figure 5.1 Under the translation of the field line frozen to the particles 1 and 2, the component normal to the planes $z = $ constant is preserved. When the velocity in the plane $z = z_1$ is different from that in the plane $z = z_2$, tangential components of field appear, and change as indicated by the figure.

$a + \delta a$, $(\delta x^i)_n = \delta a$ at $x = 0$ and $(\delta x^i)_n = 2\delta a$ at $y = 0$. But, evidently, this is just a geometrical trick and the change in H_n is limited. It may be shown that, for compact surfaces, the tangential component of the field can only grow linearly with time and may experience oscillations (Arnold, 1972). One may think that an exponential solution could be obtained for surfaces of negative curvature, because in that case the trajectories of the fluid elements diverge exponentially. But it is a geometric fact (Efimov, 1964) that such surfaces cannot be embedded into a three-dimensional Euclidean space. Exponential growth of the field is possible in a fluid whose flow paths do not lie on a surface and where there is a space-filling flow line i.e. a line which passes arbitrarily close to every point of a spatial region (see Chapters 6 and 7).

This discussion concerns the ideal case of completely frozen-in fields, but the inclusion of even a very small diffusivity may qualitatively change the result: the frozen-in condition may be violated in the manner described in the previous section. As will be shown in Section V, however, the above restrictions imposed on the topology of the flow have force even when magnetic field diffusion is taken into account. The behavior of field in a turbulent flow is considered in Chapter 8.

IV The Kinematic Dynamo

Let us consider the precise formulation of the dynamo problem. According to Faraday's induction law, the change of the magnetic field is proportional to the curl of the electric field:

$$\partial \mathbf{H}/\partial t = -c \nabla \times \mathbf{E}.$$

In a moving medium, instead of the usual Ohm's law, relating an electric field to the current $\mathbf{j}$, we have

$$\mathbf{E} = \sigma^{-1}\mathbf{j} - c^{-1}\mathbf{v} \times \mathbf{H},$$

where $\mathbf{v}$ is the flow velocity and σ is the conductivity of the fluid. In the co-moving frame of reference Ohm's law is always valid; the additional term $-c^{-1}\mathbf{v} \times \mathbf{H}$ arises from the transformation to the frame of reference in which the fluid is moving. We shall consider only nonrelativistic motions $v \ll c$. In this approximation one may also ignore the displacement currents, i.e.

$$\mathbf{j} = (c/4\pi)\nabla \times \mathbf{H}.$$

From these three relations, known as far back as pre-Maxwellian times, we obtain, after the elimination of $\mathbf{j}$ and $\mathbf{E}$, the basic equation of dynamo theory (the induction equation)

$$\partial \mathbf{H}/\partial t = \nabla \times (\mathbf{v} \times \mathbf{H}) - \nabla \times (\nu_m \nabla \times \mathbf{H}), \qquad (15)$$

which clearly shows that, if the magnetic field was solenoidal initially, it is solenoidal for all time,

$$\nabla \cdot \mathbf{H} = 0.$$

Thus, given the initial and boundary conditions, the evolution of the field is entirely determined by the velocity $\mathbf{v}(\mathbf{r}, t)$ and the magnetic diffusion coefficient

$$\nu_m = c^2/4\pi\sigma.$$

It is remarkable that only one hydrodynamic parameter, the fluid velocity, enters the induction equation while the density, pressure, etc. do not appear explicitly. This allows us to formulate the classical "kinematic dynamo problem" for the evolution of magnetic field for a given fluid motion. In this formulation it is natural to assume that there are no external e.m.f.s, i.e. the magnetic field vanishes at infinity. If the motion is confined to a finite volume which is surrounded by a nonconducting medium ($\sigma \to 0$) then outside the volume

$$\nabla \times \mathbf{B} = \mathbf{0}, \qquad \nabla \cdot \mathbf{B} = 0, \qquad (16)$$

and the field decreases as r^{-3} as $r \to \infty$. If the surroundings are motionless but have finite (non-zero) conductivity, then the spatial part of the field still decreases as r^{-3}, or as r^{-2} if the ohmic decay of the field is included. Solutions of equations (15) and (16) must be matched at the boundary dividing the moving fluid from its stationary surroundings.

Sometimes spatially periodic velocity fields are considered that do not decrease at infinity. In this case it is natural to require, as a boundary condition, that the magnetic field also be limited (for example, be periodic) at infinity.

The aim of the dynamo problem is to answer the question: is an unlimited growth of the magnetic field (or its maintenance) by fluid motions possible in the presence of magnetic diffusion? In the kinematic formulation most attention is paid to the growth of field since stationary solutions (static or oscillatory) are special, and possible only with certain restrictions (having the form of equalities) imposed on the velocity field. The necessity of taking into account magnetic diffusion, even when very small, is important. In completely frozen-in conditions, it is easy to obtain very slow, power law types of magnetic field amplification, for example, through a differential rotation or the Couette flow considered earlier. This growth is, however, replaced asymptotically $(t \to \infty)$ by exponential field decay when $\nu_m \neq 0$. Therefore, flows with an exponential field growth are of interest. Note that, in principle, flows are possible in which the magnetic energy becomes infinite in a finite time (Shukurov *et al.*, 1983). It is clear also that, in any situation where the field grows, the magnetic forces $\mathbf{j} \times \mathbf{H}$ will eventually start to have an effect on the flow, at which time the kinematic formulation should be replaced by a fully dynamic one. Some aspects of this problem will be considered in the context of the turbulent dynamo in Chapter 10.

Note that the induction equation is linear with respect to $\mathbf{H}$, i.e. if $\mathbf{H} = \mathbf{0}$ initially, then $\mathbf{H} = \mathbf{0}$ for all time. We therefore speak in dynamo theory only of the amplification of an initial field. To generate a field one must add to the induction equation an additional e.m.f. independent of $\mathbf{H}$. Usually such e.m.f.s of a thermal or plasma nature are used to explain the appearance of the initial distribution of field. Drobyshevsky (1977) considers models, so-called "semi-dynamos", where the additional e.m.f.s act continuously, in conjunction with hydrodynamic flows, to play a major part in field generation.

V   A Necessary Condition for Dynamo Action

The dynamo process succeeds when the motions amplifying a magnetic field triumph over diffusion and dissipation. It is clear then that the vigor of these motions must be sufficiently great. Estimates were made by Backus (1958); see also Moffatt (1978).

Consider a body, say a star or a galaxy, surrounded by a nonconducting medium, in which the motion of an incompressible fluid is maintained at a velocity **v**. From (15) it is easy to obtain an equation governing the evolution of the total magnetic energy (multiplied by 4π).

$$(d/dt) \int \tfrac{1}{2}H^2 \, d^3\mathbf{r} = \int_V \mathbf{H} \cdot (\mathbf{H} \cdot \nabla)\mathbf{v} \, d^3\mathbf{r} - \int_V \nu_m (\nabla \times \mathbf{H})^2 \, d^3\mathbf{r}. \quad (17)$$

The second integral on the right-hand side of Eq. (17) describes the dissipation of magnetic energy according to the Joule-Lenz law. Evidently this integral is a minimum for the lowest main harmonic which describes the mode of slowest field decay in the absence of motion, i.e.

$$\int \nu_m (\nabla \times \mathbf{H})^2 \, d^3\mathbf{r} \geq \nu_m L^{-2} |\gamma_1| \int H^2 \, d^3\mathbf{r},$$

where L is a characteristic dimension of the region and γ_1 is the dimensionless decay rate of that harmonic.

We estimate the first integral in (17) which alone can be responsible for growth of the magnetic energy. Evidently

$$\mathbf{H} \cdot (\mathbf{H} \cdot \nabla)\mathbf{v} \equiv H_i H_k \nabla_k v_i = \tfrac{1}{2} H_i H_k (\nabla_k v_i + \nabla_i v_k),$$

i.e. only the symmetric part of the velocity shear can regenerate. Thus, first, a velocity constant in space (though not necessarily in time) does not lead to field generation. Second, for a solid body rotation one may also easily obtain $\nabla_k v_i + \nabla_i v_k = 0$. Since

$$\int H_i H_k (\nabla_k v_i + \nabla_i v_k) \, d^3\mathbf{r} \leq |\nabla_k v_i + \nabla_i v_k|_{\max} \int H^2 d^3\mathbf{r},$$

the magnetic energy can grow only if

$$(Lv/\nu_m)(L/l) \geq |\gamma_1|, \quad (18)$$

where l and v are respectively the characteristic length and magnitude of the fluid velocity, where $|\nabla_k v_i|_{\max} = v/l$. The dimensionless magnitude on the left-hand side of inequality (18) is the magnetic Reynolds number and is determined by the ratio of the two terms on the right-hand side of the basic Eq. (15). Usually the magnetic Reynolds number is defined using the maximum scale, say

$$R_m = Lv/\nu_m.$$

Note that condition (18) is satisfied more easily the smaller the scale of the motion, but then the field scale also decreases as a result of field line tangling, i.e. the inequality is correct only at the initial stage. When $l \sim L$, it is reduced to the simple relation $R_m \gtrsim |\gamma_1|$ with the value depending on the geometry. For a spherical region $|\gamma_1| = \pi^2 \approx 10$ and for a thin disk $|\gamma_1| = \pi^2/4 \approx 2.5$. Thus, to initiate magnetic field growth, the magnetic Reynolds number must be sufficiently large. Let us recall (Chapter 3) that in astrophysics R_m is very large, i.e. the criterion is satisfied. From this, however, it does not follow that a dynamo is possible: the criterion is only necessary. In particular, it may be satisfied in simple Couette-type flows in which dynamo action is obviously impossible. The dynamo actually imposes graver demands on the structure of the velocity field than on its magnitude.

VI   Non-Existence Theorems

To understand the requirements imposed on the structure of the velocity field, we consider first flows which do not produce a dynamo whatever the amplitude of the velocity field. Statements distinguishing such flows are called "non-existence theorems."

Consider a plane, incompressible conducting fluid flow, bounded at infinity, and of the general type

$$v_x = v_x(x, y, z, t), \qquad v_y = v_y(x, y, z, t), \qquad v_z = 0,$$

$$\nabla \cdot v = \partial v_x/\partial x + \partial v_y/\partial y = 0, \qquad \nu_m = \text{constant}. \qquad (19)$$

All instantaneous flow lines lie in the planes $z = \text{constant}$, however they change from plane to plane. The initial magnetic field is arbitrary and may depend on all three coordinates.

In Section II we have shown that under the frozen-in condition the z-component of the field is transported independently of all other components. The introduction of magnetic diffusion leads to the decay of this component. Indeed, the corresponding projection of the induction equation is

$$\partial H_z/\partial t + (v_x\,\partial/\partial x + v_y\,\partial/\partial y)H_z = \nu_m \nabla^2 H_z.$$

Thus H_z behaves like a passive scalar (a temperature or smoke density) and obeys the same equation. On integrating this equation after multiplying by H_z, and recognizing that the magnetic field vanishes at infinity, we obtain

$$(\mathrm{d}/\mathrm{d}t)\int H_z^2\,\mathrm{d}^3\mathbf{r} = -\int (\nabla H_z)^2\,\mathrm{d}^3\mathbf{r}. \tag{20}$$

From this it immediately follows that $H_z^2 \to 0$ as $t \to \infty$. This integral consequence of the induction equation we call the dissipation theorem.† It was first derived by Zeldovich in 1956 for the magnetic dynamo problem.

The behavior of the two-dimensional field $\mathbf{H}_2 = (H_x, H_y)$ is more complicated. In contrast to H_z, it may grow due to the tangling of the field lines by the motion over the planes $z = $ constant. Moreover, when the H_z component is a function of z, its transport can also change the two-dimensional field (Figure 5.1). Even if H_z depends only on x, y and not on z at an initial time, the z-dependence will appear if the fluid motion depends on z, i.e. if $\partial v_x/\partial z \neq 0$ and/or $\partial v_y/\partial z \neq 0$. Indeed, if initially $\partial H_z/\partial z = 0$, then even if $\nu_m = 0$

$$\frac{\partial}{\partial t}\frac{\partial H_z}{\partial z} = \frac{\partial v_x}{\partial z}\frac{\partial H_z}{\partial x} + \frac{\partial v_y}{\partial z}\frac{\partial H_z}{\partial y},$$

and is in general non-zero. The two-dimensional vector $\mathbf{H}_2$ may be represented as the sum of solenoidal and potential parts

$$H_x = \frac{\partial\Phi}{\partial y} + \frac{\partial\phi}{\partial x}, \qquad H_y = -\frac{\partial\Phi}{\partial x} + \frac{\partial\phi}{\partial y}.$$

The potential part is completely determined by the z-dependence of H_z since, from the solenoidal condition

$$\partial H_x/\partial x + \partial H_y/\partial y = -\partial H_z/\partial z,$$

it follows

$$\nabla_2^2\phi = -\frac{\partial H_z}{\partial z}, \qquad \nabla_2^2 = \frac{\partial^2}{\partial x^2} + \frac{\partial^2}{\partial y^2}.$$

---

† The theorem can be generalized to the case $\nu_m = \nu_m(z)$.

Taking into account the boundary conditions one can easily integrate this last equation to give ϕ for any time in any layer. In this way we determine the function $\phi(x, y, z, t)$. During the decay of H_z, the gradients, ∇H_z, may temporarily grow. Due to the dissipation theorem (20) this growth accelerates the field dissipation. Therefore ϕ will also decay, and after a sufficient time has elapsed the two-dimensional field will take the form

$$\mathbf{H}_2 = \nabla \times \mathbf{n}\Phi, \qquad \mathbf{n} = (0, 0, 1),$$

i.e. the two components, H_x and H_y, will be described by one variable $\Phi(x, y, z, t)$. From the basic Eq. (15) it immediately follows that this variable is transported like a passive scalar, i.e. it too decays asymptotically. Its derivatives H_x and H_y can temporarily grow due to the decrease in scale, but from the dissipation theorem for Φ it follows that as $t \to \infty$ the two-dimensional field also decays.

For clarity, we will write down an equation for Φ before the decay of H_z and ϕ sets in. To do this we shall take the z-component of the curl of Eq. (15) and make use of the fact that $\nabla_2^2\Phi = -(\nabla \times \mathbf{H})_z$. We obtain

$$\partial\Phi/\partial t + \mathbf{v} \cdot \nabla\Phi = \nu_m \nabla^2\Phi + S, \qquad (21)$$

where the source S is linear in H_z,

$$S = (\mathbf{v} \times \nabla)_z\phi + \nabla_2^{-2}(\partial/\partial z)[\partial(v_x H_z)/\partial y - \partial(v_y H_z)/\partial x].$$

An arbitrary function of z which may appear on the right-hand side of this equation is inessential for calculating the integral Φ^2, because Φ can be chosen so that its mean over every plane $z = $ constant vanishes.

In the above discussion we have followed the proof given by Zeldovich and Ruzmaikin (1980). The non-existence theorem for a planar flow depending only on x and y was first proved by Zeldovich (1956). A similar result for motion over spherical surfaces was obtained earlier by Bullard and Gellman (1954).

It is natural to try to extend the theorem to the case of motion over stationary surfaces of arbitrary form. In Section II we noted that under the frozen-in condition the component of magnetic field normal to such surfaces is bounded. We now prove this by setting $\nu_m = 0$ in the induction equation.

Let a fluid be moving over a system of stationary surfaces $\psi(r) =$ constant. Then the field component normal to these surfaces, $H_\psi = \mathbf{H} \cdot \nabla \psi$, satisfies

$$\mathrm{d}H_\psi/\mathrm{d}t = 0, \qquad \mathrm{d}/\mathrm{d}t = \partial/\partial t + \mathbf{v} \cdot \nabla,$$

i.e. for a fixed particle, H_ψ is conserved. By multiplying this equation by H_ψ and taking into account $H_\psi(\mathbf{v} \cdot \nabla)H_\psi = \frac{1}{2}\nabla \cdot (\mathbf{v}H_\psi^2)$, we obtain

$$(\mathrm{d}/\mathrm{d}t) \int H_\psi^2 \, \mathrm{d}^3\mathbf{r} = \int_S H_\psi^2 \mathbf{v} \cdot \mathrm{d}\mathbf{S}.$$

The integral on the right-hand side vanishes if we require that the normal velocity be equal to zero on a remote surface S or on the boundary of a finite body. From this it follows that, for smooth, exponentially-growing, form-preserving solutions, $H_\psi = 0$.

The degeneracy into lines or points of some individual surfaces $\psi =$ constant does not alter this result if the field and velocity remain smooth. Then a sum of integrals over these singular surfaces of zero area will appear on the right-hand side of the last equation. Each of these integrals is evidently zero. For the remaining components of the magnetic field, which lie on the surfaces $\psi =$ constant, we introduce a vector potential $\mathbf{H}_2 = \nabla \times \mathbf{n}\Phi$ with a single normal component Φ. As in the case considered above, this satisfies the transport equation for a scalar. Thus asymptotic growth of field is impossible.

VII   The Slow Dynamo

The inclusion of the magnetic diffusion qualitatively changes our last conclusion: dynamo action becomes possible. This is a consequence of the fact that diffusive transport of a vector [described by the term $-\nu_m \nabla \times (\nabla \times \mathbf{H})$ in Eq. (15)] differs from the diffusive transport of a scalar ($\kappa \nabla^2 T$). The diffusion of the component H_ψ cannot generally occur independently of the other components. Indeed, from (15) we have

$$\mathrm{d}H_\psi/\mathrm{d}t = \nu_m \nabla^2 H_\psi - \nu_m \nabla \cdot [\nabla_\psi \mathbf{H} + (\mathbf{H} \cdot \nabla)\nabla\psi], \qquad (22)$$

where $\nabla_\psi = \nabla_k \psi \cdot \nabla_k$. In the second summation there are in general, in addition to H_ψ, tangential field components lying in the surface.

The exceptions are, as one may easily see, the planar ($\psi = z$) and spherical ($\psi = r^2$) cases.

Multiplying Eq. (22) by H_ψ and integrating the result over the entire volume using Gauss' theorem, we obtain the generalization of the dissipation theorem (Ruzmaikin and Sokoloff, 1980)

$$(d/dt) \int \tfrac{1}{2} H_\psi^2 \, d^3\mathbf{r}$$

$$= -\nu_m \int [(\nabla H_\psi)^2 + 2H_\psi \nabla_k H_i (\nabla_i \nabla_k \psi) + H_\psi H_k \nabla_i (\nabla_i \nabla_k \psi)] \, d^3\mathbf{r}. \quad (23)$$

The first term in the integral on the right-hand side of (23) makes a purely negative contribution to the field growth rate. The second and the third terms can, in principle, make a positive contribution, their importance depending on how the normal curvatures are distributed over the surfaces $\psi = $ constant. When the tensor $\nabla_i \nabla_k \psi$ is proportional to δ_{ik} or when it is zero, these terms vanish. Evidently, only planes or spherical surfaces satisfy this condition; for motion over cylindrical surfaces $\psi = \tfrac{1}{2}(x^2 + y^2) = \tfrac{1}{2} r^2$, the normal curvature is not isotropic. The tensor $\nabla_i \nabla_k \psi$ is diagonal and has only one component $\nabla_r \nabla_r \psi = 1$, the corresponding contribution to (23) being

$$-2\nu_m \int H_r (\partial H_\phi / \partial \phi) \, d^3\mathbf{r}.$$

It is due to this term, which represents the tangling of r- and ϕ-components, that a dynamo is possible. An example of such a dynamo was described in detail in Chapter 4 (Ponomarenko, 1973; Gailitis and Freiberg, 1976; for a toroidal geometry, see also Tverskoy, 1965). We emphasize here the necessity of motion to generate H_ϕ (and H_z) from H_r.

Since the relevant term is proportional to ν_m it is natural to expect that the growth rate for this kind of dynamo will depend critically on the magnetic viscosity. More precisely, if $H \sim \exp \gamma t$, we expect $\gamma \sim (v/L) f(R_m)$, where R_m is the magnetic Reynolds number. In astrophysics, where R_m is large (Chapter 3), the growth rate is small compared with v/L, so such dynamos are called slow. In the Earth's core, R_m is not large, and so the geodynamo is necessarily slow (Golitsyn, 1981).

If the field scale were the same for $R_m \to \infty$, then evidently $\gamma \sim (v/L) R_m^{-1}$. In reality field evolution leads, as a rule, to a decrease in the characteristic scales of the field. This, in its turn, results in an increase of the growth rate but this still tends to zero as the magnetic Reynolds number increases for fixed L and v. The trouble is that the interaction terms in Eqs. (22) and (23) contain only first derivatives of the transverse field components while the first integral on the right-hand side of (23) involves second derivatives. Therefore the occurrence of skin-layers or ropes of field does not yield growth rates independent of R_m (Ruzmaikin and Sokoloff, 1980). We will explain this using the simple example of motion over cylinders. It may be assumed that, due to a differential rotation, the ϕ-component of the field is generated from the r-component (see Chapter 4), i.e. crudely $H_\phi \sim \Omega \gamma^{-1} H_r$, and $H_\Phi \sim \exp(im\,\phi)$. By an order of magnitude argument we have, from (23),

$$\gamma = C_1 \Omega^2 m / \gamma R_m - C_2 \Omega m^2 / R_m,$$

where C_1 and C_2 are dimensionless constants. The growth rate is maximal when $m = C_1 \Omega / (2C_2 \gamma_{max}) \sim R_m^{1/3}$ and then equals $\gamma_m \sim R_m^{-1/3} \Omega$. The corresponding solution (compare Chapter 4) as $R_m \to \infty$ has the form of a surface wave concentrated in a layer of thickness $\Delta r \sim R_m^{-1/2} L$ near the surface $\psi = \text{constant}$, and with a wavelength along that surface of $\Delta \phi \sim R_m^{-1/3}$.

CHAPTER 6

The Topology of Flows in Brief

The purpose of this chapter is to give some idea of the topological concepts which already permeate modern dynamo theory. The presentation will be restricted to a consideration mainly of knots and linkages, and we shall demonstrate their relationship to the non-existence theorems. The knots and linkages are close to the problem of the construction of non-degenerate surfaces on which the velocity field or the magnetic field may be placed. The covering of a spatial domain by a streamline or a magnetic field line is an alternative to motion over the surfaces. We shall avoid here such topological refinements as the "surface" completing the space.

In the previous chapter we have shown that for a given velocity field the non-existence theorems establish two facts: (1) no generation of magnetic field is possible when the velocity field has planar or spherical symmetry; (2) in a general two-dimensional flow, generation is possible, in principle, though it will necessarily be weak in the sense that the growth rate, γ, will tend to zero as $R_m \to \infty$ (alternatively, for a given velocity and spatial scale, $\gamma \to 0$ as $\nu_m \to 0$).

At first sight, two-dimensional flows seem to be a very narrow class since a small disturbance in the velocity field would seem to be enough to make them three-dimensional. This is not, however, quite the case: the two-dimensional condition, i.e. that the flow lines lie on the surfaces $\psi = $ constant, requires that the equation

$$\mathbf{v} \cdot \nabla \psi = 0, \tag{1}$$

possesses a first integral.†

---

† Clearly Eq. (1) has a solution in a small neighborhood of any point in any smooth velocity field. This means that the dimensionality of the velocity field is a global, and not a local, topological property.

The origin of this condition can be understood by studying the induction equation for a stationary velocity field $\mathbf{v}(\mathbf{r})$ with $\nu_m = 0$

$$\partial\mathbf{H}/\partial t + (\mathbf{v} \cdot \nabla)\mathbf{H} = (\mathbf{H} \cdot \nabla)\mathbf{v}. \tag{2}$$

When one seeks a solution of this first order equation by the method of characteristics he arrives at the equations governing the streamlines:

$$v_x^{-1} \, dx = v_y^{-1} \, dy = v_z^{-1} \, dz. \tag{3}$$

The first integral ψ of this system must satisfy Eq. (1). When (3) has two first integrals $\psi = $ constant and $\eta = $ constant the streamlines are the intersections of these two families of surfaces. It is known that when streamlines lie on a surface, the infinitesimal distance between two neighboring fluid particles can grow only as a power of the time (see, for instance, Arnold, 1972). Hence follows the power law for the frozen-in magnetic field.

When the system (3) has no first integrals, i.e. a streamline fills a spacial domain, the exponential separation of two adjacent fluid particles is possible (Henon, 1966; Arnold, 1972, 1980). This results in the exponential growth of the frozen-in magnetic field.

The situation, however, will change crucially when one takes into account even a small magnetic diffusivity, i.e. when one adds to (2) the term $\nu_m \nabla^2 \mathbf{H}$. For motions over surfaces, exponential growth of magnetic field can arise but only with a growth rate that tends to zero as $R_m^{-1} \to 0$ (the slow dynamo, see Chapter 7). For a stationary motion in which streamlines fill entire spatial domains, a sharp decrease in scale, resulting in a strong increase in the processes of magnetic diffusion, takes place. So the problem of the fast dynamo, whose growth rate tends to a finite limit as $R_m \to \infty$, is not simple. An example of a fast dynamo is the mean field dynamo operating in a non-reflectionaly invariant turbulent flow, for which

$$\alpha \equiv \int \langle \mathbf{v}(t, \mathbf{r}) \cdot \nabla \times \mathbf{v}(t + s, \mathbf{r}) \rangle \, ds \neq 0,$$

(see Chapter 8). At large magnetic Reynolds numbers this flow necessarily possesses two features. First it is non-stationary (only stationary on average). Second, it is topologically complicated. The integral α has a demonstrable topological interpretation (Moffatt, 1969). This is a generalization of the topological invariant that determines the linkage of two closed vortex lines in a statistically stationary continuous fluid.

In this chapter we consider this particular topological property of the flow in detail, and set aside the problem of non-stationarity. In particular, we point out how this property relates to the non-existence theorems. It is nevertheless clear that this invariant, and also some other invariants (say the invariant determining the linkage of stream-lines), are neither sufficient nor necessary for dynamo action. For instance, $\mathbf{v} \cdot (\nabla \times \mathbf{v}) \neq 0$ for the general planar flow $v_z = 0$, $v_{x,y}(x, y, z, t)$ and yet dynamo action is impossible; conversely, some dynamos work in turbulent flows of zero α (see Chapter 8). We believe, however, that some topological compliance of the flow is necessary for any fast dynamo. The problem is to find the appropriate topological interpretation.

I Knots, Linkages and Non-Existence Theorem

We first consider situations in which all the flow lines are closed. Then one can easily understand how to find a three-dimensional flow which essentially differs from the two-dimensional flows. The point is that flow lines which can be placed on two-dimensional, non-degenerate surfaces (i.e. are not lines or points) are not linked to one another. The notion of "linked lines" belongs to topology [see, for example, the book "Topology" by Seifert and Threlfall (1934)]. We call the non-intersecting and non-self-intersecting closed curves C_1 and C_2 "linked" if they cannot be continuously deformed (without cutting) into two circumferences lying in, say, two parallel planes (Figure 6.1). One understands the linkage of many curves similarly (Figure 6.2).

It is remarkable that to obtain linkage one does not require two different curves: one can achieve it with only one curve by tying "a knot". This, the simplest type of linkage, will be studied separately. We shall often illustrate the facts of linkage theory by knots (Crowell and Fox, 1963).

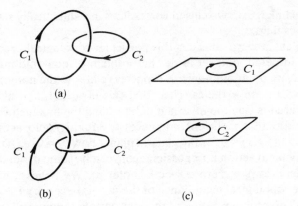

Figure 6.1   Linkage pictures: (a) and (b) show linked curves, in (c) the curves are unlinked.

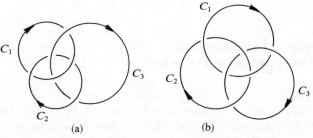

Figure 6.2   Various types of linkages for three curves: In (a) the curves (C_1, C_2) and (C_2, C_3) are linked in pairs. In (b), the curves are not linked in pairs; there is, however, a triple link.

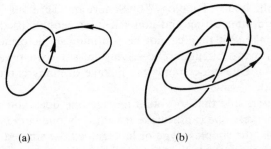

Figure 6.3   Various linkages of two closed circuits: in (a), the coefficient of the linkage is 1; in (b) it is 2.

In Figure 6.2 we have drawn linkages for three circumferences, and as can be seen these may be of several different types. Even two circumferences may be linked in different ways (see Figure 6.3). More precisely, two linkages are called "topologically equivalent" if they can be continuously deformed into each other (without breaking). We note that, from the viewpoint of its own geometry, each of these curves is a normal circumference. Linkages are, however, the subject not of the one-dimensional topology of curves but of the three-dimensional topology of curve positions in space.

In defining equivalent linkages some delicate questions arise as to the use of reflections. It may happen that two linkages can be transformed into each other only by the use of reflections. Such linkages are said to be equivalent but not isotopic. An example of a mirror-asymmetric knot is given in Figures 6.4a and b. The simplest knot is the trifolium. Another simple knot, the figure "eight" (mirror-symmetric), is shown in Figure 6.4c.

It is important to recognise that flow lines are not simply curves, they incorporate a direction. In particular, besides the operation of knotting, or of linkage mirror-reflection, one may consider the operation of reversing the direction in which a circuit round the knot or linkage is described. For a flow line, this corresponds to time reversal. It often happens that the successive operations of spatial reflection and time reversal return the knot to its initial position (see, for example, Figures 6.4a and b). There are also irreversible knots (Figure 6.5). Unlike velocity, magnetic field (and vorticity) is a pseudovector, i.e. it does not change direction under reflection. Therefore the arrow

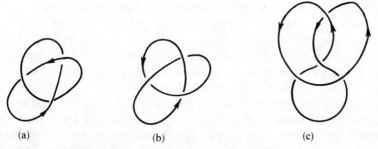

(a)   (b)   (c)

Figure 6.4   A trifolium (a) and its mirror reflection (b)—different (nonisotopic) knots; (c) shows a figure eight, an example of a mirror-symmetric knot.

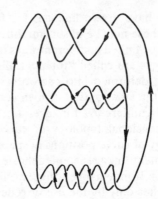

Figure 6.5 An example of an irreversible knot (Trotter, 1963).

indicating the direction of a single field line has only a conventional meaning. It is only needed in order to show the relative directions of two field lines. Note also that the equations governing the magneto-hydrodynamics of perfect conductors are invariant under the simultaneous transformations $\mathbf{r} \to -\mathbf{r}$ and $t \to -t$. But, of course, not all solutions of these equations will have this symmetry.

How can one know that two given knots are equivalent or not? The solution to the first question is rather simple: one knot is deformed into the other. If this cannot be done the second question will have to be answered. This is the central point of the theory of knots and linkages. In topology many invariants, i.e. values and objects that do not change under knot deformation, can be constructed. If these invariants are different for two knots then the knots themselves are topologically different. The invariants are often not numbers at all. For example, the most popular topological knot invariant is a polynomial named "the polynomial of Alexander" (Crowell and Fox, 1963). Such polynomials have already been used in the physics of long molecules (Frank-Kamenetsky and Vologodsky, 1981), but have not yet been applied to magnetohydrodynamics.

Here we restrict ourselves to a consideration of the simplest problem in which we do not know *a priori* what kind of knots the fluid flow lines tie; we are required first to determine how many such linkages exist in a space filled by the liquid. Let us return to Figures 6.3.

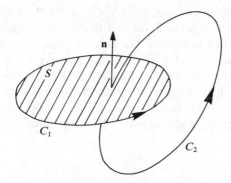

Figure 6.6 Illustration of the notion of the coefficient of linkage: **n** is the normal to the surface S determined by the right-hand screw rule. A negative piercing is shown.

It is seen that the curves are linked only once in Figure 6.3a and twice in Figure 6.3b. How does one determine the number of linkages in the general case? Consider again two linked curves and let us assume that they have no knots (Figure 6.6). Then we can span one of these curves, C_1 say, by a simply-connected surface S. The second curve C_2 pierces the surface S several times. It is clear that the number of piercings depends on the surface S we have chosen to span the contour C_1. If, however, we attach a sign to each piercing, the algebraic sum of the piercings will be independent of the choice of the surface S. In order to ascribe a sign, we recall that the curve C_1 has a direction. Then by the right-hand screw rule we can determine the direction of the normal to the surface S; we consider the piercing to be positive if the vector tangent to the curve C_2 (which also has a direction) points in the same direction as the normal, and negative if it points in the opposite direction. The algebraic sum of piercings is the linkage coefficient. Evidently the linkage coefficient is a topological invariant of linkage, i.e. it cannot change under a continuous deformation of the linkage. Moreover, it does not depend on which curve we consider, the first or the second. The linkage coefficient of two curves can be computed from the Gauss formula

$$G(C_1, C_2) = (4\pi)^{-1} \oint_{C_1} \oint_{C_2} r_{12}^{-3} \mathbf{r}_{12} \cdot (d\mathbf{r}_1 \times d\mathbf{r}_2).$$

If a number of curves, say C_1, C_2, C_3, are linked we can define a total linkage coefficient. To obtain this, one determines separately the coefficients of linkage of the curve C_1 with C_2 and C_3, then of the curve C_2 with C_1 and C_3, then of the curve C_3 with C_1 and C_2. After that, the results are summed and divided by two. Obviously, it is possible for the linkage coefficient to be zero in a case of nontrivial linkage (Figure 6.7). Such a case should be described by a more complex invariant, e.g. the previously mentioned Alexander polynomial.

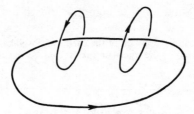

Figure 6.7   A nontrivial linkage with a zero total coefficient of linkage.

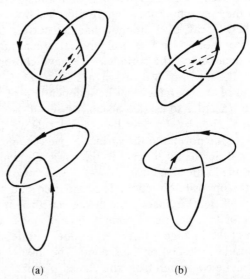

(a)   (b)

Figure 6.8   Calculation of the coefficient of linkage: (a) for a right-handed trifolium (+1); (b) for the left-handed trifolium (−1). The dashed lines complement the knots to the linkage.

One should also take into account the fact that the various curves C_i can be, so to speak, of differing intensity. For example, let the curve C_1 be a tube carrying a flux F_1 and the curve C_2 a tube with flux F_2. Then, naturally, the linkage coefficient (a dimensionless number) should be normalized (divided) by the product $F_1 F_2$. In the case of a knotted curve the calculation of the linkage coefficient is somewhat more delicate. We cannot span this curve by a simply-connected surface. Therefore we have to complement the knot by pairs of oppositely directed segments so that the knot could unravel into a linkage of unknotted curves. How to do this for the simplest knots—a trifolium and a figure eight—is illustrated in Figure 6.8. It should be pointed out that before using Gauss' formula one should first break the knot into linkages.

Now, our problem is to learn how to calculate the coefficients of linkage for flow lines and magnetic field lines. When speaking of flow lines we shall be referring to stationary flows, although flow lines can be defined instantaneously for unsteady motions.

II Linkages and Helicities

The linkages of two curves is a nonlocal concept, so clearly a knowledge of the local value of the velocity or magnetic field does not answer the question of linkages. On the other hand, the idea of linkage is connected with the solenoidal (divergenceless) nature of the field. Otherwise lines would begin somewhere and could be deformed back to that point. The vanishing of the divergence of the field implies the existence of a vector potential, with the aid of which one can determine the linkage coefficient. The vector potential at a point cannot be determined by the field at that point alone; *the field is required at other points too, and in this sense the vector potential is nonlocal.* In addition, it is a gauge-invariant, i.e. we can add any function ∇f to it. In hydro- and electro-dynamics, vector potentials are used as auxiliary variables useful in performing calculations. If, however, we are interested in the topological features of a flow or magnetic field the vector potential becomes quite essential. Without it one cannot decide, for example, whether the flow or field lines are linked or not, and hence one cannot see in an invariant way whether the flow is two- or three-dimensional. Experiments by Aharonov and Bohm (1959) showed that the vector potential has a physical meaning in quantum mechanics. We should like to emphasize the importance of this quantity in hydro- and magnetohydro-dynamics.

Consider an incompressible fluid flow $\nabla \cdot \mathbf{v} = 0.$[†] Then the vector potential of the velocity field is given by

$$\mathbf{v} = \nabla \times \mathbf{a}, \tag{4}$$

from which it follows that $\mathbf{a}$ is determined uniquely up to the addition of the gradient of an arbitrary function ∇f. In the usual gauge transformations $\nabla \cdot \mathbf{a} = 0$, and the vector potential can be extracted from (4) by solving

$$\mathbf{a}(\mathbf{r}) = (4\pi)^{-1} \int |\mathbf{r} - \mathbf{r}'|^{-3} \mathbf{v}(\mathbf{r}') \times (\mathbf{r} - \mathbf{r}') \, d^3\mathbf{r}'. \tag{5}$$

To obtain this relation, it is sufficient to take the curl of (4) and write down the standard solution of the resulting Poisson equation, $\nabla^2 \mathbf{a} = -\nabla \times \mathbf{v}$.

In order to understand what quantity describes the coefficient of linkage of the flow lines one may consider the flow contained in two closed flow tubes (see Figure 6.1). Each flow tube is orientated in the direction of the velocity and has an infinitesimal cross-section. Evidently, the velocity component normal to its surface vanishes; $v_n = 0$. From the continuity condition, $\nabla \cdot \mathbf{v} = 0$, it follows that each flow tube is characterized by a constant intensity $\mathscr{I} = \int \mathbf{v} \cdot d\boldsymbol{\sigma}$, where the integral is taken over the cross-section of the tube.

Consider one of the situations shown in Figures 6.1a, b, and c. According to the definition of the linkage coefficient $\Gamma = 1$ in case (a), $\Gamma = -1$ in case (b) and $\Gamma = 0$ in case (c). Let

$$\chi_v \equiv \Gamma \mathscr{I}_1 \mathscr{I}_2 = \mathscr{I}_1 \int \mathbf{v}_2 \cdot d\mathbf{S},$$

where $\mathscr{I}_1$ and $\mathscr{I}_2$ are the intensities of the first and second tubes respectively, S is the surface spanned by the first tube (Figure 6.6)

† The arguments below may also be applied to a stationary compressible flow, for which $\nabla \cdot (\rho \mathbf{v}) = 0$, ρ being the density. Instead of the velocity one must then consider the linear momentum density $\rho \mathbf{v}$.

and v_2 is the velocity in the second tube. The use of Stokes's theorem and the substitution of the value of the integral $\mathscr{I}_1$ yields

$$\chi_v = \mathscr{I}_1 \int_{C_1} \mathbf{a} \cdot d\mathbf{l} = \int_\sigma \oint_{C_1} v_l a_l \, dl \, d\sigma.$$

We have used the fact that $\mathscr{I}_1$ is constant and that across the tube the component of the potential a_l does not change. This last property follows from (5) and from the fact that the cross-section of the tube is small. Since $dl \, d\sigma = d^3\mathbf{r}$ and $v_l a_l = \mathbf{v} \cdot \mathbf{a}$, it follows that

$$\chi_v = \int \mathbf{v} \cdot \mathbf{a} \, d^3\mathbf{r}. \qquad (6)$$

On calculating the coefficient of linkage of the first tube by the second we obtain the same value. It is evident that for n tubes $\chi_v = \Sigma_{k=1}^n \chi_v^{(k-1)}$. By assuming that the tubes have equal intensities we obtain the coefficient of linkage

$$\Gamma = \chi_v / 2\mathscr{I}^2.$$

It is easy to show that the integral χ_v is a generalization of the Gauss integral G. The integral in (6) is gauge-invariant. Indeed, the substitution $\mathbf{a} \to \mathbf{a} + \nabla f$ adds to χ_v an integral which may be transformed to one over the curved surface of the tube:

$$\int \mathbf{v} \cdot \nabla f \, d^3\mathbf{r} = \int \nabla \cdot (f\mathbf{v}) \, d^3\mathbf{r} = \int f v_n \, dS = 0.$$

It should be pointed out that these arguments can be applied only to flows in a simply-connected volume. For example, if we confine each tube inside a hard sheath, the linkage of such "hard" tubes will make no contribution to the flow helicity χ_v. The formal reason for this is the impossibility of then using Stokes's theorem.

If the flow is defined in a multiply-connected volume then to distinguish the vector potential of the velocity uniquely one has to give, besides the usual gauge transformation, its circulation, $\int \mathbf{a} \cdot d\mathbf{l}$, over the contours which cannot be deformed into a point. The helicity of the flow also depends on this condition. In other words, if a closed flow tube exists, its helicity will depend on the full flow and will evidently be gauge-invariant, because at the boundary $v_n = 0$. We mention here an important feature of flows with closed lines. In this case two flow lines cannot form an arbitrary linkage because they have to bound the strip filled by the flow lines. Here the Gauss integral describes the linkages completely (Frank-Kamenetsky and Vologodsky, 1981).

Strictly speaking the concept of knotted or linked lines refers only to closed lines. Incompressible flow lines, do not, however, have to close despite the divergenceless character of such flows. They can come from, and go to, infinity and it can happen that one flow line will completely fill an entire spatial domain or surface.

Two types of topologically distinct flows are possible. One of them corresponds to motion over a system of stationary surfaces. In fact, these surfaces are topologically equivalent to tori and can be constructed from them. The tori can be enclosed in and linked to each other. In some specific cases tori degenerate into planes (both radii of the torus tending to infinity), cylinders (one of the radii tending to infinity), spheres (one of the radii being equal to the other) and rings or ropes (a way of reducing one of the radii to zero). The flow lines on these surfaces can be either closed or fill the surface everywhere as in the well known example of the close winding of a torus where the ratio of the number of turns of the flow line along one circumference to that along the other is irrational. In another type of flow, the lines completely fill entire spatial domains. Such flows have properties similar to those of turbulent flows, and they are therefore called "stochastic". In the case of inviscid incompressible fluid flows moving under conservative forces Arnold (1966) has shown that only Beltrami flows can be of this type, i.e. flows for which $\nabla \times v \propto v \propto a$ $[v \times (\nabla \times v) = 0]$. In other words, the flow can be stochastic only when the conservative forces vanish. Note that such flows always imply an element of over-idealization since they cannot decline smoothly at infinity. On the other hand, as we shall see later (Chapter 7), it is under the action of stationary flows of this kind that accelerated diffusion of a magnetic field occurs.

One often has to consider problems with spatially periodic or axisymmetric velocity fields. When the knottiness of such flows, which do not vanish at infinity, is calculated there is further arbitrariness in the vector potential of the velocity field. In this case, to determine the flow helicity correctly, one should remember that such flows are an idealization of quasi-periodic or toroidal flows, that do vanish at infinity. Accordingly, when calculating the helicity of the flow one should choose that gauge transformation of the velocity field that is the appropriate limit of a sequence of such realistic flows. This is the method we shall select below. Of course, we can use gauge transformations of the potential of, say, a cylindrical flow which do not correspond to the gauge transformations of the toroidal flow potential.

The topological invariant (6) can also be introduced for an arbitrary stationary incompressible flow. Invariants of this type, as well as those for flows with open lines, preserve the meaning of the linkage coefficient. The essence of the proof, given by Arnold, lies in the construction of short paths closing long open lines and in proving that the result does not depend on the choice of these paths. Roughly speaking, a topological meaning for the invariant (6) for open flow lines can be understood through the example of flow lines winding over a torus. In fact, using the cylindrical coordinates (r, ϕ, z), let

$$\mathbf{v} = v_0[0, J_1(r/r_0), J_0(r/r_0)],$$

where J is the Bessel function of the first kind. For small r and r_0, i.e. near the cylinder's z-axis, $v = \omega r$ and $\omega = v_0/r$. It is easy to see from the relation $\mathbf{v} = \nabla \times \mathbf{a}$ that in this case the potential $\mathbf{a}$ coincides with $r_0 \mathbf{v}$, where r_0 is a fixed length, so that

$$\mathbf{v} \cdot \mathbf{a} = r_0 v_0^2 (J_1^2 + J_0^2)_{r<r_0} \simeq r_0 v_0^2 J_0^2.$$

Let us calculate the number of flow line linkages per unit length along the cylinder's axis. The intensities of the flow tubes of radius $r_\varepsilon \ll r_0$ orientated along the axis and in the azimuthal direction are $F_z = \int v_z \, dS_z$ and $F_\phi = \int v_\phi \, dS_\phi$. The number of linkages in a volume of radius r, is

$$(F_z F_\phi)^{-1} \int \mathbf{v} \cdot \mathbf{a} \, d^3 \mathbf{r} = v_z \omega \pi r^2 / F_z (\tfrac{1}{2} F_\phi r_0^{-1}).$$

Zero helicity of flow does not mean that linkages are absent; the net effect may be zero if they have different signs. If, however, a suitable choice of gauge for the potential $\mathbf{a}$ can reduce the density of the flow helicity in the entire volume to zero

$$\mathbf{v} \cdot \mathbf{a} \equiv 0, \tag{7}$$

then, as is well known, the vector potential $\mathbf{a}$ may be expressed in the form $\mathbf{a} = \eta \nabla \psi$, where η and ψ are coordinate functions, and the velocity $\mathbf{v} = \nabla \eta \times \nabla \psi$ is directed along the lines of intersection of the surfaces $\psi = $ constant and $\eta = $ constant. Hence, condition (7) is equivalent to condition (1) for motion over two-dimensional non-degenerate stationary surfaces. So identity (7) is a topologically invariant condition for two-dimensional flow.

Strictly speaking, a set of the surfaces $\psi = $ constant can degenerate into lines or into stagnation points (where $\mathbf{v} = \mathbf{0}$), but in the dynamo problem these degeneracies prove to be incidental. Note that by a suitable choice of gauge transformation the value of $\mathbf{v} \cdot \mathbf{a}$ can always be made zero in the neighborhood of any given point. In other words, a set of surfaces on which the velocity field lies can always be constructed near that point. Beyond this local region, however, the equality $\mathbf{v} \cdot \mathbf{a} = 0$ does not hold when the flow has linkage. Degenerate lines arise where the potential $\mathbf{a}$ develops peculiarities. For example, in the case of the cylinder just considered, for $r \ll r_0$ one may choose $a_\phi = a_z = 0$ and $a_r = \frac{1}{2} v_0 rz - v_0 r_\phi$, so that $\mathbf{v} \cdot \mathbf{a} = 0$. But then the potential $\mathbf{a}$ is not unique: the line $r = 0$ is a branch line of the potential. Certainly, continuing the calculation formally (by cutting), we find that the number of linkages differs from zero. For the surfaces $\psi = $ constant (in this case $r = $ constant) on which the velocity lies, the peculiarity of the potential corresponds to their degeneracy into lines. Note that on these branch lines the velocity does not vanish. If the `gion where we construct the velocity potential is expanded and only `.inite number of branch lines occur in this expanded region, then the linkages (though present) cannot produce fast generation of the field (see Chapter 5). In a stochastic flow a branch line of the potential can fill a complete spatial domain—the stochastic region—in which the magnetic field behavior is more complicated (Chapter 7).

III Other Helicities

Apart from the flow helicity characterizing flow line linkages, it is often expedient to consider other helicities characterizing other types of linkages. These helicities were considered earlier than the flow helicities. There is a vortex helicity (Moffatt, 1978)

$$\chi_\omega = \int \mathbf{v} \cdot \boldsymbol{\omega} \, d^3\mathbf{r} = \int \mathbf{v} \cdot (\nabla \times \mathbf{v}) \, d^3\mathbf{r},$$

which describes the linkages of vortex lines, i.e. the curves parallel to field $\boldsymbol{\omega} = \nabla \times \mathbf{v}$. Interest in the vortex helicity centres on the fact that a similar quantity determines the generation of the mean field in reflectionally asymmetric turbulence. This generation is character-ized by the pseudoscalar (Chapter 8)

$$\alpha \sim \langle \mathbf{v} \cdot \boldsymbol{\omega} \rangle,$$

where averaging is carried out over the ensemble of turbulent cells. In order to identify α with the number of linkages one has to obey certain rules. The difficulty is that $\int \mathbf{v} \cdot \boldsymbol{\omega} \, d^3\mathbf{r}$ is proportional to the number of linkages of the vortex lines only when the normal com-ponent of the vorticity, ω_n, vanishes at the boundary of the integration volume. Otherwise nothing would prevent us from choosing an integration volume so small that it would contain no linkages whatever (linkages are nonlocal objects!).

It is important that α can be expressed through the mean flow helicity which is determined in the same way as the mean vortex helicity. Note that for a turbulent cell this value

$$\chi_v = \int \mathbf{v}_T \cdot \mathbf{a}_T \, d^3\mathbf{r},$$

is evidently gauge-invariant.

Let us find a connection between the mean flow and vortex helicities. Consider a homogeneous isotropic, reflectionally-asym-metric turbulence. Applying symmetry arguments, we find that the correlation tensor for the vector potential of the velocity field is, in Fourier representation,

$$\langle a_i(\mathbf{k}_1) a_j(\mathbf{k}_2) \rangle = (c\delta_{ij} + dk_ik_j + h\varepsilon_{ijl}k_l)\delta(\mathbf{k}_1 + \mathbf{k}_2); \qquad \mathbf{k} = \mathbf{k}_1,$$

where h is proportional to the helicity $\mathbf{v} \cdot \mathbf{a}$. Using this expression to calculate the correlation tensor of the velocity field, $\mathbf{v} = i\mathbf{k} \times \mathbf{a}$, we obtain

$$\langle v_i v_j \rangle = \{ck^2(\delta_{ij} - k^{-2}k_ik_j) - hk^2 e_{ijl}k_l\}\delta(\mathbf{k}_1 + \mathbf{k}_2).$$

From this we find

$$\langle \mathbf{v} \cdot \boldsymbol{\omega} \rangle = -k^2 \langle \mathbf{v} \cdot \mathbf{a} \rangle.$$

In the coordinate representation, this is

$$\langle \mathbf{v} \cdot \boldsymbol{\omega} \rangle = \nabla^2 \langle \mathbf{a} \cdot \mathbf{v} \rangle. \tag{8}$$

Since $\mathbf{a} \cdot \mathbf{v} = 0$ at the boundary (which is equivalent to the usual condition $v_n = 0$), Eq. (8) establishes a one-to-one correspondence between the two helicities.

In general, $\mathbf{a} \cdot \mathbf{v}$ and $\mathbf{v} \cdot \boldsymbol{\omega}$ are not related so simply. It is easy to give examples where flow lines lie on planes parallel to the xy-plane and depend on z, and for which $\mathbf{v} \cdot \boldsymbol{\omega} = -v_x(\partial v_y/\partial z) + v_y(\partial v_y/\partial z) \neq 0$. Therefore the density of vortex helicity cannot be used to distinguish between two- and three-dimensional flows. The vanishing of the density of the vortex helicity is a stronger restriction on the flow than the vanishing of the flow helicity. In any case, if a flow has no stagnation points ($\mathbf{v} = \mathbf{0}$), it immediately follows from $\mathbf{v} \cdot \boldsymbol{\omega} \equiv 0$ that $\mathbf{v} \cdot \mathbf{a} \equiv 0$.

The helicity χ_v introduced is analogous to the well known magnetic helicity $\int \mathbf{H} \cdot \mathbf{A} \, d^3\mathbf{r} = \chi_H$, where $\mathbf{A}$ is a vector potential for the magnetic field $\mathbf{H}$ (Woltjer, 1958; Moffatt, 1978). This determines the number of linkages of the magnetic field lines, and reflects the well-known analogy between the induction equation and the equation governing the infinitesimal vector separation of two neighboring fluid elements. This last analogy has been earlier related to the frozen field property. It is natural to compare the vortex helicity χ_ω with the helicity $\int \mathbf{H} \cdot (\nabla \times \mathbf{H}) \, d^3\mathbf{r} \sim \int \mathbf{H} \cdot \mathbf{j} \, d^3\mathbf{r}$ describing the linkage of electric currents. In addition, by analogy with the cross-helicity $\chi_{H\omega} = \int \mathbf{H} \cdot \mathbf{v} \, d^3\mathbf{r}$ (a magneto-vortex helicity) which determines the linkages of the magnetic field lines with the vortex lines (Pouquet et al., 1976; Moffatt, 1978), it is natural to introduce

$$\mathcal{T}_{Hv} = \int \mathbf{H} \cdot \mathbf{a} \, d^3\mathbf{r}.$$

Evidently this quantity characterizes the number of magnetic field lines linked to the flow lines $\mathbf{v}$; it is gauge-invariant when the volume is infinite or when $H_n = 0$ at the boundary of a finite fluid volume. In particular, $\mathcal{T}_{Hv}$ is a gauge-invariant for a magnetic flux tube. Note that $\mathcal{T}_{Hv}$ is not a helicity since it is a scalar, as distinct from a pseudoscalar. Another well known scalar is the energy $\int H^2 \, d^3\mathbf{r}$ determining the linkages of the magnetic lines with the electric current lines. In hydrodynamics the linkage of flow lines with vortex lines is determined by the integral $\int v^2 \, d^3\mathbf{r}$. It must be emphasized that this interpretation is suitable only for fluids and fields which occupy finite, simply-connected volumes, or which vanish sufficiently rapidly at infinity. In particular, it cannot be applied to potential flows for which $\boldsymbol{\omega} \equiv 0$, but $\int v^2 \, d^3\mathbf{r} \neq 0$.

Generally speaking, the topological invariants, characterizing the linkages of various lines, need not be integrals of the motion. Nevertheless, it turns out that many such invariants are conserved. For example, the vortex helicity is an integral of the motion for an ideal fluid with $v_n = 0$ at the boundary, the motion being subjected to the action of conservative forces only (Moffatt, 1978). Also conserved are the magnetic helicity $\int \mathbf{H} \cdot \mathbf{A} \, d^3\mathbf{r}$, the cross-helicity $\int \mathbf{H} \cdot \mathbf{v} \, d^3\mathbf{r}$, the energy, etc. The topological invariants χ_v and $\mathscr{T}_{Hv}$ connected with the vector potential of the velocity field are, however, not necessarily conserved. Indeed, let us multiply the Euler equation

$$\partial \mathbf{v}/\partial t = \mathbf{f}, \tag{9}$$

scalarly by $\mathbf{a}$, where $\mathbf{f} = \nabla w - (\mathbf{v} \cdot \nabla)\mathbf{v}$, and ∇w includes all the conservative forces; then, adopting some gauge condition ($\nabla \cdot \mathbf{a} = 0$, say), integrate (9) in the form

$$\partial \mathbf{a}/\partial t = \nabla^{-1} \times \mathbf{f},$$

subsequently multiplying this equation scalarly by $\mathbf{v}$. Afer summing we obtain

$$\frac{d(\mathbf{v} \cdot \mathbf{a})}{dt} = 2\mathbf{a} \cdot (\mathbf{v} \times \boldsymbol{\omega}) + \nabla \cdot \mathbf{v}\{(\mathbf{v} \cdot \mathbf{a})\mathbf{v} + 2(w - \tfrac{1}{2}v^2)\mathbf{a} + \mathbf{a} \cdot \nabla^{-1} \times \mathbf{f}\}.$$

Integrating this over the whole volume occupied by the fluid and applying Gauss' theorem, we obtain

$$\partial \chi_v/\partial t = 2 \int \mathbf{a} \cdot (\mathbf{v} \times \boldsymbol{\omega}) \, d^3\mathbf{r}. \tag{10}$$

The surface integral vanishes when $\mathbf{v} = O(r^{-2})$ as $r \to \infty$, or when $\mathbf{a} \cdot \mathbf{n} = 0$ at the boundary. The integral on the right-hand side of (10) is exactly the number of linkages of the flowlines with the lines of Coriolis force.

A similar calculation shows that the rate of change of the magneto-flux helicity,

$$d\mathcal{T}_{Hv}/dt = \int \mathbf{A} \cdot (\mathbf{v} \times \boldsymbol{\omega})\, d^3\mathbf{r},$$

is determined by the linkages of the magnetic force lines also with the lines of Coriolis force. It is important to note that the magneto-flux helicity is also conserved in stationary two-dimensional flow. In this case

$$d(\mathbf{H} \cdot \mathbf{a})/dt = (\mathbf{H} \cdot \boldsymbol{\nabla})(\mathbf{v} \cdot \mathbf{a}) = \boldsymbol{\nabla} \cdot \{\mathbf{H}(\mathbf{v} \cdot \mathbf{a})\},$$

so that $\mathcal{T}_{Hv}$ is conserved.

CHAPTER 7

Magnetic Fields in Stationary Flows

I The Fast Dynamo

We have already noted several times the important rôle played by the dimensionless magnetic Reynolds number R_m in the evolution of a magnetic field. Since, under astrophysical conditions, this number is very large, a basic theoretical problem is to determine the growth rate and corresponding spatial field distributions in the limit $R_m \to \infty$. It should be pointed out that this limit must be taken after obtaining the solution (first $t \to \infty$, and then $R_m \to \infty$). The field is therefore generally not frozen to the plasma. Due to the decrease of some field scales, diffusion works efficiently in this asymptotic regime.

In Chapters 5–6 we have looked at the rôle of flow topology in the generation of magnetic fields. In the case of two-dimensional flows, only slow dynamos are possible, with growth rates $\gamma = \dot{H}/H$ that tend to zero in the limit $R_m \to \infty$. We cannot explain by such dynamos the relatively quick growth and variation of magnetic fields in most astrophysical objects of large magnetic Reynolds number. Therefore the question of whether fast dynamos exist, in which the growth rate of field tends to a finite positive value as $R_m \to \infty$, becomes of fundamental importance.

The theory gives a positive answer to this question. In Chapter 4 we referred to a simple but very significant example of a fast dynamo, the so-called "rope" dynamo. From the topological point of view, the requisite transformations (the twisting of the magnetic loop into an "eight" and its subsequent folding onto itself) are not trivial. First, they are essentially three-dimensional. Second, a single complete circuit of the closed curve becomes, after folding, a double journey round the loop. Dissipation must be invoked to convert the old curve into two curves topologically equivalent to the new circumference. Also, dissipation is needed to prevent the cross-section of the closed bundle of flow lines from becoming infinitely thin as $t \to \infty$ due to repeated transformations of this kind. All these peculiarities have been discussed in Chapter 4 and we recall them now only to ask the question whether such mechanisms can operate in hydrodynamic flows of conducting fluid. Though we can readily imagine such flows, it is not quite clear *a priori* whether they are stationary, $\mathbf{v} = \mathbf{v}(\mathbf{r})$, and it is important to find an analytical or numerical description of the process that is convenient in practical calculations.

It is worth noting that topological complexity and stochastic regions may be implied by a stationary velocity field described by smooth functions. The behavior of fluid particle trajectories is rather intricate. Two neighboring trajectories diverge with exponential rapidity. Under frozen-in conditions this would lead to an exponential growth in the magnetic field strengths but, simultaneously, an exponential decrease of some field scales occurs, the field being concentrated in ropes (threads) or sheets in which the frozen-in picture is invalid for any conductivity. A picture of spatial intermittency arises. As an example we note that regions of stochastic trajectories (along with regions of laminar flow) occur in the velocity field proposed by Arnold (1980)

$$v_x = A \sin z + C \cos y,$$

$$v_y = B \sin x + A \cos z,$$

$$v_z = C \sin y + B \cos x,$$

as was shown in the numerical experiments of Henon (1966) for the case $A = 3^{1/2}$, $B = 2^{1/2}$, $C = 1$.

In Section III we shall construct an artificial example of a stationary flow which resembles the model of the fast rope dynamo. The velocity field describes a flow that is one of extension in one direction but compression in the other two directions. The growing solution is, however, constructed in a curved Riemannian space. There are no known examples of fast dynamos in a stationary flow in Euclidean space (for slow dynamos, see Roberts, 1970). As we have mentioned in Chapter 6, to obtain a fast dynamo in Euclidean space the flow must apparently be non-stationary (at least at large magnetic Reynolds numbers) and must also have topological complexity.

The important elements of a fast dynamo are, however, already present in a stationary flow. The simplest is the exponential extension of the distance separating neighboring fluid particles. We shall study this in Section II by considering a flow with uniform strain. The other important element, which will be a theme running through this Chapter, is the finiteness of the magnetic diffusivity. The magnetic diffusivity is the main obstacle to the construction of a fast dynamo in a stationary flow. In fact, the magnetic field in a topologically complicated flow is transformed [by the term $(\mathbf{v} \cdot \nabla)\mathbf{H}$ in the induction equation] to small scales where it disappears through the action of ohmic dissipation.

In the last section of this chapter the problem of magnetic flux expulsion from a stationary flow is discussed. This aspect is important for the study of magnetic field distributions at large magnetic Reynolds numbers when ropes and sheets of field appear.

II   Flow with Uniform Strain

The evolution of a magnetic field in a moving conducting medium is governed by

$$\partial\mathbf{H}/\partial t + (\mathbf{v} \cdot \nabla)\mathbf{H} = (\mathbf{H} \cdot \nabla)\mathbf{v} + \nu_m \nabla^2 \mathbf{H}, \tag{1}$$

and is due to the effects of the velocity strain or, more exactly, the symmetric part of the velocity strain tensor (Section 5.V). Consider the simple case of a uniform strain in an incompressible flow,

$$v_i = c_{ik}x_k + v_0, \tag{2}$$

where c_{ik} is a constant traceless matrix. The theory can be generalized to the case in which this matrix depends on time and, yet further, to the case when it is a random function of time. Batchelor et al. (1959) and Novikov (1961) adopted this approach to describe the properties of uniform turbulence at small scales, with Eq. (2) representing the Taylor expansion of the velocity field in the vicinity of a given point.

In this section, we shall restrict attention to the case of a constant matrix c_{ik} (see also the remark at the end of the section).

The second constant term in expansion (2) may be omitted because of the Galilean invariance of the magnetic field in slowly moving media. Furthermore, for the sake of simplicity, we shall assume that c_{ik} is diagonal.[†] We denote its eigenvalues by c_1, c_2, c_3 and for definiteness set $c_1 > 0$. Since the fluid is incompressible it is clear that $c_1 + c_2 + c_3 = 0$. Therefore at least one of the constants c_2 and c_3 is negative. Thus, the fluid elements, initially spherical, will stretch out along the x-axis and shrink along the y and/or z axes, becoming a thread if $c_2, c_3 < 0$, or a "pancake" if either c_2 or c_3 is positive.

Stationary solutions of the induction equation in the velocity field (1) were first considered by Moffatt (1963); two-dimensional flows were studied by Clarke (1965) [see also Section 3.IV of the monograph by Moffatt (1978)]. We shall study in detail the behavior of a magnetic field in the velocity field

$$\mathbf{v} = (c_1 x, c_2 y, c_3 z), \qquad c_1 + c_2 + c_3 = 0. \tag{3}$$

The linearity of the velocity field evidently implies that the infinitesimal vector δx_i will stretch exponentially fast. By virtue of the incompressibility, $c_1 + c_2 + c_3 = 0$, a spherical body of fluid is elongated along one axis (say, the first; $c_1 > 0$), and either is compressed along both the other axes $(c_2, c_3 < 0)$ into a rope, or is squeezed out along two axes $(c_1, c_2 > 0)$ and compressed in the third into a pancake. A degenerate stretching, involving algebraic growth, is possible in the case of unsymmetric matrices, $c_{ik} \neq c_{ki}$. An example is plane Couette flow where only $c_{12} \neq 0$.[‡] Within the context of the kinematic dynamo, however, we are interested only in exponential solutions, so that degenerate cases will be excluded from our discussion.

It is known (Chapter 3) that the evolution of a magnetic line of force embedded in a perfectly conducting fluid resembles that of δx_i, providing the field was initially aligned along this vector (a frozen-in

---

† The exponential extensions of interest arise from the symmetric part of c_{ik}. A purely antisymmetric matrix results in a linear growth of the magnetic field with time.

‡ The diagonalization of this matrix by a transformation to a rotating frame of reference gives exponential stretchings and squeezings. The matrix c_{ik} is, however, then periodic, so that stretching is followed by squeezing, resulting in only a linear growth of the magnetic field.

condition). One might think that this is a solution to our problem, at least in the high conductivity limit of greatest practical interest. For example, when $c_1 > 0$, and c_2, $c_3 < 0$, the H_x-component grows exponentially, the other two field components decaying exponentially. The situation, however, is in reality completely different; even for a very small (but finite) magnetic diffusivity, the conclusion is reversed.

For the purpose of making estimates, it is convenient to Fourier analyse the initial magnetic field. One should distinguish two types of situations:

(a) $c_1 > 0$, $c_3 < c_2 < 0$ (the rope)

In this case the scales along the k_{02} and k_{03} axes decrease exponentially with time with k_{02} and k_{03} evolving as $k_{02} \exp(|c_2|t)$ and $k_{03} \exp(|c_3|t)$. Hence almost all field harmonics, except the $\mathbf{k}_0 = \mathbf{0}$ harmonic and those contained in an exponentially narrow cone parallel to the k_{01}-axis (Figure 7.1a), decay dramatically at a rate proportional to $\exp[-\nu_m k^2(t)]$, i.e. like the exponential of an exponential! A significant contribution to the magnetic field arises just from the harmonics within the cones $k^2(t) = O(1)$. This cone has an elliptical cross-section with semi-axes proportional to $\exp(-|c_2|t)$ and $\exp(-|c_3|t)$. In virtue of the divergencelessness of the field ($\mathbf{H}_k \cdot \mathbf{k} = 0$), the directions of the harmonics with wave vectors within this cone are close to being orthogonal to the cone's axis; more exactly, they form an "orthogonal cone". The projections of the harmonics contained by the cone onto the k_{10}-axis are of order k_{10}/k_{20}, so that they grow in time like $\exp(c_1 t) \times \exp(-|c_2|t)$ where the first factor is due to streching along the k_{10}-axis. The magnetic field in $\mathbf{r}$-space may be estimated to be the product of the amplitude of the growing harmonic in a volume of the cone proportional to $\exp[-(|c_2| + |c_3|)t]$. Hence

$$\mathbf{H}(\mathbf{r}, t) \sim \int \mathbf{H}_k \, d^3\mathbf{k}_0 \sim \exp(-|c_2|t).$$

Thus, magnetic field stretched into a rope parallel to the 1-axis decays asymptotically.

The domain occupied by the magnetic field grows exponentially, however, due to the stretching along the 1-axis. As a result, the total magnetic energy is asymptotically

$$\int H^2 \, d^3\mathbf{r} \sim \exp(c_1 - 2|c_2|)t,$$

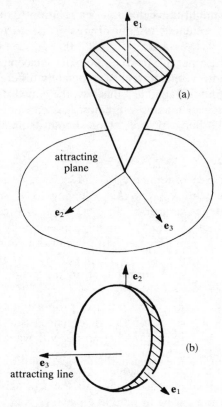

Figure 7.1 The cone of wave vectors corresponding to magnetic field harmonics which do not decay through magnetic diffusion; (a) $c_1>0>c_2>c_3$, (b) $c_1>c_2>0>c_3$.

and increases when $c_1>2|c_2|$. We must emphasize how important the magnetic diffusivity is here. It stabilizes the scales along the second and third axes but does not prevent extension along the first axis.

(b) $c_1>c_2>0$, $c_3<0$ (the pancake)

In this case the cone is constructed from wave vectors close to the (k_{01}, k_{02})-plane; see Figure 7.1b. Its aperture decreases like $\exp(-|c_3|t)$, since the cone is determined by the condition $k^2(t) \sim k_{03}\exp(|c_3|t) \sim O(1)$. It is evident that the harmonic of maximal growth is directed along the k_{01}-axis, and its wave vector is close to the k_{03}-axis. Its amplitude increases as $\exp{(c_1 t)}$, so that the magnetic field in $\mathbf{r}$-space vanishes asymptotically like

$$\mathbf{H}(\mathbf{r}, t) \sim \exp{(c_1 t)} \times \exp{(-|c_3|t)} = \exp{(-c_2 t)}.$$

To estimate the total magnetic energy one should multiply $H^2(\mathbf{r}, t)$ by the volume occupied by the magnetic field. This volume increases as $\exp{(c_1 + c_2)t}$ because the stretching now takes place in the (k_{01}, k_{02})-plane. Again, the total energy is asymptotically

$$\int H^2 \, d^3\mathbf{r} \sim \exp{\{(|c_3| - 2c_2)t\}},$$

and increases when $|c_3| > 2c_2$.

Notice that $c_2 = 0$ corresponds to planar motion: $v_y = 0$. In this case, stabilization of the field with exponentially growing magnetic energy may occur. The result does not contradict the non-existence theorem for the plane three-dimensional motion $v_x(x, y, z)$ and $v_z(x, y, z)$ (see Chapter 5). This theorem, in contrast to (2), implies that the velocity field also vanishes at infinity. In the present case the exponential decrease in the H_y-component is independent of the other components because $v_y = 0$ [see Eq. (1)]. This decaying field provides, however, a non-decaying source for the two-dimensional field (H_x, H_z) due to the exponential growth of the domain occupied by H_y.

Let us now outline the strict mathematical solution (Zeldovich *et al.*, 1983). The problem is solved in the following way. First, particular solutions

$$\mathbf{H}(\mathbf{r}, t) = \mathbf{h}(\mathbf{k}_0, t) \exp{[i\mathbf{k}(\mathbf{k}_0, t) \cdot \mathbf{r}]} \tag{4}$$

of plane wave type, with time dependent amplitudes and wave-number vectors, are studied. Substituting (4) into Eq. (1) and comparing terms of like power in r, we find that

$$d\mathbf{k}/dt = -C\mathbf{k}, \tag{5}$$

where C is the matrix c_{ik} (in general non-diagonal and time-dependent), and

$$d\mathbf{h}/dt + \nu_m k^2 \mathbf{h} = C\mathbf{h}. \tag{6}$$

The divergencelessness of the field (4) demands that

$$\mathbf{h} \cdot \mathbf{k} = 0,$$

which according to Eqs. (5) and (6) holds for all time.

Let $T_t \equiv T(t_0, t)$ be the fundamental matrix of equation (5) in the interval (t_0, t). It is evident that $T(t_0, t_0) = E \equiv \delta_{ij}$. The matrix T_t can be written symbolically in the form

$$T_t = \prod_0^t (E - C^*(u)\, du). \tag{7}$$

Since $\operatorname{tr} C^* = \operatorname{tr} C = 0$, we have $\det T_t = 1$. The solution of Eqs. (5) and (6) is easily found in terms of the matrices T_t

$$\mathbf{k}(\mathbf{k}_0, t) = T_t \mathbf{k}_0,$$

$$\mathbf{h}(\mathbf{k}_0, t) = (T_t^*)^{-1} \mathbf{h}_0 \exp\left[-\nu_m \int_0^t k^2(\mathbf{k}_0, s)\, ds \right], \tag{8}$$

where $\mathbf{h}_0 = \mathbf{h}(\mathbf{k}_0, 0)$ is the Fourier spectrum of the initial magnetic field.

Thus problem (1) has been solved by the plane wave solutions (4) and (8). At this point we note that

$$\mathbf{H}_0(\mathbf{r}) = \int \mathbf{h}_0\, e^{i\mathbf{k}_0 \cdot \mathbf{r}}\, d^3\mathbf{k}_0.$$

The linear character of the problem implies that

$$\mathbf{H}(\mathbf{r}, t) = \int \mathbf{h}(\mathbf{k}_0, t)\, e^{i\mathbf{k}(\mathbf{k}_0, t) \cdot \mathbf{r}}\, d^3\mathbf{k}_0,$$

or, after substituting (8),

$$\mathbf{H}(\mathbf{r}, t) = \int \exp\left[i(T_t \mathbf{k}_0 \cdot \mathbf{r}) - \nu_m \int_0^t (T_s \mathbf{k}_0)^2\, ds \right] (T_t^*)^{-1} \mathbf{h}_0\, d^3\mathbf{k}_0. \tag{9}$$

By making the change of variable $\mathbf{k} \to T_t \mathbf{k}$, using Parseval's theorem, and then applying the inverse transformation, we obtain an expression for the magnetic energy, namely $(1/8\pi)$ times

$$\int H^2\, d^3\mathbf{r} = \int \exp\left[-2\nu_m \int_0^t (T_s \mathbf{k}_0)^2\, ds \right] [(T_t^*)^{-1} \mathbf{h}_0]^2\, d^3\mathbf{k}_0. \tag{9'}$$

The customary asymptotic analysis of (9) leads to the conclusions described earlier (Zeldovich *et al.*, 1983). We note now an additional point. In the simplest case (3) the solution (9) takes the form

$$H_x(\mathbf{k}, t) = H_x(\mathbf{k}, 0) \exp\left\{c_1 t + \frac{\nu_m}{2}\left[\frac{k_x^2}{c_1}(e^{-2c_1 t} - 1)\right.\right.$$
$$\left.\left. + \frac{k_y^2}{c_2}(e^{-2c_2 t} - 1) + \frac{k_z^2}{c_3}(e^{-2c_3 t} - 1)\right]\right\},$$

with similar forms for H_y and H_z. Let $c_3 < c_2 < 0$, $c_1 > 0$. In the asymptotic phase, the field first grows at the rate c_1 independently of the magnetic viscosity. The characteristic duration of the phase is

$$t_* \simeq \tfrac{1}{2}|c_3|^{-1} \ln\left(|c_3|/\nu_m k_z^2\right) \equiv \tfrac{1}{2}|c_3|^{-1} \ln R_m,$$

i.e. the maximal growth of $H_x(\mathbf{k}, t)$ occurs when the scales in the z-direction are large. As $t \to \infty$, only the constant magnetic field will grow exponentially and the x-dependence with also vanish exponentially. It is easily seen that at $t \sim t_*$

$$H_*^2/H_0^2 \sim R_m^{c_1/|c_3|},$$

where, obviously, $1 \le c_1/|c_3| \le 2$.

It appears that the results obtained for this simple case of a constant diagonal matrix C are crudely valid when $C_\xi(t)$ is the matrix for a stationary random process satisfying the natural demands of ergodicity. A simple example is $C_\xi(t)$ constant during time intervals Δt, during which it takes the independent values $C_0, C_1, \ldots, C_n$.

The reduction of the stochastic case to the simple one is based on the Furstenberg–Tutubalin theorem (see Zeldovich *et al.*, 1983) which states that there exists a basis (e_1, e_2, e_3) and non-random constants $\gamma_1 > \gamma_2 > \gamma_3$ where $\gamma_1 + \gamma_2 + \gamma_3 = 0$ (the so-called Lyapunov indices) such that, asymptotically for large t,

$$(T_t \mathbf{k})^2 \sim k_1^2 e^{2\gamma_1 t} + k_2^2 e^{2\gamma_2 t} + k_3^2 e^{2\gamma_3 t},$$

where k_1, k_2, k_3 are the coordinates of $\mathbf{k}$ in the basis. The basis (e_1, e_2, e_3) depends on the complete path of the process $C_\xi(t)$ and cannot be defined at the initial moment $t = 0$.

The main idea of such reductions is very simple. Consider the deformations of a spherical domain into an ellipsoid by matrices with unit determinants (as for the flow of an incompressible fluid). Then

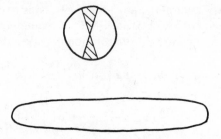

Figure 7.2   When a sphere is stretched into an ellipsoid by an unimodular matrix, the measure in the squeezing directions (shaded regions) is less than one-half.

the measure of points of the sphere which are stretching is greater than one-half (Figure 7.2) so that, under a random rotation of the stretching–squeezing directions, the effects of stretching will predominate. As a result the initial sphere will be stretched in some direction. Thus, by considering the problem (2) with a random matrix $C_\xi(t)$ in the basis (e_1, e_2, e_3) we reduce it to the earlier case of a constant diagonal matrix $C = (-\gamma_1, -\gamma_2, -\gamma_3)$. Hence the main conclusions of this section are valid also for the stochastic case; viz:

The magnetic field decays exponentially in time [as $\exp(-\gamma_2 t)$] and uniformly in $\mathbf{r}$. When $H_0 = O(r^{-3})$ as $r \to \infty$ and $\gamma_2 \neq 0$ the total magnetic energy $\int H^2 \, d^3 \mathbf{r}$ grows exponentially.

In two-dimensional stochastic flows ($\gamma_2 = 0$), which correspond to reversible stochastic processes, the magnetic field decays through random corrections $\eta (\sim t^{1/2})$ to the growth rate [compare this with the non-random case (a)].

The spatial distribution of the magnetic field when subject to a uniform strain (2) in the velocity field is ultimately a set of thin ropes and (or) sheets. The mean field of such an intermittent distribution is, evidently, considerably less than in the field concentrations. For instance, in the case of ropes, $\langle H \rangle \sim R_m^{-1} Hn$, where n is the number of ropes in the volume.

The spectrum of the magnetic field is anisotropic. The magnetic ropes are associated in k-space with the pancake configurations, and sheets are associated with k-ropes.

Note that a linear approximation to the velocity field such as (2) is incapable of treating the turbulent dynamo problem properly. To deal with this it is necessary to consider non-local properties of the velocity field (for instance, a finite correlation length).

III   Straining Motion in a Three-Dimensional Compact Manifold

The flow with uniform strain discussed in the previous sections has a major defect: it is spatially infinite. Here we shall consider an artificial straining motion confined in a compact domain (Arnold, 1972; Arnold *et al.*, 1981). The flow is stationary and simulates the key property of stochastic flow—the exponentially increasing separation of adjacent fluid particles.

The behavior of the magnetic field for a given stationary flow of an incompressible conducting fluid is governed by the induction equation

$$\partial \mathbf{H}/\partial t + (\mathbf{v} \cdot \nabla)\mathbf{H} = (\mathbf{H} \cdot \nabla)\mathbf{v} + R_m^{-1} \nabla^2 \mathbf{H},$$
$$\nabla \cdot \mathbf{v} = 0, \qquad \nabla \cdot \mathbf{H} = 0. \tag{10}$$

The flow region is a three-dimensional compact manifold constructed in Cartesian coordinates as the product of the two-dimensional torus $\{(xy) \bmod 1\}$ and a segment $0 \le z \le 1$ with end points identified according to the law

$$(x, y, 0) \equiv (2x + y, x + y, 1).$$

In this manifold one can introduce as Riemannian metric the metric of $\mathbb{R}^3$, invariant with respect to the transformations

$$(x, y, z) \to (x + 1, y, z),$$
$$(x, y, z) \to (x, y + 1, z), \tag{11}$$
$$(x, y, z) \to (2x + y, x + y, z + 1).$$

The last transformation is accomplished by the matrix

$$A = \begin{pmatrix} 2 & 1 \\ 1 & 1 \end{pmatrix},$$

which has eigenvalues

$$\lambda_{1,2} = \tfrac{1}{2}(3 \pm 5^{1/2}),$$

$$\lambda_1 \lambda_2 = 1, \qquad \lambda_1 \simeq 2.62 > 1, \qquad \lambda_2 \simeq 0.38 < 1. \tag{12}$$

The transition from the coordinates (x, y, z) to the coordinates (p, q, z), where p is directed along the eigenvector with $\lambda_2 < 1$ and q along the eigenvector with $\lambda_1 > 1$, transforms the metric into

$$ds^2 = e^{-2\mu z}\, dp^2 + e^{2\mu z}\, dq^2 + dz^2, \qquad \mu = \ln \lambda_1 \simeq 0.96, \tag{13}$$

which is invariant with respect to the transformations (11), and hence determines an analytic Riemannian structure throughout this three-dimensional compact manifold. In this Riemannian space we consider the stationary velocity field

$$\mathbf{v} = (0, 0, v), \tag{14}$$

where $v = $ constant, so that $\boldsymbol{\nabla} \cdot \mathbf{v} = 0$ and $\boldsymbol{\nabla} \times \mathbf{v} = \mathbf{0}$. In this motion the interval separating adjacent fluid elements is exponentially extended in the q-direction and compressed in the p-direction.

In the space with metric (13), the differential operators are

$$\boldsymbol{\nabla} = (e^{\mu z}\, \partial/\partial p,\ e^{-\mu z}\, \partial/\partial q,\ \partial/\partial z),$$

$$\boldsymbol{\nabla} \cdot \mathbf{H} = e^{\mu z}\, \partial H_p/\partial p + e^{-\mu z}\, \partial H_q/\partial q + \partial H_z/\partial z,$$

$$(\boldsymbol{\nabla} \times \mathbf{H})_p = e^{-\mu z}[\partial H_z/\partial q - \partial(e^{\mu z} H_q)/\partial z],$$

$$(\boldsymbol{\nabla} \times \mathbf{H})_q = e^{\mu z}[\partial(e^{-\mu z} H_p)/\partial z - \partial H_z/\partial p],$$

$$(\boldsymbol{\nabla} \times \mathbf{H})_z = e^{\mu z}\, \partial H_q/\partial p - e^{-\mu z}\, \partial H_p/\partial q,$$

$$\nabla^2 = e^{2\mu z}\, \partial^2/\partial p^2 + e^{-2\mu z}\, \partial^2/\partial q^2 + \partial^2/\partial z^2;$$

$\nabla^2 \mathbf{H}$ is calculated as $-\boldsymbol{\nabla} \times \boldsymbol{\nabla} \times \mathbf{H}$.

For velocity (14), the projection of equation (10) onto the directions $\exp\{-\mu z\}\boldsymbol{\nabla}_p$, $\exp\{\mu z\}\boldsymbol{\nabla}_q$ and $\boldsymbol{\nabla}_z$ yields

$$\frac{\partial H_p}{\partial t} + v\frac{\partial H_p}{\partial z} = -\mu v H_p + R_m^{-1}\left[(\nabla^2 - \mu^2)H_p - 2\mu\, e^{\mu z}\frac{\partial H_z}{\partial p}\right],$$

$$\frac{\partial H_q}{\partial t} + v\frac{\partial H_q}{\partial z} = \mu v H_q + R_m^{-1}\left[(\nabla^2 - \mu^2)H_q + 2\mu\, e^{-\mu z}\frac{\partial H_z}{\partial q}\right], \tag{15}$$

$$\frac{\partial H_z}{\partial t} + v\frac{\partial H_z}{\partial z} = R_m^{-1}\left(\nabla^2 - 2\mu\frac{\partial}{\partial z}\right)H_z.$$

The equation for the z-component of the field does not involve the other two components, so that asymptotically H_z attenuates as $t \to \infty$. Indeed, we can multiply the last equation by H_z and integrate over the volume enclosed between the planes $z = 0$ and $z = 1$. Since the integrals $\oint H_z^2 \, dp \, dq$ coincide in these planes, we have

$$(d/dt) \int H_z^2 \, dp \, dq \, dz = -2R_m^{-1} \int (\nabla H_z)^2 \, dp \, dq \, dz.$$

The negative sign on the right-hand side confirms our claim. In consequence we may now assume, for the sake of simplicity, that the H_z-component of the field is zero. The equations for the p and q-components differ only in the sign of μ ($\mu \to -\mu$), so it is sufficient to consider just the q-component which we denote by H with $\mu > 0$. Then

$$\partial H/\partial t + v \, \partial H/\partial z = \mu v H + R_m^{-1} (\nabla^2 - \mu^2) H. \qquad (16)$$

Let us formulate the boundary conditions. The easiest way to do this is to revert to the initial coordinates system (x, y, z). The symmetry of (11) means that the function H is periodic in x, y, and can be written as

$$H(x, y, z, t) = \sum_{n,m} H_{n,m}(z, t) \exp [2\pi i (nx + my)]$$

$$= \sum_{\alpha,\beta} h_{\alpha,\beta}(z, t) \exp [i(\alpha p + \beta q)], \qquad (17)$$

where n, m are integers; α, β are connected with $2\pi n$ and $2\pi m$ through the linear transformation relating (x, y) to (p, q) (which is equivalent to the rotation of the Cartesian axes x, y through an angle $\tan^{-1}(2 - \lambda_2) \simeq 58°$). The symmetry property with respect to translation along the z-axis, $H(x, y, z, t) = H(2x + y, x + y, z + 1, t)$, allows us to impose restrictions on the Fourier-amplitudes. By substituting this relation into (17), we obtain

$$H_n(z + 1) = H_{A'n}(z),$$

where $\mathbf{n} = (n, m)$ and A' is the transpose of the matrix A. In the case under consideration $A' = A$ and the translation along z is equivalent to the transformation from Fourier-components with indices $(n, m) = \mathbf{n}$ to those with indices $A\mathbf{n}$. The case $n = m = 0$ (when the magnetic field does not depend on x, y or p, q) is exceptional; then $A\mathbf{0} = \mathbf{0}$. In

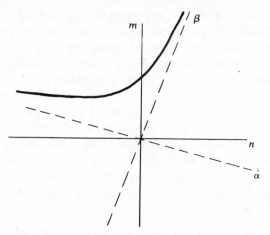

Figure 7.3 The dashed axes indicate the directions of eigenvectors of the matrix A with the eigenvalues $\lambda_1 > 1 > \lambda_2$. Since $\lambda_1 \lambda_2 = 1$, the product mn is preserved under the action of A on the vector with components n, m, i.e. a translation along a hyperbola occurs. Such a hyperbola is shown in the figure.

general the application of the matrix A to the vector **n** moves the point (n, m) along a hyperbola in the (n, m)-plane (Figure 7.3).

If the function $H(x, y, z, t)$ is analytic then its Fourier components $h_{\alpha,\beta}(z, t)$ must decrease exponentially with increasing α and β. Then according to (11) the function $h_{\alpha,\beta}(z, t)$ must decrease not more slowly than a double exponential. If the function is k times differentiable then it will decrease as a polynomial.

Thus, the solutions of Eq. (16) must be periodic in x and y. If the solution does not depend on p and q then $H(z, t)$ must be periodic in z. When there is a dependence on p or q, the Fourier components $h_{\alpha,\beta}(z, t)$ must quickly decrease with increasing $|z|$.

We consider first the case $R_m^{-1} = 0$. In the Lagrangian frame of reference, the solution of the Cauchy problem is immediate. Returning to Euler coordinates we have

$$H(p, q, z, t) = e^{\mu v t} H(p, q, z - vt, 0). \qquad (18)$$

When the initial field does not depend on p and q, the solution is represented by a superposition of the eigenfunctions $\exp(2\pi imz)$ with eigenvalues

$$\gamma_m = \mu v - 2\pi imv, \qquad m = 0, \pm 1, \pm 2, \ldots.$$

This can easily be seen by expressing (18) as a Fourier series in z. If the initial field does depend on p and q (being a periodic function of x, y) then, as noted above, the Fourier components of the expansion in p and q must diminish with increasing $|z|$. Hence the above set of functions does not describe a solution satisfying the boundary conditions, a situation which arises, in fact, because the operator of the translation along the z-axis in (18) has a continuous spectrum.

Let us consider the more general case, $R_m^{-1} \neq 0$. As before, Eq. (16) possesses solutions, periodic in z and independent of p and q, with eigenvalues

$$\gamma_m = \mu v + R_m^{-1}(4\pi^2 m^2 - \mu^2) - 2\pi i m v. \tag{19}$$

When the initial field depends on p and q, the nature of the solution is quite different from (18). The translation $z - vt$ along the z-axis is equivalent to a displacement (along a hyperbola) of the harmonic numbers of $h_{\alpha\beta}(z, t)$ at fixed z. Therefore, any given harmonic eventually moves into the region of large wave numbers where dissipation becomes crucial. Asymptotically, as $t \to \infty$, it decays, regardless of R_m. Let us describe this process.

When we seek solutions having the form (17), i.e. $h_{\alpha\beta}(z, t) \exp i(\alpha p + \beta q)$, Eq. (16) gives us

$$\frac{\partial h_{\alpha\beta}}{\partial t} + v \frac{\partial h_{\alpha\beta}}{\partial z} + R_m^{-1}[\mu^2 - R_m \mu v + \alpha^2 e^{2\mu z} + \beta^2 e^{-2\mu z}] h_{\alpha\beta} = R_m^{-1} \frac{\partial^2 h_{\alpha\beta}}{\partial z^2}. \tag{20}$$

It is natural to assume that as $R_m^{-1} \to 0$ the terms containing exponents will play the central rôle. So, we consider an abbreviated version of the equation, namely

$$\partial h_{\alpha\beta}/\partial t + v \, \partial h_{\alpha\beta}/\partial z + [R_m^{-1}(\alpha^2 e^{2\mu z} + \beta^2 e^{-2\mu z}) - \mu v] h_{\alpha\beta} = 0.$$

Its two first integrals are

$$I_1 = z - vt,$$

$$I_2 = h_{\alpha\beta}(z, t) \exp[-\mu z + (R_m^{-1}/2\mu v)(\alpha^2 e^{2\mu z} - \beta^2 e^{-2\mu z})].$$

With their aid it is easy to construct the solution of the Cauchy problem for the initial field $h_{\alpha\beta}(z, 0)$:

$$h_{\alpha\beta}(z, t) = h_{\alpha\beta}(z - vt, 0) \exp\{\mu vt - (R_m^{-1}/2\mu v)$$

$$\times [\alpha^2 e^{2\mu z}(1 - e^{-2\mu vt}) - \beta^2 e^{-2\mu z}(1 - e^{2\mu vt})]\}$$

$$\simeq h_{\alpha\beta}(z - vt, 0) \exp[\mu vt + (R_m^{-1}/2\mu v)\beta^2 e^{-2\mu(z - vt)}].$$

We see that each $\alpha\beta$-harmonic, $h_{\alpha\beta}(z, 0)$, of the initial (z-limited) field $H(z, 0)$ grows at first in proportion to $\exp(\mu v t)$, then as it shifts to the right along the z-axis with the velocity v it sharply attenuates, like $\exp\{-(R_m^{-1}/2\mu v)\beta^2 \exp[-2\mu(z - vt)]\}$, in a characteristic time of

$$t_* \simeq v^{-1}z + (2\mu v)^{-1} \ln (R_m\beta^{-2}).$$

The field scale in z begins to change quickly [at about the same time but with $\ln (R_m^{-1}\beta^2)^{-1}$ replaced by $\ln (R_m^{-1}\mu^2)^{-1}$], i.e. the use of the truncated equation then becomes unreasonable.

To find the asymptotic solution as $t \to \infty$ we switch from (20) to the Schrödinger-type equation

$$d^2\psi/dz^2 + (k^2 - U)\psi = 0,$$

where

$$\psi = H \exp\left(-\tfrac{1}{2}vR_mz - \gamma t\right),$$

$$k^2 = -\gamma R_m - \tfrac{1}{4}v^2 R_m^2 + \mu v R_m - \mu^2,$$

$$U(z) = \alpha^2 e^{2\mu z} + \beta^2 e^{-2\mu z} = 2|\alpha\beta| \cosh (2\mu Z),$$

$$Z = z + \mu^{-1} \ln (|\alpha/\beta|^{1/2}).$$

The potential U has a minimum $U_{\min} = 2|\alpha\beta|$ at the point $Z = 0$, and increases rapidly in an exponential manner on both sides of the minimum. We may crudely estimate that the lowest "energy" level is of order $U_{\min}$. This gives

$$\gamma \simeq -\tfrac{1}{4}v^2 R_m + \mu v - \mu^2 R_m^{-1} - U_{\min} \xrightarrow[\varepsilon \to 0]{} -\tfrac{1}{4}v^2 R_m.$$

One may show that as $z \to \infty$, the corresponding eigenfunction is of the form

$$h_{\alpha\beta}(z, t) \simeq \exp\left[(-\tfrac{1}{4}v^2 R_m + \mu v)t + \tfrac{1}{2}vR_mz - 2|\alpha\beta| \cosh Z\right].$$

Thus we conclude that, asymptotically as $t \to \infty$, only the solution independent of p and q survives.

Note the essentially three-dimensional nature of this problem. The stretching takes place in the p, q-plane, and the flow velocity is directed along the z-axis. It is this translation along the z-axis that provides the amplification factor $\exp(\mu v t)$ for a field independent of p and q. On the other hand, the same translation along z produces the continuous spectrum which results in the dramatic damping of the harmonics that depend on p or q.

A numerical study of a realistic velocity field in Euclidean space (Section I) made by E. Korkina (unpublished) shows that the growth rate $\gamma(R_m)$ behaves similarly, but becomes positive in a certain finite interval of R_m.

IV Magnetic Flux Expulsion

A rather common feature of moving conducting fluids is the nonuniform distribution of magnetic flux within it; the field concentrates in separate spatial regions. This shows up even in flows of simple topology, for example in a differential rotation $\omega_z(x, y)$, and is a general characteristic of stationary flows with closed streamlines. This subject is discussed in detail in the monograph by Moffatt (1978), some additional aspects, associated with the redistribution of the magnetic fields in compressible fluids and with turbulent diamagnetism, being considered by Parker (1979) and Vainshtein *et al.* (1980). So we discuss the basic idea only briefly, and mention new results.

Zeldovich (1956) was the first to call attention to the expulsion of magnetic flux from two-dimensional flows. Of great importance are the subsequent solutions obtained by Weiss (1966), which demonstrate this phenomenon in both a single vortex and in a grid of vortices. Flux expulsion from a region of two-dimensional flow arises from the fact that a stationary magnetic field must ultimately vanish on the streamlines. Indeed, let (Moffatt, 1978)

$$\mathbf{v} = (\partial\psi/\partial y, -\partial\psi/\partial x, 0),$$

where $\psi(x, y)$ is the streamfunction. In a stationary state the induction equation (10) reduces to

$$\mathbf{v}\nabla \cdot \mathbf{A} = R_m^{-1}\nabla^2\mathbf{A}.$$

By integrating this equation over the area bounded by a closed flow line $\psi = $ constant, we find that in an incompressible fluid the left-hand side vanishes while the right-hand side is proportional to the product of the velocity circulation and $\partial\mathbf{A}/\partial\psi$. It follows that $\mathbf{A}$ must be constant, i.e. $\mathbf{H} = \mathbf{0}$.

On the other hand, from Chapter 5 we know that a spatially limited magnetic field in a two-dimensional flow vanishes asymptotically everywhere. Therefore, the existence of stationary magnetic configurations is due to special boundary conditions that allow nonzero magnetic fields at infinity. In the stationary state the field is in fact always concentrated into ropes, or into sheets whose thickness is proportional to $R_m^{-1/2}$ as in the simplest flow of Section II.

Spatial intermittency seems to be the reason for the temporary growth of the magnetic energy in a two-dimensional flow. The trouble is that initially the formation of ropes or sheets results in a strong increase in the gradient of the vector potential (i.e. in the field). Then, due to the growth of field gradients, the field dissipates (under the zero boundary condition at infinity) or tends to a steady state (if the fluid is allowed to extend to infinity in some directions).

An interesting result was obtained by Childress (1979) who calculated the asymptotic $(R_m \to \infty)$ behavior of the helicity in two-dimensional periodic flows; see also Anufriev and Fishman (1982). Childress divided the field into mean (constant) and fluctuating parts $\mathbf{H} = \langle \mathbf{H} \rangle + \mathbf{H}'$, and evaluated the helicity as the symmetric matrix α_{ij} in the relation

$$\alpha_{ij}\langle \mathbf{H}_j \rangle = \langle \mathbf{v} \times \mathbf{H} \rangle_i.$$

By taking advantage of delicate boundary layer techniques, Childress showed that, in the case of plane periodic flow $\psi = \sin kx \sin ly$, the helicity decreases as $R_m^{-1/2}$ as the magnetic Reynolds number increases, and a stationary magnetic field becomes concentrated near the boundaries. Consider the axisymmetric velocity field (in cylindrical coordinates z, r, ϕ)

$$\mathbf{v} = (r^{-1}\,\partial\psi/\partial r, \; -r^{-1}\,\partial\psi/\partial z, \; w(\psi)),$$

where $w(\psi)$ is an arbitrary function of ψ, $\psi = 0$ at the boundary and

$$-(\nabla^2 - 2r^{-1}\,\partial/\partial r)\psi > 0, \qquad 0 < z < 1, \qquad 0 < r < r_0.$$

The helicity (and accordingly the mean e.m.f.) proves to be different from zero in the limit $R_m \to \infty$, and the strongest magnetic flux concentration occurs near the axis of the cylinder. This result indicates once more the topological difference between plane and cylindrical flows (see Chapters 5 and 6).

Childress considered an axisymmetric magnetic field for which dynamo action is, *a priori*, impossible. It is interesting to invoke a dependence on the angle ϕ and to construct a class of dynamo-solutions for the above velocity field. In accordance with the results of Chapter 5 these must be slow (i.e. have growth rates that tend to zero as $R_m \to \infty$) or simply be stationary.

Today, much effort is expended in investigating, in the context of solar and stellar convection, the magnetic field behaviors of stationary convective lattices of three-dimensional flows. Drobyshevsky and Yufreev (1974) recognized the topological asymmetry of stationary thermal convection. Usually fluid rises at the centers of cells and falls at their peripheries, so that the ascending fluid elements, unlike the descending ones, are disconnected from one another. This produces a peculiar valve effect (topological pumping) allowing downward

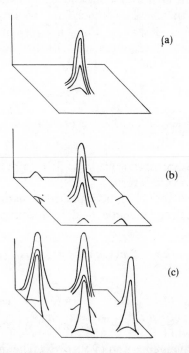

(a)

(b)

(c)

Figure 7.4 Perspective plots of steady vertical field strength ($R_m \geq 200$) (a poor schematic representation of the fine computer drawings by Galloway and Proctor, 1981); (a) $z = 0$, (b) $z = 0.5$, (c) $z = 1$.

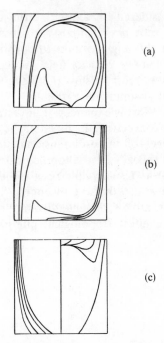

(a)

(b)

(c)

Figure 7.5 Field line plots in vertical planes for steady solution: (a) through the cell center and a side mid-point, (b) through cell center and a vertex, (c) across a side (after Galloway & Proctor, 1981).

transport of mean horizontal magnetic field to the bottom of a cell but impeding its upward return.

Galloway and Proctor (1981) investigated numerically the evolution of an initially vertical magnetic field in a convective lattice with hexagonal cells

$$\mathbf{v} = \{u(x, y) \cos \pi z, v(x, y) \cos \pi z, w(x, y) \sin \pi z\},$$

where

$$u = (3\lambda^2/4\pi) \, \partial w/\partial x, \qquad v = (3\lambda^2/4\pi) \, \partial w/\partial y,$$

$$w = \tfrac{1}{3} \cos (2\pi x/3^{1/2}\lambda) + \tfrac{2}{3} \cos (\pi x/3^{1/2}\lambda) \cos (\pi y/\lambda).$$

This flow has no divergence and $(\nabla \times \mathbf{v})_z = 0$. The induction equation (10) was integrated for this velocity field for a sequence of magnetic Reynolds numbers: $R_m = 20, 40, 80, 120, 200$ and 400. At small R_m

no topological changes in the field lines were observed and the results were similar to those for two-dimensional axisymmetric flows. Changes appeared for $R_m \geq 80$. Figure 7.4 shows the distribution of the vertical component of magnetic field in the final steady state at three different levels $z = 0$, 0.5, 1 and for $R_m = 200$. At the bottom of the cell H_z is strongest in the center; at the top of the cell the situation is different—the magnetic flux is concentrated into the six angles of the hexagon. Magnetic field architecture makes a strong aesthetic impression!

Figure 7.5, also based on their work, shows the final steady state magnetic field lines for $R_m = 200$. Not illustrated but displayed in their paper is the prior evolution of the field, which develops a neutral point within the cell which gradually moves towards the boundary: reconnections of field lines take place resulting in flux redistribution and concentration.

CHAPTER 8
Kinematic Turbulent Dynamos

Cosmic magnetic fields are, as a rule, evolutionary in a turbulent medium (see Chapter 3). A turbulent dynamo is therefore of major interest in astrophysical applications. It is natural to assume that in a well-mixed, stationary turbulence the magnetic fields become tangled and of small scale. How can one explain, then, the prevalence of large-scale astrophysical magnetic fields (see Chapter 2)?

Cosmic turbulence has two special features uncharacteristic of laboratory turbulence. First, it is usually stratified, i.e. a density gradient $\nabla\rho$ is present; second, it occurs in rotating bodies (planets, stars, galaxies). This results in a violation of the mirror-symmetry of the mean statistical parameters of the turbulence. Even from dimensional arguments, it is clear that a pseudoscalar, namely $\boldsymbol{\Omega} \cdot \nabla\rho$ where $\boldsymbol{\Omega}$ is the angular velocity, can be defined which breaks the parity. To clarify what violation of reflection symmetry $(\mathbf{r} \to -\mathbf{r})$ means, imagine an ensemble of screws, such that the numbers of screws with opposite directions of thread (say, left-hand and right-hand) are unequal. On being reflected in a mirror, a left-hand screw transforms into a right-hand screw and vice versa. Since, therefore, the number of screws with one thread becomes on reflection the number of screws with the other thread, similarly turbulence that on average is left-handed will on reflection become on average right-handed, i.e. the flow will possess helicity.

The idea that the mean magnetic field, in a randomly moving medium whose flow is not mirror-symmetric, could grow, was first proposed by Parker (1955). Parker considered a set of screw-like vortices similar to cyclones in the Earth's atmosphere. Mirror symmetry is violated and a mean helicity arises when the number of right-handed cyclones is unequal to the number of left-handed cyclones. It was, however, a rather special model that did not capture many theorists' attention. Non-mirror-invariant mean velocity fields as generators of large-scale magnetic fields also appeared in well-known papers by Braginsky (1964). Only after the work of Steenbeck, Krause and Rädler (1966), however, has it become clear how general and how important the violation of mirror symmetry in random motions is for the generation of mean magnetic field. The works by Parker and Braginsky were only forewarnings of a storm that was to break in the theory of the turbulent dynamo and its applications following the work of Steenbeck and his colleagues. An excellent review of the theory of mean magnetic field generation by a mirror asymmetric flow is given in the monographs of Moffatt (1978) and Krause and Rädler (1980). These authors use a two-scale approach: the velocity and the magnetic field are decomposed into mean (large-scale) and fluctuating (small-scale) components. For the mean magnetic field one can obtain with this approach a closed linear differential equation whose coefficients are determined by the mean velocity and the mean statistical parameters of the fluctuating velocity components. In a homogeneous isotropic turbulence these parameters describe two main effects: a turbulent diffusion of the mean magnetic field and a mean helicity. Inhomogeneity in the turbulence results in a special "turbulent diamagnetism" first recorded by Zeldovich (1956).

The theory of mean magnetic fields in this approach excels in simplicity and physical clarity, and it has found many useful applications in astrophysics. But the penalty for this is an almost total ignorance of the fluctuating magnetic fields. Nevertheless, at large Reynolds numbers, these fields exceed the mean fields, and play a very important rôle in turbulence dynamics. Theorists are also still anxious about the classical question (Batchelor, 1950) of whether or not small-scale magnetic fields can be generated in the absence of a mean helicity and a mean magnetic field. Therefore we will depart from the traditional exposition of the subject, and try to conduct the discussion in a way that does not leave the small-scale fields in the background. Our purpose, however, is to clarify the basic physical features of the theory, and so we shall focus our attention on illustrative aspects. This will explain the first section below where peculiarities in the evolution of a magnetic field in a random medium are discussed.

In this chapter we shall restrict ourselves to kinematic models. Some nonlinear aspects of turbulent dynamo theory are discussed in the next chapter.

I   Random Walk of Fluid Particles

When dissipation is neglected, the magnetic field behavior is similar to that of the vector connecting two neighboring (infinitesimally separated) particles in a fluid flow (Chapter 3). Therefore, the evolution of the magnetic field in a random turbulent motion is reduced to understanding the behavior of a wandering pair of fluid particles.

It is very easy to calculate the basic statistical properties of fluid particles in a flow when it is modeled by the generalization to the continuous case of the discrete process of random walk, in which each step is independent of the past history. In probability theory such a process is called "Markovian". The main quantitative feature of the flow which we shall use is its short memory time, or in other words the smallness of the characteristic correlation time, t_c, when compared with other characteristic times arising in the problem. If $\Delta t \gg t_c$, the displacement, $\Delta X_i = X_i(t + \Delta t) - X_i(t)$, of a particle moving along a trajectory is uncorrelated to the particle's position at time t. Therefore if we divide the time axis into intervals of length Δt, then the particle displacements during these time intervals will be statistically independent, i.e. $X_i(t)$ will be a random variable governed by a Gaussian distribution. It is well known that due to the diffusion of the particle the standard deviation $\langle X_i^2(t) \rangle$ will grow linearly in time. We are interested in the behavior of a pair of particles moving along neighboring trajectories.

Let us assume that the flow is incompressible ($\nabla_k v_k = 0$), homogeneous and isotropic. If, in addition, we suppose that it is mirror-symmetric, i.e. it does not change under the transformation $X_i \to -X_i$, then, as it is well known from hydrodynamics (Batchelor, 1953; Monin and Yaglom, 1967), the correlation of velocities at the points X_1 and X_2 depends only on their vector separation, $r = X_1 - X_2$, and is uniquely determined by an even function $f(r)$. More precisely,

$$\langle v_i(X_1, t)v_j(X_2, t') \rangle = \tfrac{2}{3}v_0^2[f(r)\delta_{ij} + \tfrac{1}{2}r(\delta_{ij} - r_i r_j r^{-2}) \, df/dr]\psi(t - t'),$$

$$(1)$$

in which we see the constant factor v_0^2, which is the mean square of the velocity fluctuations at $r = 0$. The general properties of the function $f(r)$ are well known. Near the point $r = 0$

$$f(r) \simeq 1 - r^2/2\lambda^2, \qquad \lambda \equiv [-f''_{(0)}]^{1/2}. \qquad (1a)$$

In the case of short correlation times considered here, the function $\psi(t-t')$ may be approximated by $\tau\delta(t-t')$. In other words, the velocity field is a limit of the velocities $v^\Delta(\mathbf{x}, t) \sim (\tau/\Delta t)^{1/2}$ as $\Delta t \to 0$. Here the characteristic correlation time of the velocity field,

$$\tau = v_0^{-2} \int_0^\infty \langle \mathbf{v}(t) \cdot \mathbf{v}(t+s) \rangle \, ds,$$

is introduced.

Eq. (1) determines the most general tensor of the second rank, depending only on $\mathbf{r}$, that satisfies the requirements detailed earlier. When mirror-symmetry is absent Eq. (1) is supplemented by the additional term $\frac{2}{3}v_0\chi(r)\varepsilon_{ijk}r_k\psi$, where the pseudoscalar $\chi(0)$ describes the correlation (the helicity density)

$$\chi(0) = (\tau/v_0)\langle \mathbf{v} \cdot \nabla \times \mathbf{v} \rangle.$$

This quantity will play an important rôle in the generation of the mean magnetic field.

For the neighboring pairs the one-point correlations of velocity derivatives, which are determined by the second derivative of f, are important [see (5), (6) below]. We write down two of them:

$$\langle \nabla_l v_i \nabla_l v_i \rangle = \langle \omega_i \omega_l \rangle = -5v^2\tau f''(0) = 5v^2\tau/\lambda^2,$$

$$\langle \nabla_l v_i \nabla_i v_l \rangle = 0, \tag{2}$$

where $\boldsymbol{\omega} = \nabla \times \mathbf{v}$.

It is also convenient to introduce the spectral components of the velocity field

$$\mathbf{v}(\mathbf{r}, t) = \int \mathbf{v}(\mathbf{k}, t) \, e^{i\mathbf{k}\cdot\mathbf{r}} \, d^3\mathbf{k}.$$

In Fourier-representation with the non-mirror-invariant term included, the correlation (1) takes the form

$$\langle v_i(\mathbf{k}, t)v_j(\mathbf{k}, t) \rangle = \left\{ \frac{E(k)}{4\pi k^2} \left(\delta_{ij} - \frac{k_i k_j}{k^2} \right) \delta(\mathbf{k} + \mathbf{k}') \right.$$

$$\left. + i\frac{\partial \chi(k)}{\partial k} e_{ijl}k_l \right\} \psi(t-t'), \quad \langle v_i(\mathbf{k}, t) \rangle = 0, \tag{3}$$

where the spectral density $E(k)$ determines $\frac{1}{2}\langle v^2 \rangle = \int E(k)\, dk$, has the dimensions $[v^2] \cdot [L]$, and is related to $f(r)$ by

$$E(k) = \frac{1}{3} v^2 k^2 \pi^{-1} \int_0^\infty f(r) r^2 [(\sin kr)/kr - \cos kr]\, dr.$$

The flow under consideration is three-dimensional. It is interesting to compare it with the corresponding two-dimensional (homogeneous and isotropic) turbulence. Let all the quantities depend only on two coordinates, say x and y, and let $v_z = 0$. Then instead of (3) we have

$$\langle v_i v_j \rangle = v_0^2 [f(r)\delta_{ij} + r(\delta_{ij} - r^{-2} r_i r_j)\, df/dr]\psi, \qquad (i, j = 1, 2) \qquad (3')$$

where $\mathbf{r}$ is now the two-dimensional vector (x, y), and we have kept the same notation for f and its derivative. The difference between this form and (3) is formally mainly due to the two-dimensionality of the Kroneker delta δ_{ij}: here $\delta_{ii} = 2$ while in the three-dimensional case $\delta_{ii} = 3$. But two-dimensional turbulence actually has qualitatively distinct features. First and foremost, the mirror-non-invariant term is absent. This is natural because one cannot tie a knot without moving out of the plane. Also, in the absence of dissipation, a law of conservation for the square of the vorticity exists. For a detailed discussion of the properties of two-dimensional turbulence we refer the reader to the review by Mirabelle and Monin (1979). For our problem, two-dimensional flows are interesting because they do not act as dynamos (see Chapter 5). Therefore they may be used as a test against which to assess arguments which, when applied to the three-dimensional case, predict dynamo action. Nevertheless (Zeldovich, 1956), two-dimensional magnetohydrodynamic turbulence possesses diamagnetism, and there is a possibility of strong but temporary growth in the magnetic energy.

Let us express the statistical characteristics of a three-dimensional random walk of particles in terms of the correlations we have defined. We first consider one particle. By scalarly multiplying the equation $d\mathbf{r}/dt = \mathbf{v}$ by $\mathbf{r}$ and averaging, we obtain

$$d(\langle r^2 \rangle)/dt = 2\nu_T,$$

where

$$\nu_T = \int_0^\infty \langle \mathbf{v}(t') \cdot \mathbf{v}(t - t') \rangle\, dt' = \frac{1}{3} v^2 \tau \qquad (4)$$

is the coefficient of turbulent diffusion.

The distance between a pair of particles, $\boldsymbol{\xi} = \mathbf{r}_1 - \mathbf{r}_2$, is also a random Gaussian variable. Clearly, $\langle \xi_i \rangle = 0$. It is easy to see how the mean square $\langle \xi^2 \rangle$ changes by using the relation

$$\langle \exp i\mathbf{k} \cdot \boldsymbol{\xi} \rangle = -\exp\left(-\tfrac{1}{2}k^2 \langle \xi^2 \rangle\right)$$

and the equation obtained above with ξ^2 calculated from $r_1^2 + r_2^2 - 2\mathbf{r}_1 \cdot \mathbf{r}_2$. Summing we obtain (Kazantsev, 1967)

$$d\langle \xi^2 \rangle / dt = (8\tau/3) \int \left(1 - e^{-\frac{1}{2}k\langle \xi^2 \rangle}\right) E(k) \, dk.$$

It is natural that distant particles $(k^2 \langle \xi^2 \rangle \gg 1)$ move away from each other, at a rate determined by twice the coefficient of diffusion ν_T. For a pair of close particles the exponent can be expanded. In the first non-zero approximation

$$d\langle \xi^2 \rangle / dt = (4\tau/3)\langle \xi^2 \rangle \tau \int E(k)k^2 \, dk$$

$$\sim \langle \omega^2 \rangle \tau \langle \xi^2 \rangle. \tag{5}$$

Thus neighboring particles also move away from each other, exponentially fast if the vorticity is non-zero. The characteristic timescale during which they separate is of order

$$\tau_2 = (l/v)^2 \hat{\tau}^{-1}, \tag{6}$$

where l is a length characteristic of the turbulence. This conclusion, on the separation of the randomly wandering pair, is important for the dynamics of small-scale magnetic fields (Section III) because the frozen-in field behaves like the vector connecting the pair of particles.

On the face of it, the scattering law is exponential [see Eq. (5)]. This formula, however, is valid only for small time intervals. Actually expansion (5) presupposes that the magnitude of the spectral function $E(k)$ is significant for small k, i.e. on large-scales. This fact is of great importance. It turns out that on small-scales the behavior of the pair (and of the magnetic field also, see Section III) is affected by the motion on the large-scales.

In the two-dimensional case also, the particles move apart but at a much slower rate $(\sim t)$. This is evidently because the distance between the particles, along the perpendicular to the surface of motion, is limited. (It is constant for a plane; see also Chapter 5.)

The particle trajectories are level lines of a vector potential which here has only a single non-zero component. The trajectories do not intersect or tangle, and therefore neighboring particles, moving along adjacent trajectories, remain close to each other; only a linear growth of their separation with time is possible (Arnold, 1972). It is interesting that on surfaces of constant negative curvature, exponential separation of a pair of particles is possible. Such surfaces cannot, however, be imbedded in three-dimensional Euclidean space (see Chapter 5).

II   The Mean Field Dynamo

The behavior of the magnetic field is described by the induction equation. In frozen field conditions, when one can ignore the dissipation, this equation is, for an incompressible fluid, similar to the transport equation for the distance between two neighboring particles

$$\partial H_i / \partial t = -v_k \nabla_k H_i + H_k \nabla_k v_i, \qquad \nabla_k \equiv \partial / \partial x_k. \qquad (7)$$

We shall assume that the motion is random with zero mean velocity. In this and subsequent paragraphs we shall consider the behavior of the mean field $\langle H_i \rangle \equiv B_i$ and the mean squared field $\langle H_i^2 \rangle$. Averaging is carried out over the ensemble of possible velocity fields or, in more physical terms, over scales and times exceeding the characteristic correlation scale, l, and the correlation time, τ, of the velocity field.

The averaging of equations like

$$\partial H_i / \partial t = \hat{L}_{il} H_l,$$

to which Eq. (7) with the stochastic operator $\hat{L}$ belongs, is a non-trivial mathematical operation considered in a number of papers (see, for example, van Kampen, 1974; Knobloch, 1978). The result is a closed equation for the mean field. We prefer another, less strict but more illuminating way of solving Eq. (7), which is also in useful (see, for example, Kraichnan, 1975, 1976) and describes strictly the behavior of the mean field when the velocity correlation time is small and the velocity field is sufficiently well described by its second moments.

We integrate Eq. (7) formally as

$$H_i(\mathbf{r}, t) = \int (-v_k \nabla_k H_i + H_k \nabla_k v_i) \, dt' + H_i(\mathbf{r}, 0),$$

and we assume that the initial magnetic field is uncorrelated to the velocity field. By substituting this expression into the right-hand side of Eq. (7) and averaging, we obtain

$$\partial B_i / \partial t = \int_0^t \hat{P}_{il} B_l \, dt', \tag{8}$$

where the operator in the integrand is of the form

$$\hat{P}_{il} = \delta_{il} \langle v_k v'_m \rangle \nabla_k \nabla_m - 2 \langle v_k \nabla_l v'_i \rangle \nabla_k$$
$$+ \delta_{il} \langle v_k \nabla_k v'_m \rangle \nabla_m - \langle v_k \nabla_k \nabla_l v'_i \rangle + \langle \nabla_k v_i \nabla_l v'_k \rangle. \tag{8'}$$

The prime means that the velocity is taken at the time t' and the correlations in angular brackets are calculated at the spatial point x_i. The approach is not rigorous because of the assumed factorization of the correlation of magnetic and velocity fields.

The factorization can be justified for a velocity field with a short correlation time (Molchanov et al., 1982), when the averaging over the time interval $(0, t)$ mainly depends on the magnetic field over its previous history. The correlation does not factorize only for the short time interval $(t - \Delta t, t)$, where Δt is the correlation time. The magnetic field can however then be expressed in terms of its earlier values by a Taylor expansion.

Magnetic field generation is determined mainly by the tensor $v_k \nabla_l v_i$. It is important to emphasize that this tensor does not generally coincide with the helicity and has a more general significance. For example, for a two-dimensional velocity field (where $v_z \equiv 0$ but v_x and v_y depend on all three coordinates), one can easily see that the mean field component B_z is not affected by the other components and decays independently. After that, one can prove in the standard way that dynamo action is impossible. The helicity $\langle \mathbf{v} \cdot (\nabla \times \mathbf{v}) \rangle$ may be non-zero because $\partial / \partial z \neq 0$ but this is irrelevant (Chapter 5).

In the two-scale approximation $L \gg l$, where l is the correlation scale of the velocity field and L is the scale over which mean values change, the third-rank tensor $\langle v_k \nabla_l v_i \rangle \equiv a_{ikl}$ may be reduced to a pseudotensor of second-rank α_{ik}. In fact, for any t

$$\langle v_k(\mathbf{x}) \nabla_l v_i(\mathbf{x}) \rangle = \partial \langle v_k(\mathbf{y}) v_i(\mathbf{x}) \rangle / \partial x_l |_{\mathbf{x}=\mathbf{y}}.$$

Let us introduce new variables

$$\mathbf{r} = \tfrac{1}{2}(\mathbf{x} - \mathbf{y}), \qquad \mathbf{R} = \tfrac{1}{2}(\mathbf{x} + \mathbf{y}).$$

For completely homogeneous anisotropic turbulence the correlation depends only on $\mathbf{r}$. The symmetry

$$\partial \langle v_k(\mathbf{y}) v_i(\mathbf{x}) \rangle / \partial x_l |_{\mathbf{x}=\mathbf{y}} = -\partial \langle v_i(\mathbf{y}) v_k(\mathbf{x}) \rangle / \partial y_l |_{\mathbf{x}=\mathbf{y}}$$

is obvious even when this tensor lacks reflectional symmetry ($\mathbf{r} \to -\mathbf{r}$). Hence, a_{ikl} is antisymmetric in the subscripts i, k and can therefore be written as $\varepsilon_{ikm} \alpha_{lm}$; inverting, we have

$$\alpha_{ik} = \varepsilon_{klm} \langle v_m \nabla_i v_l \rangle = \langle (\nabla_i \mathbf{v}) \times \mathbf{v} \rangle_k.$$

In a weakly inhomogeneous two-scale turbulence, an $\mathbf{R}$-dependence appears. The third-rank corrections to the pseudotensor α_{ik} are, however, clearly of order l/L. The third-rank tensor a_{ikl} must be used when the scale of magnetic field is of order l (i.e. $L \sim l$).

In a homogeneous isotropic velocity field only the first terms of (8') are non-zero. Then

$$\int \langle v_i v'_k \rangle \, \mathrm{d}t' = \tfrac{1}{3} \langle v^2 \rangle \delta_{ik} \tau,$$

$$\int \langle v_i \nabla_k v'_l \rangle \, \mathrm{d}t' = -\tfrac{1}{6} \varepsilon_{ikl} \tau \langle \mathbf{v} \cdot (\nabla \times \mathbf{v}) \rangle,$$

with constant diffusion coefficient and helicity. Here generation is decided by the mean helicity, and

$$\partial \mathbf{B} / \partial t = \alpha \nabla \times \mathbf{B} + \nu_T \nabla^2 \mathbf{B}, \tag{9}$$

where

$$\alpha = -\tfrac{1}{3} \tau \langle \mathbf{v} \cdot (\nabla \times \mathbf{v}) \rangle. \tag{10}$$

It is interesting that the inclusion of the correlations described by the last two terms of Eq. (8') is equivalent to introducing spatial dependence into the helicity. But then, as can easily be seen, instead of $\alpha \nabla \times \mathbf{B}$ as in (9), one has $\nabla \times (\alpha \mathbf{B})$.

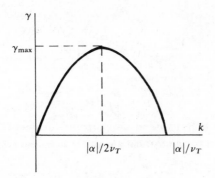

Figure 8.1 Growth rate of the mean magnetic field versus wavenumber for a random medium with a nonzero mean helicity.

Equation (9) implies that exponential growth of mean magnetic field is possible. It is easy to construct fast dynamo solutions of the form

$$\mathbf{B} = (\pm \sin kz, \cos kz, 0)\, e^{\gamma t}, \qquad \gamma = -\nu_T k^2 \pm \alpha k, \qquad (11)$$

which grow exponentially for a helicity with one of the signs and for sufficiently small values of $k < |\alpha|\nu_T$,. i.e. sufficiently large scales. The maximal growth rate, $\gamma_{max} = \alpha^2/4\nu_T$, is attained when $k_{max} = |\alpha|/2\nu_T$ (Figure 8.1).

It has been shown (Moffatt, 1978; Krause and Rädler, 1980) that the generation term in Eq. (9) has a simple electrodynamic meaning, namely, as an electric field and consequently a mean electric current directed along the magnetic field:

$$\mathbf{j} = -\sigma\alpha\,\mathbf{B} + \cdots.$$

This current means that an ensemble of screw-like vortices with non-zero mean helicity is capable of generating loops in the magnetic field in a plane perpendicular to that of the initial field (Figure 8.2).

We will conclude this section with a few remarks. Our first remark is concerned with the Ohmic diffusivity, ν_m, which we ignored in deriving Eq. (9) for the mean field. This Lagrangian procedure can be generalized to the case of finite magnetic diffusivity (Molchanov *et al.*, 1982). The idea is to describe magnetic diffusion by a Wiener process, as has been done in the theory of Brownian motion. We recall that Brownian motion can be thought of as a viscous process

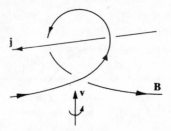

Figure 8.2 Creation of a current (anti-) parallel to a magnetic field due to the motion with $\alpha \neq 0$.

in which the idea of the path of a fluid element is not invoked (*a la* Langevin) but is described by an (inviscid) motion along random paths (as for Einstein-Smoluchousky–Wiener). The application of this method to the case of short-time correlations results in the same Eqs. (8) and (9), but with ν_T replaced by $\beta = \nu_T + \nu_m$.

The second remark refers to the assumption of instantaneous correlations which seems to be a serious restriction to the theory, since in real turbulence the correlation time $\Delta t \sim \tau = l/v_0$. It is possible to calculate corrections to Eqs. (8) and (9) which are proportional to powers of the small parameter $\Delta t/\tau$. The form of the first term on the right-hand side of (9) does not change; only corrections are added to α. In the third-order approximation a qualitatively new effect arises, a negative contribution to the diffusion coefficient (Knobloch, 1977):

$$(\Delta t/\tau)^3 \langle \nabla_j v_i \cdot \nabla_m v_j \cdot v_r v_s \rangle \nabla_r \nabla_s B_m$$

$$\approx (\Delta t/\tau)^3 [4\langle \nabla_l v_p v_s \rangle \langle \nabla_p v_l v_s \rangle - \langle \nabla_l v_p v_s \rangle \langle \nabla_s v_l v_p \rangle] \Delta B_i/15$$

$$= -(\langle \alpha^2 \rangle/18)(\Delta t/\tau)^3 \nabla^2 B_i. \tag{12}$$

Here $\langle \alpha^2 \rangle$ denotes the mean square helicity fluctuation. This result was first obtained from heuristic arguments by Kraichnan (1976) [see also Moffatt (1978)]. Thus, the growth of the mean magnetic field can be influenced both by the mean helicity and (even in the absence of a mean α) by its fluctuations.

In the next approximation, however, along with the corrections to the coefficients in Eqs. (8) and (9) and the equation governing the second moments, there already appear terms in the third space derivatives. The order of the moment equations will rise in the next approximations. This implies that the proper moment equations for the magnetic field are integral equations. Apparently one should regard Eqs. (8) and (9) as the first approximation in the expansion of the exact equations in some asymptotic series, and take the corrections very carefully into account. The approximation is rather good as long as the characteristic growth times of the magnetic field are greater than τ. For instance, in the isotropic, reflectionally non-symmetric case, the characteristic growth time for the mean magnetic field is [see (11)]

$$2\beta/\alpha^2 \sim \tau/\chi,$$

which exceeds τ because in real situations $\chi \ll 1$. The asymptotic properties of (9) have been discussed in detail by Isakov *et al.* (1981). The minimum growth time for the second moment of the field is $5\tau/4$ (see Section IV), i.e. 25% greater than τ.

Next we turn to a topological point. Mean field generation is possible in the isotropic case only if the mean helicity, $\langle \mathbf{v} \cdot (\nabla \times \mathbf{v}) \rangle$, is non-zero. In general this condition is not a necessary one for dynamo action. For example, in homogeneous anisotropic turbulence, it is possible to construct the solutions $B \sim \exp \gamma t$, Re $\gamma > 0$ when $\langle \mathbf{v} \cdot (\nabla \times \mathbf{v}) \rangle = 0$ but $v_i \omega_k \neq 0$. In fact, it is easy to show that in the long-wave approximation

$$\gamma^2 \simeq k_x^2 \alpha_2 \alpha_3 + k_y^2 \alpha_1 \alpha_3 + k_z^2 \alpha_1 \alpha_2, \qquad \alpha_1 + \alpha_2 + \alpha_3 = 0,$$

where α_1, α_2 and α_3 are the eigenvalues of the matrix (8') and $\mathbf{k}$ is the wave vector. When $\alpha_1, \alpha_2, \alpha_3 \neq 0$ (i.e. for situations that are not two-dimensional) then clearly solutions exist with Re $\gamma > 0$.

In conclusion we note the importance of time dependence for the velocity field even though it is on average stationary. At first sight it might seem possible to introduce the same transport coefficients, α and β, for a purely stationary flow, $\mathbf{v}(\mathbf{r})$. Suppose one constructs a stationary flow with $\mathbf{v} \cdot (\nabla \times \mathbf{v}) \sim \alpha_s$. The α_s so introduced is time invariant $(t \rightarrow -t)$. It should be emphasized that, in the turbulent (non-stationary) case, time reversal results in a change of sign of α (and β). So we do not see how α_s (and β_s) can be constructed so that the equation analogous to (9) gives fast dynamos in the limit of large magnetic Reynolds number. Of course, one cannot exclude slow dynamos when α and β depend on ν_m.

III   The Spectrum of Small-Scale Fields

A large-scale magnetic field in a turbulent dynamo is inevitably accompanied by small-scale fields. (We shall denote by **b** the small-scale random component of the magnetic field with zero mean.) Moreover, it is impossible to generate a large-scale field without small-scale ones, for the source in Eq. (9) is the mean value

$$\langle \mathbf{v} \times \mathbf{b} \rangle = \alpha \mathbf{B} - \beta \nabla \times \mathbf{B}, \qquad \mathbf{b} \equiv \mathbf{H} - \mathbf{B}. \tag{13}$$

In this section we shall make some crude estimates of the spectral distribution of this component.

By subtracting Eq. (9) from the induction equation for the total field **B** + **b**, we obtain the equation governing **b** (see, for example, Moffatt, 1978)

$$\partial \mathbf{b} / \partial t = \nabla \times (\mathbf{v} \times \mathbf{B}) + \nabla \times \mathbf{G} + \nu_m \nabla^2 \mathbf{b}, \tag{14}$$

where $\mathbf{G} = \mathbf{v} \times \mathbf{b} - \langle \mathbf{v} \times \mathbf{b} \rangle$. In deriving Ohms law (13) one usually ignores this term (the first-order smoothing approximation). Its presence is necessary, however, when determining the spectrum of the small-scale field; it is through this G-term that energy transfer from the large to the small scales takes place. In this sense its rôle is similar to that of the nonlinear term $(\mathbf{v} \cdot \nabla)\mathbf{v}$ in the Navier–Stokes equations. In our case, however, the transfer is accomplished by the interaction of **v** with the field **b**. Small-scale magnetic fields can grow not only due to distortions of the mean field but also independently due to self-excitation. The dynamo problem for the small-scale magnetic field when **B** = 0 will be studied in the next section. Here we consider crudely the action of the mean magnetic field, neglecting the self-excitation of small-scale fields. This problem was first considered by Golitsyn (1960) and was developed by Moffatt (1978).

The long-term rate of change of **b** is obviously the same as that of the large-scale field $B \sim \exp(\gamma t)$. It is clear that the characteristic growth time of the mean field γ^{-1} exceeds the overturning time of a turbulent cell, as well as the characteristic time of magnetic energy transfer down the spectrum. Therefore, in the period $t < \gamma^{-1}$, the small-scale field may be assumed stationary on average, i.e. one can write $\partial \langle b^2 \rangle / \partial t = 0$.

To obtain order of magnitude estimates, we introduce an approximation: turbulent viscosity. We replace the term $\langle \mathbf{b} \cdot \nabla \times \mathbf{G} \rangle_k$, which causes the energy cascade in wave number ($\mathbf{k}$) space, by $\nu_T(k)k^2 b_k^2$, where $\nu_T \sim (\tau k^2)^{-1}$. Here $\tau(k)$ is the characteristic time of energy transfer for the given scale. Instead of field and velocity amplitudes we consider the spectral functions

$$\langle b^2 \rangle / 8\pi = \int M(k)\,\mathrm{d}k, \qquad \langle v^2 \rangle / 2 = \int E(k)\,\mathrm{d}k.$$

In the kinematic dynamo problem, i.e. when the back reaction of the magnetic field on the motion is not taken into account, the function $E(k)$ is assumed given, for example, by the Kolmogorov form $E(k) \sim k^{-5/3}$. From the conservation of energy flux along the spectrum,

$$\left(\int_k E\,\mathrm{d}k \right) \Big/ \tau(k) = \text{constant}, \tag{15}$$

one immediately deduces $\tau(k)(\sim k^{-2/3}$ for the Kolmogorov spectrum). We fix the external scale for the turbulence, i.e. an upper bound for the inertial range, k_0. Then the lower (dissipation) bound for the inertial range is obtained by equating $\tau(k)$ to the characteristic dissipation time, i.e. $\tau_d = (\nu_m k^2)^{-1}$, so that

$$k_m = k_0 R_m^{3/4}.$$

The magnetic energy input at the scale k_0 takes place through the large-scale field $\mathbf{B}$ [the first term on the right-hand side of equation (14)]. Equating, in order of magnitude, this input flux to the flux transported to large wave numbers in the region where $\tau(k) \ll \tau_d$, we immediately obtain

$$M(k) \approx B^2 E(k)/3\nu_T^2(k)k^2 \approx \tfrac{1}{3}\tau^2(k)k^2 E(k^2)B^2, \tag{16}$$

the analogue of Golitsyn's relation (see Moffatt, 1978, p. 290). For the Kolmogorov spectrum and k close to k_0, this gives (Ruzmaikin and Shukurov, 1982)

$$M(k) \sim k^{-1}. \tag{17}$$

After integration over all k in the inertial range, (16) yields, in fact, an important relationship between the mean energy of the small-scale magnetic field and the energy of the large-scale field. The result depends in an essential way on the spectrum of the "background" hydrodynamic turbulence.

For the Kolmogorov spectrum,

$$\langle b^2 \rangle \sim (\ln R_m) B^2. \tag{18}$$

For another spectrum, say $E(k) \sim k^{-q}$ where $q < \frac{5}{3}$,

$$\langle b^2 \rangle \sim R_m^{(5/3-q)/(1-q/3)} B^2. \tag{18a}$$

Using such crude arguments, we certainly cannot estimate the coefficient of proportionality in these relations, but the dependence on the magnetic Reynolds number is found simply. It is interesting that, when $R_m \gg 1$, the small-scale fields are always stronger than the large-scale ones. To estimate the energy contained in the small-scale fields, the relation $\langle b^2 \rangle = R_m B^2$ is commonly used. Such a relation is valid only for two-dimensional turbulence (Zeldovich, 1956; see also Chapter 5). It is obtained from (18a) with $q = 1$. This spectrum is typical for homogeneous isotropic two-dimensional turbulence (Mirabelle and Monin, 1979). The spectrum of three-dimensional turbulence is steeper, and the exponent of the magnetic Reynolds number is therefore smaller. The Kolmogorov spectrum proves to be exceptional, a logarithm appearing instead of an exponent. In this case, the characteristic time of energy transfer at scale k, and the time of overturning of a turbulent cell are of the same order of magnitude, $\tau^{-1} \simeq v_k \cdot k$. It may be shown that, if the spectrum is steeper than that of Kolmogorov ($q > \frac{5}{3}$), then the energy maximum for the small-scale fields moves to k_0 so that $\tau \sim \tau(k_0)$ and $\langle b^2 \rangle \sim B^2$.

In the region where the Ohmic dissipation is considerable, order of magnitude arguments give the spectrum (Golitsyn, 1960)

$$M(k) \sim k^{-11/3}.$$

The qualitative picture of the spectrum for the kinematic dynamo is given in Figure 8.3a.

An exponential growth in the large-scale magnetic field, at the rate γ, leads to growth of spectral density of magnetic energy at the rate 2γ, for all scales. The magnetic energy will first become of the same order as the kinetic energy near the dissipation scale k_m. From that time onwards the inertial interval becomes increasingly dominated by magnetohydrodynamic effects. The back reaction of the field on the motion becomes significant eventually, and $M(k) \simeq E(k)$. The turbulence then takes the form of random magnetohydrodynamic waves, propagating with approximately the Alfvén speed v_A, where

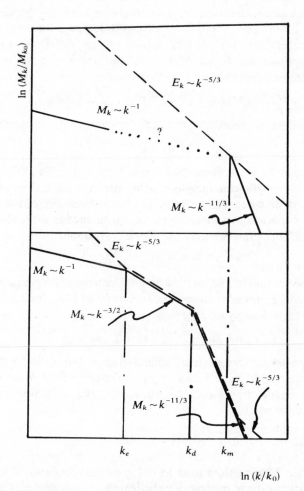

Figure 8.3 The hydrodynamical and magnetic spectra in a mirror-symmetrical turbulence, (a) kinematic regime; (b) MHD turbulence.

$v_A^2 = \langle b^2 \rangle / 4\pi\rho$ and ρ is the density. The time of the energy transfer along the spectrum is now decided by the characteristic time of the nonlinear interaction between these waves. In weak magnetohydrodynamic turbulence the triple wave interactions dominate (Sagdeev and Galeev, 1969), and $\tau^{-1}(k) = \omega_k(kM(k)/U)$, where $\omega_k = kv_A$ and U is a homogeneous function with dimensions cm^2/s^2. On writing $U = v_A^2$ we obtain $\tau^{-1}(k) = k^2 M(k)/v_A$, and from energy flux conservation,

$$\mathcal{E} = kM(k)/\tau(k) = k^3 M^2(k) v_A^{-1} = \text{constant},$$

we obtain the spectrum

$$M(k) \sim k^{-3/2}, \tag{19}$$

first derived by Iroshnikov (1963) and Kraichnan (1965). The spectrum of the small-scale magnetic field, after the inertial interval has evolved, is shown in Figure 8.3b. The lower bound k_e of the inertial range for this magnetohydrodynamic turbulence is estimated from $E(k_e) = M(k_e)$. Here we may use $E \sim k^{-5/3}$, so that

$$k_e = k_0(v_0/v_A)^3.$$

The upper bound of the inertial range is determined by equating the rate of energy transfer down the spectrum with the rate of energy dissipation, or simply by $\tau(k) = \tau_d$:

$$k_d = (\varepsilon/v_A v_m^2)^{1/3} = k_0 R_m^{2/3} (v_0/v_A)^{1/3}.$$

Note that, in the case of small-scale dynamos that act independently of the generation mechanism of the large-scale field dynamos, the spectrum in the range $k_0 < k < k_e$ (or $k_0 < k < k_m$ in the linear regime) is quite different.

IV The Magnetic Field in Mirror-Symmetric Turbulence

It is well known that a magnetic field line is subjected in turbulence of zero mean helicity to two effects. First, the turbulent tangling decreases the field scale and increases the rate of Ohmic dissipation.

Second, as is clear from Section I, the stretching of the separation between neighboring particles results in field amplification. The question of the asymptotic $(t \to \infty)$ behavior of the magnetic field by a given turbulent flow was first raised and solved by Batchelor (1950). He used the analogy between the induction equation for the field and the Helmholtz equation for the vorticity. He concluded that asymptotically an exponential growth of the field is possible if the magnetic viscosity ν_m is smaller than the kinematic viscosity ν. This analogy is valid during the initial stages of evolution of field and vorticity. In particular the law (5) for the separation of adjacent particles involves $\langle \omega^2 \rangle$. The use of analogy is, however, not justified at later times. It should be noted that the induction equation is linear, while the equation governing the vorticity $\omega = \nabla \times v$ is nonlinear, the velocity v in the former being maintained by external sources. Moreover, Batchelor's arguments are applicable to two-dimensional homogeneous turbulence, where a dynamo is known to be impossible (Zeldovich, 1956). Later, Saffman (1963), by an approximate qualitative analysis, concluded that dynamo action cannot occur in mirror-symmetric turbulence, and that the result of Zeldovich (1956), about a temporary growth of the small-scale field with a subsequent asymptotic decay, could be generalized from two- to three-dimensional turbulence.

In the years that followed a number of papers (Kazantsev, 1967; Kraichnan and Nagarajan, 1967; Knobloch, 1978; Vainshtein, 1980b) were published which proved, with the help of different mathematical methods, that a dynamo of small-scale fields can operate in homogeneous isotropic turbulence. In the papers by Knobloch and Vainshtein, the velocity correlation tensor is expanded near $r = 0$ in the form (1a). Obviously this gives the dispersion law (5) for a pair of neighboring particles and gives the characteristic time (6). If Ohmic dissipation is negligible, the magnetic field grows indefinitely and exponentially in the characteristic time determined by the second derivative of the correlation function $f''(0)$ or, equivalent, the second moment of the spectral function. Vainshtein then included the magnetic viscosity in order to deduce the correct boundary condition for the correlation function of the magnetic field at $r = 0$. Were ν_m zero, the field would decrease its scale indefinitely so that a dynamo would be impossible. But the analyses performed in the above papers does not consider the behavior of the velocity field at large distances.

Saffman (1963) drew attention to the importance of the velocity correlation for large scales. The situation is well exemplified by the flow $v_i = c_{ik}x_k$ considered in the previous chapter. If the flow is assumed to take place in infinite space [$c_{ik}(t)$ is a random function of time] then the magnetic energy grows indefinitely in time (Section 7.I). Although this example is artificial, it nevertheless clearly points to the necessity of considering velocity field correlations for large r.

Kazantsev (1967) derived an equation for the correlation function of the magnetic field with no restriction on the small-scale correlations of the velocity field. The only assumptions were the homogeneity, isotropy and Markovian nature of the process (the smallness of the memory time as compared with other characteristic times). The problem reduced to evaluating

$$\langle H_i(\mathbf{k}, t)H_j(\mathbf{k}', t')\rangle = w(k, t)(\delta_{ij} - k^{-2}k_ik_j)\delta(\mathbf{k}+\mathbf{k}')\delta(t-t'),$$

where the angle brackets denote the average over the velocity and initial magnetic field distributions. It is convenient to use the correlation function in r-space

$$w(\mathbf{r}, t) = \int w(\mathbf{k}, t)\,e^{i\mathbf{k}\cdot\mathbf{r}}\,d^3\mathbf{k}.$$

Kazantsev wrote down the solution of the induction equation as an infinite series in the small parameter R_m^{-1}. By summing this series with the aid of diagram techniques and subsequently differentiating, he obtained an integral equation for $w(\mathbf{k}, t)$. It was then shown that this equation can be simply deduced by the Wiener path method mentioned in Section II (Molchanov et al., 1982), only using the assumption of short velocity correlation time.

In r-space the equation of Kazantsev may be reduced by the substitutions

$$V(r) = \int (4\pi k^2)^{-1}E(k)\,e^{i\mathbf{k}\cdot\mathbf{r}}\,d^3\mathbf{k},$$

$$m(r) = \frac{1}{2}\left(\nu_m + \tfrac{1}{3}\nu_T - r^{-3}\int_0^r V(s)s^2\,ds\right), \tag{20}$$

$$\psi(r, t) = (2mr^2)^{-1/2}\int_0^r w(s, t)s^2\,ds,$$

[ν_T is the coefficient of turbulent diffusion (4)] to the Schrödinger-type equation

$$\partial\psi/\partial t = m^{-1}(r)\,\partial^2\psi/\partial r^2 - 2U(r)\psi$$

for a variable mass in the potential

$$U(r) = \tfrac{1}{2}r^{-1}\,df/dr + (mr^2)^{-1} - \tfrac{1}{8}m^{-3}(dm/dr)^2.$$

The asymptotic values of the mass and potential are

$$m(r) \simeq \begin{cases} \tfrac{1}{2}\nu_m, & r \to 0, \\ \tfrac{3}{2}\nu_T, & r \to \infty, \end{cases}$$

$$U(r) \simeq \begin{cases} 2\nu_m/r^2, & r \to 0, \\ 2\nu_T/r^2, & r \to \infty. \end{cases}$$

Evidently when $\nu_T \gg \nu_m$ there may be a potential well in the intermediate region. In analysis it is convenient to use a specific expression for the correlation function $V(r)$. We shall assume that

$$V(r) = \exp(-r^2/2\lambda^2).$$

Then V is everywhere positive, diminishes at infinity, and is regular as $r \to 0$. A sketch of the potential is shown in Figure 8.4, and of the

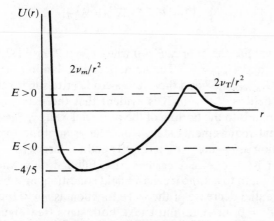

Figure 8.4   Potential function in the problem of small-scale magnetic fields in a given velocity field. The state with energy $E > 0$ is indicated by the dashed line and corresponds to a temporary growth of the magnetic field energy with subsequent asymptotic decay.

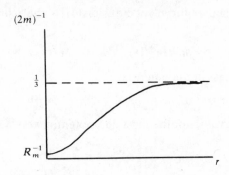

Figure 8.5 Dependence of the "effective mass" on radius.

mass in Figure 8.5. The presence of negative-energy levels is this potential implies the existence of unstable exponential solutions, for which

$$\langle H^2\rangle \sim \exp(2\gamma t), \qquad E = -\gamma < 0.$$

For at least one level to exist it is (crudely speaking) necessary that

$$\int_{r_1}^{r_2} [2m(E-U)]^{1/2}\, dr \geq \tfrac{1}{2}\pi. \qquad (21)$$

The first level has the energy $E \approx 0$ when $r_1 \approx 0.2$, $r_2 \approx 1.2$, $(2m)^{-1} \approx R_m^{-1} + 0.3r^2$ and $U_{min} \approx -\tfrac{4}{5}$. The numerical calculation of the integral (21) gives $(R_m)_{min} \approx 58$ and therefore a crude criterion for the infinite growth of field is $R_m \geq 60$. It is evident that the maximum growth rate corresponds to the bottom of the potential well: $\gamma_{max} \approx (4/5)l^{-1}v$.

Additional requirements stem from the asymptotic forms of the wave function as $r \to 0(\psi \sim r^2$, $w(0) = 1)$ and for large $r[\psi \sim \exp(-r)]$. The former is, as one may easily verify, fulfilled only when $\nu_m \neq 0$, i.e. for a dynamo to act the frozen-in field condition must be violated. The exponential decrease of the wave function as $r \to \infty$ requires that $\int_0^\infty w(x)x^2\, dx = 0$, i.e. $w(r)$ must take both signs (see Figure 8.6). It is interesting that the vorticity correlation $\langle\omega^2\rangle$, calculated directly from (1), has the same form. This partially justifies the analogy with the vorticity used by Batchelor. In the Fourier-representation the condition obtained implies that, at small wave numbers, the spectrum of the magnetic energy, $\langle H^2 \rangle / 2 = \int M(k)\, dk$, is of the form

$$M(k) \sim k^4, \tag{22}$$

since $M(0) = 4\pi k^2 \displaystyle\int_0^\infty w x^2 \, dx = 0.$

Of particular interest is the limit $R_m \gg 1$. In this case (due to the growth of the effective mass) the field concentrates on small scales $\sim R_m^{-1/2}$.

Note that, although the three-dimensional picture is statistically mirror-symmetric, a helicity is necessarily locally present in the motion. The form of the correlation in Figure 8.6 shows that the directions of the magnetic field lines are opposite for large r but the same for small r, i.e. as they approach each other, the magnetic loops must accumulate by turning themselves about. It may be imagined that locally the loops are transformed into figure eights, and subsequently merge with one another.

When the magnetic Reynolds number is less than $(R_m)_{min}$, infinite growth of magnetic field is impossible, and the spectrum of $E(=-\gamma)$ is positive and continuous. The evolution of an initial field is, however,

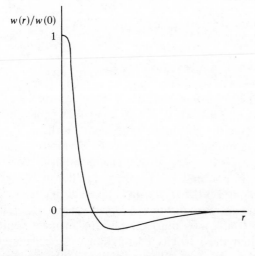

Figure 8.6 The diagonal component $w = \langle H_i H_i \rangle$ of the correlation tensor of the magnetic field at large Reynolds numbers.

not necessarily one of monotonic decay. When $1 \ll R_m < (R_m)_{min}$, a temporary amplification of field takes place. This effect is well known in two-dimensional turbulence (Zeldovich, 1956; Pouquet, 1978). Note that, in the two-dimensional case, the potential $U(r)$ is positive everywhere, and is given by $2mU = \frac{3}{4}r^{-2}$.

In conclusion, we should point out that the qualitative estimates made above have been supported by a numerical study of the Kazantsev equation to be published by V. G. Novikov and the present authors. This study also examines other facets of the problem: the dependence of γ on R_m, the speed of propagation of inhomogeneities in field, other choices of $V(r)$, and so on.

CHAPTER 9

The Turbulent Dynamo in a Disk

In the previous chapter we obtained the equations governing the average properties of the magnetic field. It was shown that in infinite space they possess solutions which grow in time. In reality we must deal with the finite volumes of planets, stars or galaxies. The simplest case is a thin turbulent rotating disk (or slab) surrounded by a vacuum. This can model a galactic disk (Chapters 12 and 13), an X-ray disk (Chapter 15) and even a stellar convection zone (Chapter 11) or the fluid parts of a planetary core. So in this chapter we shall study in detail mean field generation in a thin disk.

In the axisymmetric case the large-scale dynamo problem for a thin disk reduces to the solution of two coupled equations involving the time and the single space coordinate z which measures distance perpendicular to the plane of the disk. This is because, in the first approximation the other spatial coordinate, r, appears only parametrically. The resulting "local" eigenvalue problem has been considered in a number of works (Parker, 1971; Vainshtein and Ruzmaikin, 1971; Moffatt, 1978; Ruzmaikin et al., 1979; Ruzmaikin, Sokoloff and Turchaninoff, 1980) with a variety of different assumptions about the form of the turbulent helicity. Common features of all these models are a field generation threshold and certain symmetry properties. In other respects, the solutions are, as a rule, substantially different. The differences stem from the functional dependence on z postulated for the helicity, which permit the generation equations to admit a variety of different modes. It is therefore of particular importance to examine qualitatively the properties of the disk dynamo equations, and this is the purpose of this chapter.

The next step in disk dynamo theory is to construct a global dynamo solution by combining these local solutions i.e. the solutions that treated r as a parameter. We can find an analogy in the quantum theory of molecules where the spectrum and the electron wave func-

149

tions are first calculated for fixed nuclei and then the motion of the nuclei is added as though it were governed by the potential computed from the fixed nucleus solutions. It can be shown (see below) that the growth rate of the global solution is close to the maximum growth rate of the local solutions. Away from the radius, r_0, where the growth rate is a maximum, the field is of small amplitude.

Another approach is to take the limit in which a spheroid becomes a thin disk (Stix, 1975; Soward, 1978; White, 1978). Due to the fact that the problem is now two-dimensional, an investigation of its qualitative features is much more complicated.

I The Generation Equations and Boundary Conditions

We consider a thin (the thickness $2h$ is much less than the radius R), differentially rotating (the angular velocity is $\omega = \omega(r)$†), turbulent disk. The equations governing the generation of an (on average) axisymmetric $(\partial/\partial\phi = 0)$ magnetic field are, in cylindrical coordinates,

$$\frac{\partial B_r}{\partial t} = -\frac{\partial}{\partial z}(\alpha B_\phi) + \beta(\nabla^2 B)_r, \tag{1}$$

$$\frac{\partial B_\phi}{\partial t} = GB_r + \frac{\partial}{\partial z}(\alpha B_\phi) + \beta(\nabla^2 B)_\phi, \tag{2}$$

$$\frac{\partial B_z}{\partial t} = \frac{1}{r}\frac{\partial}{\partial r}(r\alpha B_\phi) + \beta(\nabla^2 B)_z, \tag{3}$$

where B_r, B_ϕ, B_z are the components of the average magnetic field; $G \equiv r\, d\omega/dr$ is a measure of the differential rotation, α and β are the helicity and the turbulent magnetic diffusion (see Chapter 8). Generally speaking, the mean helicity is a pseudo-tensor α_{ik}. In real astrophysical disks however, the turbulence is usually of small-scale (the mixing length is less than the disk thickness) and hence almost isotropic. Therefore, $\alpha_{ik} \approx \alpha\delta_{ik}$ approximately, where α is a pseudo-scalar. Due to the axisymmetry condition $\partial/\partial\phi = 0$, the fields B_r and

---

† For the case $\omega = \omega(z)$ see the next chapter (Section 10.I).

B_z can be replaced by a vector potential with only a ϕ-component, $A_\phi \equiv A$, i.e.

$$B_r = -\partial A/\partial z, \qquad B_z = r^{-1}\partial(rA)/\partial r, \qquad (4)$$

automatically giving $\nabla \cdot \mathbf{B} = 0$. We now write $B_\phi = B$. In the thin disk approximation, r-derivatives of the field can be ignored in Eqs. (1)–(3), i.e. $(\partial/\partial r)/(\partial/\partial z) \to 0$. Thus we may restrict ourselves to the first two equations with $(\nabla^2 \mathbf{B})_r = \partial^2 B_r/\partial z^2$ and $(\nabla^2 \mathbf{B})_\phi = \partial^2 B_\phi/\partial z^2$. In a thin disk, the B_z-component is weak,

$$B_z/B_r = O(h/R), \qquad (5)$$

and can be omitted in the subsequent analysis. The choice of boundary conditions depends on the specific physical situation. In the case of the Galaxy, while the conductivity in $|z| > h$ may be quite large, the force-free conditions seem to be reasonable since the density of the intergalactic medium is so low and the gas pressure is negligible outside the disk. Moreover, $\mathbf{B}$ must be a potential field because force-free configurations with $\nabla \times \mathbf{B} \neq \mathbf{0}$ do not vanish smoothly at infinity. Hence, due to the axisymmetry, $B = 0$ for $|z| > h$,† and the potential A satisfies the equation

$$\frac{\partial^2 A}{\partial z^2} + \frac{\partial}{\partial r}\left(\frac{1}{r}\frac{\partial}{\partial r}(rA)\right) = 0, \qquad (6)$$

the general solution of which can be written as a linear combination of the functions $\exp\{ikz\}J_1(kr)$, where k is determined from the boundary conditions $\nabla \times \mathbf{B} = \mathbf{0}$ at $r = R$, i.e. $k \sim R^{-1}$. Hence, in the thin disk case, $\partial A/\partial z \approx kA \approx 0$ on the planes $|z| = h$. In other words, we may assume that the behavior of the solutions of interest outside the disk resembles that of an infinite disk (whose exterior fields are independent of r and ϕ). In such a solution, $B = 0$ and $B_r = 0$. (Of course, outside a disk of finite dimensions the field vanishes only at infinity.) For this field $\partial A/\partial z = 0$ in $|z| > h$. We assume that the rim of the disk makes a negligible contribution to the generation process.

---

† The field B vanishes due to the absence of the normal current j_z. In fact, $j_z \sim \int B_\phi r \, d\phi = 2\pi r B$.

The boundary conditions follow from the continuity of the azimuthal and poloidal components of the field. In the thin disk approximation $(kh \sim h/R \to 0)$

$$B(\pm h) = 0, \qquad B_r = -\partial A(\pm h)/\partial z \simeq 0. \qquad (7,8)$$

We note that the first of these conditions is a result of the axisymmetry and continuity with the exterior potential field. The second boundary condition is approximate and valid only for a thin disk. The effects of violating this condition are discussed at the end of Section IV.

In the limiting case, when the disk has infinite conductivity and diffusion through the boundaries is impossible, field generation is also impossible (see Section II). Parker (1971) was the first to note the rôle played by diffusion of field through the boundaries in the disk dynamo problem.

We shall search for axisymmetric solutions of the form

$$B(z, t) = b(z)\, e^{\gamma t}, \qquad B_r(z, t) = b_r(z)\, e^{\gamma t}. \qquad (9)$$

Also it is convenient to introduce dimensionless variables by

$$z \to z/h, \qquad t \to t/h^2\beta^{-1}, \qquad \gamma \to \gamma/\beta h^{-2},$$

$$\alpha \to \alpha_0 \alpha(z), \qquad \alpha_0 \equiv \alpha_{max}. \qquad (10)$$

With the earlier notation preserved, we obtain the generation equations

$$(\gamma - \partial^2/\partial z^2)b_r = -R_\alpha\, \partial(\alpha b)/\partial z, \qquad (11)$$

$$(\gamma - \partial^2/\partial z^2)b = R_\omega b_r - R_\alpha\, \partial(\alpha b_r)/\partial z, \qquad (12)$$

$$b(\pm 1) = b_r(\pm 1) = 0. \qquad (13)$$

Here we have introduced the dimensionless numbers

$$R_\omega = Gh^2/\beta, \qquad R_\alpha = \alpha_0 h/\beta,$$

that play an important part in the problem under consideration. Instead of (11) it is often convenient to use its integral

$$(\gamma - \partial^2/\partial z^2)a = -\alpha b. \qquad (14)$$

The form (9) has allowed us to reduce our system to a one-dimensional eigenvalue problem. A certain difficulty, however, arises because the eigenvalues γ may be complex (the solution may oscillate) since the operator defined by Eqs. (11)–(13) is not self-adjoint. The eigenfunctions obtained do not therefore, in general, form a complete set in function space. When analyzing the boundary-value problem, however, we must not consider arbitrarily high harmonics in z since the problem becomes finite-dimensional (Ruzmaikin, Sokoloff and Turchaninoff, 1980). For such problems it is known that incompleteness of the eigenfunction system arises from the degeneracy of several eigenvalues at some values of R_α and R_ω. In this case the missing solutions may, for example, have the form $t^n \exp \gamma t$; in principle, more complex situations, e.g. a continuous spectrum, may occur.

As mentioned above, the next step in the theory of a disk dynamo is to take into account the dependencies of G, α and h upon r. As a result, the γ will also depend on r, i.e. if the radial derivatives are not involved, the field growth rates will be different at different radii. The radius, r_0, at which $\gamma(r)$ reaches its maximum, γ_{max}, is crucial for the entire disk. The global-solution increases as $\exp(\Gamma t)$, where Γ is only slightly less than γ_{max}. In regions where $\gamma(r)$ is small, the field is mainly determined by diffusion from the vicinity of $r = r_0$ and therefore decays exponentially with radius. We now determine the difference between Γ and γ_{max}, obtain an approximate form for the eigenfunction in the entire disk, and estimate the time required for the global solution to be set up.

The function $\gamma(r)$, which plays the rôle of a potential in the vicinity of the maximum, can be expanded as

$$\gamma(r) \simeq \gamma_{max} + \tfrac{1}{2}(r - r_0)^2 \, d^2\gamma/dr^2.$$

This is the potential of an oscillator of frequency $(d^2\gamma/dr^2)^{1/2}$. Thus the lowest energy level is

$$\Gamma \simeq \gamma_{max} - \tfrac{1}{2}(\beta \, d^2\gamma/dr^2)^{1/2}.$$

The "wave function" $B(r)$ is approximately constant in a "potential well" $\gamma(r) > \Gamma$, but outside this region

$$B(r) \sim \exp\left(-\int [\Gamma - \gamma(r)]^{1/2} \, dr \right).$$

The time needed to reach this stationary state (with constant Γ) proves however to be large, viz.

$$\tau \sim [h \, d\gamma/dr]^{-1},$$

assuming $\gamma(r)$ does not undergo abrupt changes over distances of the order of the disk thickness. In other words, the distance between the zero and the next levels is of order h/R. As the solution decreases exponentially towards the edge of the disk, the order of magnitude of $(\partial/\partial r)/(\partial/\partial z)$ becomes less than h/R. Soward (1978) showed it to be of order $(h/R)^{2/3}$. The approximation can hardly be applied to the central region of a galactic disk (Ruzmaikin and Shukurov, 1981).

In the disk, the function $\alpha(r, z)$ is determined by the angular velocity and the gradient of the gas density (see Sections 11.III and 13.I). Since density changes are most pronounced in the direction perpendicular to the disk,

$$\alpha \simeq l^2 \omega \; \partial \ln \rho/\partial z,$$

where l is the mixing length of the turbulence. As a first approximation it is natural to consider ω to be independent of z, and $p(z)$ to be a symmetric function. Then $\alpha(r, z)$ is an antisymmetric function of z.

The generation equations with antisymmetric $\alpha(r, z)$ possess an important symmetry property: under the transformation $z \rightarrow -z$ the system (11)–(13) is invariant under

$$a(-z) = a(z), \qquad b_r(-z) = -b_r(z), \qquad b(-z) = -b(z)$$

or under

$$a(-z) = -a(z), \qquad b_r(-z) = b_r(z), \qquad b(-z) = b(z).$$

Therefore, all solutions of the generation equations may be divided into two groups depending on the form of $b(z)$: odd (dipolar in b_r and b_z) or even (quadrupolar in b_r and b_z). This symmetry property allows us to confine attention to the interval $0 < z < 1$.

Two types of generation are naturally distinguished, depending on the relative contribution of the sources on the right-hand side of Eq. (2); the $\alpha\omega$-dynamo for which $|GB_r| \gg |\partial(\alpha B_r)/\partial z]$, a condition which seems to be satisfied throughout the Galaxy, and the α^2-dynamo for which the differential rotation can be neglected. In each of these cases the system features a single dimensionless number, R_α for the α^2-dynamos and

$$D = R_\alpha R_\omega$$

for the $\alpha\omega$-dynamos, for which the equations take the form

$$(\gamma - \partial^2/\partial z^2)b_r = -\partial(\alpha b)/\partial z, \tag{15}$$

$$(\gamma - \partial^2/\partial z^2)b = Db_r, \tag{16}$$

after renormalizing the amplitude, $b_r \to R_\alpha b_r$.

II The Properties of the Principal Modes

We now demonstrate that important conclusions about the forms of the first even and odd modes may be drawn without solving the generation equations.

(a) Even solutions

It may seem that the first mode can have no zeros in the interval $0 < z < 1$, but actually the radial field component of a growing solution must necessarily change sign in this interval. To see this integrate Eqs. (1) and (2) over the interval $0 < z < 1$ in the thin disk approximation with a smooth $\alpha(z)$ function, to obtain

$$\frac{\partial}{\partial t}\int_0^1 B_r\,\mathrm{d}z = \beta\,\frac{\partial B_r}{\partial z}\bigg|_0^1, \tag{17}$$

$$\frac{\partial}{\partial t}\int_0^1 B\,\mathrm{d}z = G\int_0^1 B_r\,\mathrm{d}z + \beta\,\frac{\partial B}{\partial z}\bigg|_0^1. \tag{18}$$

Note that, due to the conditions $\alpha(0) = 0$ and $B(1) = 0$, the helicity does not appear in these equations at all. Suppose, as is natural to expect for the first excited mode, that $B(z)$ has no zeros within the interval $(0, 1)$; without loss of generality we may take $B > 0$. An important conclusion may be drawn from Eqs. (17) and (18) in the case of a growing even (quadrupole) solution. Then $\partial B_{\phi,r}(0)/\partial z = 0$, $\partial B_\phi(1)/\partial z < 0$ and hence for $G \gtrless 0$,

$$\int_0^1 B_r\,\mathrm{d}z \gtrless 0, \qquad \partial B_r(1)/\partial z \gtrless 0, \tag{19}$$

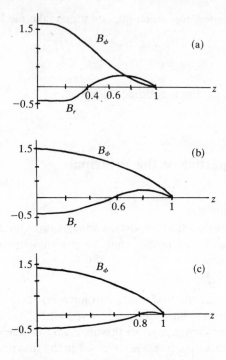

Figure 9.1 The main harmonic of the field excited by the $\alpha\omega$-dynamo in a thin disk when the dynamo number approaches a critical value: (a) $D = -100$, (b) $D = -20$, (c) $D = D_{cr} = -8$.

Since $B_r(1) = 0$, it follows that the radial magnetic field component must change sign near the surface of the disk (see Figure 9.1).

The results (17) and (18) also allow us to clarify the rôle of diffusion in the dynamo process. At first glance, it appears that diffusion may be safely neglected. If we set $\beta = 0$ in Eq. (17), however, we immediately obtain $\int_0^1 B_r \, dz = \text{constant}$, so that $\int_0^1 B_\phi dz$ increases linearly with time, i.e. a stationary dynamo is impossible for $\beta = 0$. As the second condition in (19) shows, dynamo action can occur only when the "diffusive flux" across the boundary is non-zero. The rôle of diffusion can be understood directly from the generation equations. Let us assume $G < 0$ and $B > 0$. Then the source $-\partial(\alpha B)/\partial z$ generates, near the surface of the disk, a radial field whose sign is opposite to that of $B_r(<0)$ in the central part of the disk. The radial field due to the differential rotation $-|G|B_r$, makes a negative contribution to $\partial B/\partial t$ which may only be compensated for by the diffusion term $\beta\ \partial^2 B/\partial z^2$. The latter is bound to be positive near the boundary of the disk, i.e. the field will spread out from the disk. From this it is clear that, in addition to the conditions (19), the function B must have a turning point to the left of the zero of B_r. On the other hand, it should be understood that diffusion must not be too potent for otherwise the growth rate would be negative.

The above arguments refer to the case of growing solutions. In a stationary state $(\partial/\partial t = 0)$

$$\beta\ \partial B_r/\partial z = 0, \qquad \beta\ \partial^2 B/\partial z^2 = 0 \quad \text{at } z = 1,$$

and the components B_r and B are of one sign in the disk. The qualitative transition to the stationary solution is shown in Figure 9.1a–c. The corresponding solutions for $\alpha = \sin \pi z$ and $D = -100$, -20, and -8 were obtained by numerical integration of Eqs. (15) and (16) by Ruzmaikin, Sokoloff and Turchaninoff (1980). The last of these values of D is the critical one at which generation is marginal (i.e. $\gamma = 0$).

If $\alpha(z)$ is discontinuous the condition (18) remains valid, but at a discontinuity the right-hand side of Eq. (17) must be supplemented by the addition of the jump in the source αB. For example, if $\alpha = \theta(z) - \theta(-z)$, where $\theta(z) = 1$, $z > 0$ and $\theta(z) = 0$, $z < 0$ (Parker, 1971), then we have, instead of Eq. (17),

$$(\partial/\partial t) \int_0^1 B_r\, dz = -2B(0) + \beta\ \partial B_r(1)/\partial z, \qquad (17a)$$

and, instead of the second condition in (19),

$$\beta\ \partial B_r(1)/\partial z \gtrless 2B(0). \qquad (19a)$$

The resulting solutions may differ qualitatively from those obtained with a continuous $\alpha(z)$. The dynamo is, however, impossible when $\beta = 0$ since it then follows from Eq. (5), for example, that $B(0) = 0$. Relation (17a) shows that a discontinuous $\alpha(z)$ introduces an additional source at the point of discontinuity.

A similar integral relation can be obtained for $\alpha(z)$ of δ-type; here an additional contribution to the B derivative arises at the point of source concentration.

(b) Odd solutions

We show next that there are no static odd (dipole) solutions of the $\alpha\omega$-dynamo for smooth $\alpha(z)$ and with fields which have no zeros in the interval $0 < z < 1$. Indeed, for $\gamma = 0$, we have from (15)–(16)

$$b_r = -D^{-1} \partial^2 b / \partial z^2, \qquad D\alpha(z) = -b^{-1} \partial^3 b / \partial z^3,$$

with boundary conditions

$$b(1) = b(0) = 0, \qquad b_r(1) = b_r(0) = 0.$$

It is evident that the function $\partial^3 b / \partial z^3$ [and, hence, $\alpha(z)$] must change sign within the interval $0 < z < 1$. This seems to indicate that for such $\alpha(z)$ (which are those of greatest astrophysical interest) only oscillatory (Im $\gamma \neq 0$) odd modes can be excited, and this surmise is supported by the numerical calculations of Ruzmaikin, Sokoloff and Turchaninoff (1980).

Note that, in the α^2-dynamo as well as in the full problem, a non-oscillatory even mode can be excited (see Section III) in addition to the oscillatory odd mode.

III   Some Examples

(a) Decay modes

It is easy to solve the system (11)–(14) when sources are absent ($\alpha = G = 0$). For odd modes

$$\gamma_k^d = -k^2 \pi^2, \qquad k = 0, 1, 2, \ldots, \tag{20}$$

and for even modes

$$\gamma_k^q = -(k + \tfrac{1}{2})^2 \pi^2. \tag{21}$$

The corresponding eigenfunctions are doubly degenerate. For example, both

$$b_r = \sin k\pi z, \quad b = 0, \qquad \text{and} \qquad b_r = 0, \quad b = \sin k\pi z,$$

are eigenfunctions for γ_k^d. Note that the dipole mode with $\gamma_k^d = 0$ is trivial and we have $\mathbf{B} \equiv 0$ even if $\alpha \neq 0$ and $G \neq 0$. Therefore the lowest quadrupole mode decays four times as slowly as the lowest dipole mode. This partly explains an extremely important property of Galaxy models: the lowest quadrupole mode is the first to be excited. It is also clear (and confirmed by numerical calculations) that for small α and G, i.e. small D, the eigenfunctions and eigenvalues are similar to those of the decay modes.

(b) Differential rotation

Another instructive example is the case of a non-helical flow ($R_\alpha = 0$) in differential rotation, for which

$$\frac{\partial B_r}{\partial t} = \beta \frac{\partial^2 B_r}{\partial z^2}, \qquad \frac{\partial B}{\partial t} = -GB_r + \frac{\partial^2 B}{\partial z^2}.$$

If, as earlier, solutions of the form $B = b(z) \exp \gamma t$ are sought, then surprisingly we again obtain the expressions (20) and (21) for γ! The source GB_r in the second equation removes the degeneracy, however: only one pair of eigenfunctions correspond to each γ. It is clear that the reduced set will be incomplete, since the dynamo boundary problem is not self-adjoint. The missing solutions are easily obtained directly from the equations (without the substitution $B \sim \exp \gamma t$). For example, the solutions of even symmetry satisfying the boundary conditions are

$$B_r = e^{-|\gamma_k|t} \cos (|\gamma_k|^{1/2} z), \qquad B = Gt \, e^{-|\gamma_k|t} \cos (|\gamma_k|^{1/2} z).$$

Thus, consistent with the picture we have developed for the action of differential rotation, the azimuthal field first grows linearly with time but eventually, following the decay of the radial field, decays exponentially.

(c) α^2-approximation

Let us consider another limiting case, $R_\omega = 0$, which corresponds to a disk in solid-body rotation with helical turbulence. In this case the generation equations reduce to a single Schrödinger-type equation. In fact, the system

$$\gamma b_r - b_r'' = -R_\alpha (\alpha b)', \qquad \gamma b - b'' = R_\alpha (\alpha b_r)',$$
$$b(\pm 1) = 0, \qquad b_r(\pm 1) \simeq 0, \tag{22}$$

(the prime denotes differentiation with respect to z') is invariant to the replacement of the eigenvector (b_r, b) by the vector $(b_r + b, b - b_r)$.

Since the latter must, to an arbitrary multiplicative factor, coincide with the former, we immediately see that either $b_r = ib$ or $b_r = -ib$. Therefore, it is sufficient to consider only one equation, e.g. that for $b_r = ib$:

$$b'' - \gamma b = iR_\alpha (\alpha B)'.$$

[The case $b_r = -ib$ is equivalent to replacing R_α by $-R_\alpha$.] By substituting

$$b = \psi \exp\left(\tfrac{1}{2}iR_\alpha \int \alpha \, dz\right), \tag{23}$$

we reduce the problem to one of solving a Schrödinger equation with a complex potential

$$\psi'' + (E - U)\psi = 0, \qquad \psi(\pm 1) \simeq 0,$$
$$E = -\gamma, \qquad U = -\tfrac{1}{4}R_\alpha^2 \alpha^2 + \tfrac{1}{2}iR_\alpha \alpha'. \tag{24}$$

The potential (and therefore the solutions) depends not only on the function α but also on its derivative. From this it follows that solutions with smooth and non-smooth $\alpha(z)$ must be substantially different. Note that the solutions for the $\alpha\omega$-dynamo also depend on the derivative of $\alpha(z)$ (See Section IV). This statement is non-trivial because in that case no α derivatives appear in the generation equations.

For slowly changing $\alpha(z)$ and sufficiently large R_α the potential is real, and for antisymmetric $\alpha(z)$ it has two wells (Figure 9.2) as in the well-known problem of Lifshitz (Landau and Lifshitz, 1958). The presence of energy levels indicates whether generation is possible. The eigenfunctions produced will be odd and even (dipolar and quadrupolar). In the simplest case $\alpha = 1$, there exists one potential well which, as is easily shown, yields the following solution

$$\gamma_n = \tfrac{1}{4}R_\alpha^2 - (\tfrac{1}{2}\pi n)^2, \qquad n = 0, 1, 2, \ldots$$
$$\begin{pmatrix} b_r \\ b \end{pmatrix} = \begin{pmatrix} \sin \tfrac{1}{2}R_\alpha z \\ \cos \tfrac{1}{2}R_\alpha z \end{pmatrix} \sin \tfrac{1}{2}\pi n (z - 1).$$

For discontinuous $\alpha(z)$ the potential is complex. Let us consider the example

$$\alpha(z) = \begin{cases} 1, & z > 0, \\ -1, & z < 0. \end{cases} \tag{25}$$

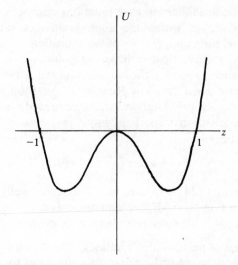

Figure 9.2 The potential function for the α^2-dynamo when $\alpha(z)$ is a smooth anti-symmetric function and R_α is large.

In addition to the boundary conditions at $z = \pm 1$, we require conditions for the point $z = 0$:

$$[\psi] = 0, \qquad [\psi'] = iR_\alpha\psi(0).$$

The square brackets denote the jump in the function they contain. The solution is

$$\gamma_n = \tfrac{1}{4}R_\alpha^2 - k^2,$$

where $k = \pi n$, $n = 0, 1, 2, \ldots$ for the odd modes ($\psi(0) = 0$), while for the even modes k is determined from the equation

$$k \cot k = \tfrac{1}{2}R_\alpha.$$

In particular, for small R_α,

$$k^2 = \pi^2(n + \tfrac{1}{2})^2 - iR_\alpha(2n + 1).$$

(d) Concentrated helicity

It is also of interest to consider generation functions of the form

$$\alpha = \delta(z - \zeta) - \delta(z + \zeta). \tag{26}$$

From the physical standpoint, such functions may occur in a rotating turbulent medium in which the density changes rather abruptly ($\alpha \sim \omega \nabla \rho$), due, for example, to a phase transition.

Moffatt (1978) investigated in detail solutions of the dynamo equations for such an $\alpha(z)$ in the $\alpha\omega$-approximation (when $R_\omega \gg R_\alpha$). The general case was treated by Ruzmaikin, Sokoloff and Shukurov (1980). In the α^2-approximation the concentrated helicity could be introduced in other ways [perhaps more realistically than in (26)]; for example, suppose

$$\alpha^2 = \delta(z - \zeta) + \delta(z + \zeta).$$

Then the potential (24) will have two δ-function wells. It is easy to solve the resulting Schrödinger equation. For example, for even symmetry and $\zeta = \frac{1}{2}$ there is one mode with the growth rate $\gamma = R_\alpha^4/64$.

(e) Application of perturbation methods

It is rather easy to solve the generation equations for small R_α and R_ω by using a perturbation theory similar to that of quantum mechanics. Again we consider two limiting cases, assuming for both that the eigenfunctions and eigenvalues reduce to their unperturbed forms when $R_\alpha = R_\omega = 0$ [see (a)]:

Case 1. $R_\alpha \ll R_\omega \ll 1$ (the $\alpha\omega$-dynamo)

We regard the right-hand sides of Eqs. (15) and (16) as perturbations. Let us renormalize the amplitude b_r so that b_r and b are of the same order of smallness: $b_r \to D^{1/2} b_r$. Then the equations take the form

$$\left(\frac{d^2}{dz^2} - \gamma - V\right)\binom{b_r}{b} = 0,$$

where

$$V = |D|^{1/2}\begin{pmatrix} 0 & \partial\alpha/\partial z \\ \pm 1 & 0 \end{pmatrix}$$

is the perturbing potential. The upper sign is for $R_\omega > 0$ and the lower for $R_\omega < 0$. By applying standard perturbation theory for a doubly degenerate level (see, for example, Landau and Lifshitz, 1958) we obtain in the first approximation

$$\Delta\gamma = \pm(V_{12}V_{21})^{1/2},$$

where V_{12} and V_{21} are the matrix elements of the perturbing potential, calculated using the eigenfunctions of the unperturbed problem. In this case the diagonal matrix elements vanish. Letting $\alpha = z$ we obtain the following first eigenvalues

$$\gamma^d = -\pi^2 \pm (\mp|D|)^{1/2} \quad \text{(an odd mode)},$$
$$\gamma^q = -\tfrac{1}{4}\pi^2 \pm (\mp|D|)^{1/2} \quad \text{(an even mode)}.$$
(27)

Thus, the even (quadrupole) field is excited first.

Case 2. $R_\omega \ll R_\alpha \ll 1$ (the α^2-dynamo)

Here we have directly the Schrödinger equations (24) with a complex potential. For the case $\alpha = z$ the result is

$$\Delta\gamma = iR_\alpha/z + R_\alpha^2/12,$$
(28)

and a crude estimate of the excitation threshold is.

$$R_{\alpha cr} = \begin{cases} 10, & \text{for the even mode,} \\ 5, & \text{for the odd mode.} \end{cases}$$

(f) Stationary solutions

We can construct stationary solutions corresponding to $\partial/\partial t = 0$ ($\gamma = 0$) for the $\alpha\omega$-models (Soward, 1978; Ruzmaikin *et al.*, 1979). In this case the system of Eqs. (15) and (16) reduces to the single equation

$$d^3B/dz^3 + D\alpha(z)B = 0.$$

It is interesting that the statement of the problem can be inverted by seeking functions $\alpha(z)$ (vanishing at $z = 0, \pm1$) which give solutions, $B(z)$, of the requisite symmetry. A class of such functions was obtained by Ruzmaikin *et al.*, (1979).

IV   Asymptotic Solutions

The behavior of dynamo solutions depends on the values of the dimensionless numbers R_α and R_ω. In particular, non-decreasing solutions occur only when these numbers exceed their critical values. In astrophysics R_ω and R_α are usually large, or at least their magnitudes exceed unity. For example, in the solar convection zone $R_\omega \sim 10^3$

and $R_\alpha \sim 10$, in the disks of double X-ray sources $R_\alpha \sim 1$–10 and $R_\omega \sim 10$–10^3, and in the gaseous Galactic disk $R_\alpha \sim 1$ and $R_\omega \sim 10$.

It is difficult, however, to determine exact values for these numbers and for the form of the function α describing the turbulent helicity. Great significance is therefore attached to the qualitative investigation of the properties of the solutions and especially to the approximate evaluation of the growth rate of field, to the determination of the localization zone of the solution, and to the elucidation of those features of the α-function that have a significant effect on the behavior of the solutions. Unlike studies made under laboratory conditions, astrophysics takes no special notice of the exact calculation of the critical R_α and R_ω, or the growth rates. In fact, it suffices to study the limiting cases of small and large R_α and $|R_\omega|$. We have already considered the former [see point (e) of Section III]; here we shall deal with strong excitation. In this case diffusion has insufficient time to spread the generated field far from the regions in which it is created, and the characteristic zone of field localization proves to be small compared with the disk thickness. Therefore one can apply the short-wave asymptotics of WKB theory (Isakov et al., 1981). We shall restrict ourselves to the one-dimensional problem, though the transition to the two- or three-dimensional cases does not present any particular difficulties.

(a) α^2–dynamos

Here we can apply the WKB-method to the Schrödinger equation (24) with the complex potential

$$U = -\tfrac{1}{4}R_\alpha^2\alpha^2 + \tfrac{1}{2}iR_\alpha\alpha'$$

(see Figure 9.2). Obviously, when R_α is large the first term of the potential dominates. The solution is localized in the vicinity of the minima of U (i.e. the α-extrema), and the first mode has the growth rate

$$\gamma = \tfrac{1}{4}R_\alpha^2 + O(R_\alpha).$$

The correction to γ and the finer features of the solution depend on the behavior of α near the extrema $z = \pm z_0$. When $\alpha''(\pm z_0) \neq 0$ (Isakov et al., 1981)

$$\gamma = \tfrac{1}{4}R_\alpha^2 - \tfrac{1}{2}R_\alpha\omega + \tfrac{1}{2}i\omega^{-1}\alpha'''(z_0) + O(R_\alpha^{-1}),$$
$$b_\pm(z) = C_\pm \exp\left[\tfrac{1}{2}iR_\alpha(z \mp z_0) - \tfrac{1}{4}R_\alpha\omega(z \mp z_0)^2\right]. \tag{29}$$

Here ω has the sense of frequency of oscillations in the well $[\omega^2 = -2\alpha(\pm z_0)\alpha''(\pm z_0)]$ the signs "$+$" and "$-$" corresponding to the different wells.[†]

The complete eigenfunction is constructed as even and odd combinations of $b_\pm$, corresponding to quadrupole and dipole modes. The difference in the growth rates for the even and odd solutions arises from the splitting in the wells (Landau and Lifshitz, 1958) and is exponentially small.

Let us evaluate the total number of levels excited for a given R_α. In the quasi-classical approximation this number is equal to the area of the one-dimensional phase space (px) for momenta in the interval $0 < p < p_{max}$, $p_{max} = (-2U)^{1/2}$ (Landau and Lifshitz, 1958), i.e. $2\pi p_{max} \, dz/4\pi R_\alpha^{-1}$. By taking into account the double degeneracy of levels in a potential having two wells, we obtain

$$N = (3 \cdot 2^{1/2}/\pi\varepsilon) \int_{-1}^{1} |\alpha| \, dz \sim R_\alpha \qquad (30)$$

(if $\alpha = \sin \pi z$ then $N = 9 \cdot 2^{1/2}/\pi^2\varepsilon \simeq 0.5 R_\alpha$). From (29) we have an estimate of the excitation threshold: $R_\alpha = 2$. Note that in the α^2-approximation

$$b_r/b = O(1). \qquad (31)$$

It is interesting to obtain a criterion for the validity of this approximation. One might at first think that it is sufficient to require $R_\alpha \gg R_\omega$. Substitution of the solution obtained back into (24) shows, however, that the principal term on the right-hand side is the one proportional to R_α^2. Thus the criterion is actually

$$R_\alpha^2 \gg R_\omega. \qquad (32)$$

This significantly widens the range of applicability of the solution because under astrophysical conditions $R_\alpha \gtrsim 1$ and (32) can be satisfied even for rather large $R_\omega (> R_\alpha)$.

[†] When the function $\alpha^2(z)$ has several maxima, e.g. the potential has many wells, one must consider generation in each well. We should also recall that the generation equations are equivalent to two independent Schrödinger equations differing in the sign of the imaginary part of the potential.

(b) $\alpha\omega$-dynamos

If the solution is localized then $b' \gg b$. We may therefore assume that for large dynamo numbers Eqs. (15) and (16) are replaced by

$$(\gamma - \partial^2/\partial z^2)b_r \approx -\alpha\, \partial b/\partial z, \qquad (\gamma - \partial^2/\partial z^2)b = Db_r. \qquad (33)$$

Equating terms of the same order shows that this is possible when $(D \gg 1)$

$$\gamma = \gamma_0 D^{2/3}, \qquad b/b_r \sim D^{1/3}, \qquad (34)$$

and the characteristic scale over which the field changes is $b/b' \sim D^{-1/3}$. Since the scale decreases as D increases, the solution becomes concentrated in a thin zone around $z = z_0$. In its vicinity the solution is found to have the form [cf. (29)]

$$\binom{b_r}{b} = \binom{D^{-1/3}\mu}{\nu} \exp\{[ik(z-z_0) - \tfrac{1}{2}p(z-z_0)^2]D^{1/3}\}, \qquad (35)$$

where μ, ν are constant amplitudes, k is real and $\operatorname{Re} p > 0$. In the vicinity of $z = z_0$, the helicity can be written as

$$\alpha \simeq \alpha_0 + \alpha_1(z - z_0) + \tfrac{1}{2}\alpha_2(z - z_0)^2. \qquad (36)$$

By substituting Eqs. (35) and (36) into Eqs. (33), we obtain a pair of linear inhomogeneous equations for μ and ν. Then k, p and γ are determined from their solvability condition

$$\{\gamma_0 - [ik - p(z - z_0)]^2\}^2 + \alpha[ik - p(z - z_0)] = 0,$$

(we consider the case $D > 0$).

To the accuracy of the expansion (36), it is sufficient to set to zero the coefficients of powers of $(z - z_0)$ up to the second. Now we have for both even and odd modes:

$$k = \pm\alpha_0^{1/3}/2, \qquad p = -\alpha_1/2\alpha_0^{2/3},$$
$$\gamma = \tfrac{1}{4}\alpha_0^{2/3}(1 + 2i)D^{2/3} + O(D^{1/3}). \qquad (37)$$

The location of the point z_0 is defined by the conditions

$$\alpha_0\alpha_2 + \alpha_1^2 = 0, \qquad \alpha_1 < 0, \qquad \alpha_0 \neq 0.$$

For example,

$$z_0 = \pm\tfrac{3}{4}, \quad \text{for } \alpha = \sin \pi z.$$

The approximation (33) is invalid near the points where $\alpha \simeq 0$. In fact numerical calculations (Ruzmaikin, Sokoloff and Turchaninoff, 1980) show that at such a point there is an additional maximum of the eigenfunction much bigger when $D \lesssim 10^3$ than the maximum at $z = z_0$. Moreover asymptotic solutions exist in which the field is localized only in the vicinity of the zero in α. In this case order of magnitude estimates of the terms in Eqs. (15) and (16) yield

$$\gamma = O(D^{1/2}), \qquad b_r/b = O(D^{-1/2}),$$

and the coordinate stretching is

$$z \to zD^{1/4}.$$

A more detailed analysis gives (Isakov *et al.*, 1980)

$$\gamma \simeq \gamma_0 |D|^{1/2}[\alpha'(0)]^{1/2}, \qquad D \gg 1, \tag{38}$$

where γ_0 is a constant independent of the form of α and the dynamo number:

$$\gamma_0 \simeq \begin{cases} 0.4, & D < 0 \\ 0.07 + 0.6i, & D > 0 \end{cases} \quad \text{for the even modes,}$$

$$\gamma_0 \simeq 0.1 + 0.6i \qquad\qquad \text{for the odd modes.}$$

The range of validity of the approximations used can be determined by substituting the solutions into the complete generation equations (1)–(3) and comparing the orders of the terms. For the solution (37) we require,

$$R_\omega \gg R_\alpha^2, \tag{39}$$

and for (38) we must have

$$R_\omega \gg \alpha'(0)R_\alpha. \tag{40}$$

To conclude this section we recall that all the equations governing the turbulent dynamo with helicity are derived by averaging over scales and times exceeding, respectively, the correlation length l and the turbulent correlation time τ. Since the characteristic scale of the excited field decreases with the growth of the relevant dynamo number, it is clear that the approximations considered cease to be valid beyond certain limiting D and R_α. We now set down the resulting estimates, using the characteristic scales of the excited magnetic fields obtained above:

for the α^2-dynamo

$$(R_\alpha)_{\lim} \sim h/l;$$

for the $\alpha\omega$-dynamo

$$D_{\lim} \sim \begin{cases} (h/l)^4, & \alpha'(0) \neq 0, \\ (h/l)^3, & \alpha(z_0)\alpha''(z_0) + \alpha'^2(z_0) = 0. \end{cases}$$

In any generation mechanism the limiting growth rate (in sec^{-1}) is

$$\gamma_{\lim} \sim \beta/l^2 \sim \tau^{-1},$$

which is also obvious from elementary physical considerations. Such a restriction is valid also for the general turbulent dynamo. Some further requirements that arise from the peculiarity of the one-dimensional problem posed by the disk dynamo can also be mentioned.

It was assumed in our one-dimensional approximation that, at the edge of the disk, b and $b_r \to 0$ as $|z| \to h$. But, in fact, b_r does not vanish there; it merges into the vacuum solution outside the disk with $b_r \sim 0$ [like $(h/R)^n$] for $|z| \to h$, where R is the radius of the disk and $n \sim 1$. It may be shown , by the use of singular perturbation theory, that when

$$(h/R)^n \geq D^{-1/3};$$

these solutions change slowly within the disk and abruptly at its edge (boundary layers). When the radius of the disk is not too large, regeneration can occur in principle, with D less than $D_{\lim}$. This does not seem to be the case under realistic astrophysical conditions (the Galaxy or the disks of the X-ray sources). Let us stress that we are interested only in nonstationary solutions (cf. Soward, 1978).

When the function α changes so rapidly that its characteristic scale, λ, is less than that of the field but greater than l, the solutions are asymptotically those obtained for discontinuous and for δ-like α. It is necessary in such cases to renormalize the dynamo numbers, e.g. for a δ-like α

$$R_\alpha \to (\lambda/h)R_\alpha.$$

CHAPTER 10

Some Topics in Non-Linear Turbulent Dynamo Theory

The origin and maintenance of the magnetic field of most astrophysical objects are explained by turbulent dynamo theory. The main applications of the theory (stellar cycles and the Galactic dynamo, amongst others) will be considered in the following chapters. In reality, magnetic fields usually evolve into states of dynamical equilibrium with the motion, and the non-linear turbulent dynamo problem therefore has a vital significance. A general approach to the problem is presented, for instance, in the review by Roberts and Soward (1975), the monograph by Moffatt (1978) and a series of papers by Frisch and his colleagues (see, for instance Meneguzzi *et al.*, 1981). For laminar non-linear dynamos, see for instance Braginsky (1967). In this chapter we shall only use simple examples to demonstrate some effects of magnetic back reaction on the sources of the field generation. Nevertheless, some general ideas and results will emerge from these special cases.

Since helicity plays such a key rôle in mean field turbulent dynamo action (Section 8.II), it is natural to expect, in accordance with the le Chatelier principle, that the magnetic field generated in helical turbulence will primarily affect the helicity. Usually the violation of reflectional symmetry in turbulence is slight and that is why in the first approximation we need not concern ourselves with the effects of magnetic back reaction on the turbulent diffusivity and the mean flow.

Such heuristic notions have been used successfully in many works dealing with the geomagnetic dynamo, the solar cycle and the Galactic dynamo. In simple terms, the problem is to replace the value "α" in Eq. (8.9) by a function $\alpha(\mathbf{B})$. The form of this function is either postulated, or it is obtained under various approximations from the Navier–Stokes equation (with the magnetic force included). Then the problem is to solve the non-linear equation for the mean magnetic field (Section II).

In this approach it is supposed that the helicity responds instantly to changes in the mean magnetic field. This approximation is in fact excellent if we confine attention only to the back reaction of the field in the hydrodynamic helicity $\langle \mathbf{v} \cdot (\nabla \times \mathbf{v}) \rangle$ which has a short relaxation time (approximately the overturning time of a large-scale vortex). But small-scale magnetic fields, $\mathbf{h}$ are generated, by the turbulent dynamo process and these create a mean magnetic helicity $\langle \mathbf{h} \cdot (\nabla \times \mathbf{h}) \rangle$. The characteristic relaxation time for this helicity may be large. In such a case, instead of the instantaneous approximation, we can describe the back reaction of the magnetic field through a differential equation (Section III). Such dynamical regimes, though more complicated, are apparently relevant to a stellar convection zone (Chapter 11), and by recognizing their existence we hope to explain, for instance, the random, prolonged period's of weak solar activity typified by the Maunder-minimum.

In the final section (Section V) of this Chapter we will describe briefly the results of direct numerical integrations of the full magneto-hydrodynamic equations governing a turbulent dynamo (Meneguzzi *et al.*, 1981). The material of the Sections I–IV is based on papers by Kleeorin and Ruzmaikin (1981, 1982).

I A Linear Model

We begin with a simplified model of mean field generation in a thin slab (or disk). Even such a crude model is of interest in application to thin spherical stellar shells, and to the gaseous disks encircling X-ray sources. Disk systems will be discussed in Chapters 12 and 14. In this chapter, using some simplifications, we will consider only stellar shells, and in particular the solar convection zone.

It will be convenient to use local Cartesian coordinates (x, y, z). For the purpose of interpreting the results in spherical geometry we note that the two coordinate systems are related by $dx = R\,d\theta$, $dy = R \sin \theta\,d\phi$ and $dz = dr$, where R is the stellar radius. We shall consider axisymmetric situations in which $\partial/\partial y = 0$, for all mean values. The mean velocity is assumed to have only one component $v = v_y(x, z)$ whose shear, ∇v, models differential rotation.

The equation governing the generation of the mean magnetic field $\langle H \rangle \equiv B$,

$$\partial B / \partial t = \nabla \times (v \times B + \alpha B - \beta \nabla \times B),$$

has an especially simple form when one uses the y-components of the field $B_y \equiv B$ and the vector potential $A_y \equiv A$ which determines the other field components ($B_x = -\partial A / \partial z$, $B_z = \partial A / \partial x$)

$$\frac{\partial A}{\partial t} = \alpha B + \beta \left(\frac{\partial^2}{\partial z^2} + \frac{\partial^2}{\partial x^2} \right) A, \tag{1}$$

$$\frac{\partial B}{\partial t} = \frac{\partial v}{\partial z} \frac{\partial A}{\partial x} - \frac{\partial v}{\partial x} \frac{\partial A}{\partial z} + \beta \left(\frac{\partial^2}{\partial z^2} + \frac{\partial^2}{\partial x^2} \right) B. \tag{2}$$

Here β is the magnetic diffusivity including both the turbulent and ohmic contributions. The function α describes the mean helicity of the turbulence. For simplicity we shall not examine any other effects at work in the dynamo, for instance, turbulent diamagnetism or other pumping effects. Moreover, it will be supposed, in addition, that α is a function of z only and that $\partial v / \partial x = 0$. The latter assumption is appropriate for stellar applications, but for galactic or X-ray disks (Chapter 16) we will suppose instead that $\partial v / \partial z = 0$. After making these simplifications, a system of two simple equations for the mean magnetic field results. Along with the diffusion terms on their right-hand sides, there are two sources: helicity and velocity shear (differential rotation).

It is convenient to use dimensionless variables. We use the slab thickness h as the unit length, h^2 / β as the unit of time and (preserving the previous notation) introduce the following dimensionless variables:

$$\alpha \to \alpha_0 \alpha, \qquad \partial v / \partial z \to (v_0 / h) v, \qquad A \to A(\alpha_0 h^2 / \beta).$$

Upon writing

$$(A, B) = [a(z), b(z)] \exp (pt - ikx),$$

we obtain from (1) and (2) the equations

$$(q^2 - d^2 / dz^2)a = \alpha b, \tag{3}$$

$$(q^2 - d^2 / dz^2)b = -ikdva, \tag{4}$$

where

$$q^2 = p + k^2, \qquad \mathrm{Re}\, q > 0, \qquad p = \gamma + i\omega.$$

We introduce here the dimensionless dynamo number,

$$d = \alpha_0 v_0 h^2 / \beta^2, \tag{5}$$

which is positive by definition. It should be noted that the dynamo number $D = \alpha_0 v_0 R^2 / \beta^2$, usually used in spherical geometry, differs from d by the factor $(h/R)^2$. We also assume that $a(z)$ and $b(z)$ decrease rapidly with distance from the generation region, thus providing an obvious boundary condition.

To solve (3) and (4) we need explicit forms of $\alpha(z)$ and $v(z)$. In solar dynamo models (Chapter 11), it is usually assumed that $v < 0$ (i.e. the angular velocity increases with depth into the interior), or, more precisely, that $\alpha v < 0$. On the basis of these assumptions we shall examine two simple models for which Eqs. (3) and (4) can be solved analytically. In both models the regions of non-zero velocity shear and non-zero helicity do not overlap. In the first model these regions are highly concentrated and in the second they are extremely unlocalized. It is apparent that, since the sources do not overlap, it is possible to obtain an analytical solution of the non-linear problem. It is also interesting to note that the assumption of non-overlapping sources is not entirely unrealistic for stars (see Chapter 11).

As a first model we suppose that the helicity and shear are concentrated in two parallel layers, distant h apart (Moffatt, 1978)

$$\alpha = \delta(z - 1), \qquad v = -\delta(z). \tag{6}$$

Then, after applying continuity conditions at $z = 0$ and $z = 1$ and requiring the fields to vanish at infinity, we find that Eqs. (3) and (4) give

$$a(z) = (\tilde{b}/2q)\, e^{-q|z-1|}, \qquad b(z) = \tilde{b}\, e^{q(1-|z|)}, \tag{7}$$

where q is complex and

$$q^2 = \tfrac{1}{4} ikd\, e^{-2q}. \tag{8}$$

In the solar context, Eqs. (6)–(8) correspond to the northern hemisphere; for the southern hemisphere $q^2 = -\tfrac{1}{4} ikd \exp(-2q)$.

To solve (8) we write $q = q_1 + iq_2$; the two roots $(n = 0, 1)$ are then given by

$$q_1 = -q_2 \cot (q_2 \mp \pi/4),$$

$$q_2 = -\tfrac{1}{2}(|k|d)^{1/2} \exp \{q_2 \cot (q_2 \mp \tfrac{1}{4}\pi)\} \sin (q_2 \mp \tfrac{1}{4}\pi + \pi n). \tag{8a}$$

Two properties of the solution are important:

(1) For large dynamo numbers the solution has several branches having different field geometries at a given latitude (x) (for a given k), different radial (z) distributions, and different complex frequencies p. The conditions for the existence of the branch labeled m is

$$d > d_*(m, k) = \pi^2 (2m - 3)^2/k.$$

The upper sign in (8a) corresponds to the odd branches $m = 1, 3, \ldots$ the lower sign to the even ones $m = 2, 4, \ldots$. Every branch has its own critical dynamo number $d_{cr}(m, k) \geq d_*(m, k)$, where

$$d_{cr}(1, k) \leq d_{cr}(2, k) \leq d_{cr}(3, k) \leq \cdots,$$

so that in the solar context, only the first branch is of interest. On replacing k by $-k$ we have $q \to q^*$ and $p \to p^*$. This should be borne in mind when making a formal transition in the dispersion relation from the northern hemisphere to the southern $(k \to -k)$. In the northern hemisphere ω and k have the same sign for the odd branch and opposite signs for the even one.

(2) Dynamo action is possible when

$$\gamma = q_2^2 [\cot^2 (q \mp \tfrac{1}{4}\pi) - 1] - k^2 \geq 0. \tag{9}$$

The corresponding frequency is $\omega = -2q_2^2 \cot (q_2 \mp \tfrac{1}{4}\pi)$. Condition (9) is satisfied for m branches if $d > d_{cr}(m, k)$. For the first branch $d_{cr}(1, k) = (d_{cr}^{(k)})_{min}$. It is easy to see that, because $|\omega|$ increases monotonically with k, the group velocity of the associated migratory dynamo waves is directed towards the equator both for $k > 0$ and $k < 0$, i.e. for both hemispheres. In this respect the model is completely consistent with the observed facts of sunspot migration. It is interesting to note that the magnetic energy tends to be transferred towards the equator for all odd modes and away from it for all even modes.

The results of the numerical solution of (8) for the basic branch $m = 1$ are shown in Figures 10.1, 10.2 and 10.3. The k-dependence of the critical dynamo number, the dimensionless period T, and the inverse characteristic scale q_1 are shown in Table 10.1. A realistic

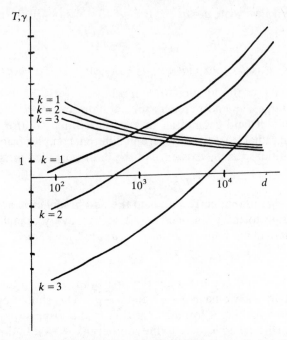

Figure 10.1   The non-dimensional period of oscillations $T = 2\pi/\omega$ and the growth rate γ of the magnetic field as a function of the dynamo number d. At large d the period does not depend on k and tends to a universal asymptotic form: $T \sim (\ln d)^{-1}$.

value of k corresponds to the latitude scale $\lambda_x = 2\pi h/k \sim R_\odot$. The dimensional period can be obtained by multiplying T by $h^2/\beta \simeq 4$ years (for $\beta = 3 \cdot 10^{12}$ cm^2/s and $h = 2 \cdot 10^{10}$ cm $\simeq 0.3 R_\odot$). It should be recalled that the quantity usually employed is not d but $D = d(R_\odot/h)^2$ which implies that $D_{cr} \sim 10^4$, in rough agreement with the results of more detailed studies (Deinzer and Stix, 1971) and numerical calculations (Ivanova and Ruzmaikin, 1977). However, one must bear in mind that the period obtained is less than that observed. It is well known (Ivanova and Ruzmaikin, 1976) that this defect is removed when the diamagnetism of the turbulent convection zone is taken into account.

Model 2 corresponds to the step-functions (Kleeorin and Ruzmaikin, 1981)

$$\alpha = \theta(z - 1), \qquad v = \theta(z) - 1, \qquad (10)$$

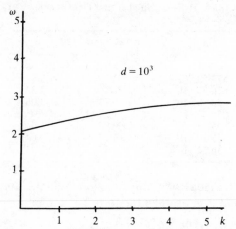

Figure 10.2 The monotonic increase of the frequency of magnetic field oscillations with increasing k. In the solar context, such a slope of the curve means that the magnetic energy is transported towards the equator.

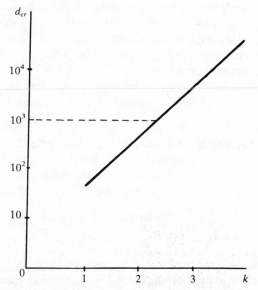

Figure 10.3 The critical dynamo number d_{cr} versus k. For realistic values of $d \sim 10^3$, the wavenumber $k \simeq 2-3$. The dependence of d_{cr} on k is exponential.

Table 10.1   The Critical Dynamo Number and Dimensionless Period
as a Function of k.

k	d_{cr}	$T = 2\pi/\Omega$	q_1	$\lambda x/R_\odot$
1	47.11	6.94	1.08	1.88
2	573.53	2.85	2.07	0.94
2.25	$10.37 \cdot 10^2$	2.46	2.32	0.84
3	$58.64 \cdot 10^2$	1.73	3.06	0.63
4	$55.27 \cdot 10^3$	1.23	4.05	0.47
5	$49.64 \cdot 10^4$	0.95	5.04	0.38
6	$43.10 \cdot 10^5$	0.77	6.04	0.31
7	$36.71 \cdot 10^6$	0.65	7.03	0.27

where $\theta(z)$ is the Heaviside unit function. Solutions of the dynamo equations (3) and (4) now take the forms

$$a(z) = \tilde{b}/4q^2 \begin{cases} \exp[q(z-1)], & z < 1; \\ [1 + 2q(z-1)]\exp[-q(z-1)], & z > 1; \end{cases}$$

$$b(z) = \tilde{b}e^q \begin{cases} (1 - 2qz)\exp(qz), & z < 0; \\ \exp(-qz), & z > 0; \end{cases}$$

(11)

where

$$q^4 = (ikd/16)\exp(-2q). \tag{12}$$

The general properties of the solutions of Eq. (12) are similar to those of the corresponding relation (8) for Model 1, although the critical dynamo numbers are lower.

Finally, we note that, according to Eq. (9), the dynamo waves existing at moderate dynamo numbers and moderate k have

$$q_1 \approx |k|, \qquad q_1 \gg q_2.$$

This means that $\partial A/\partial z \sim \partial A/\partial x$ whereas $|\partial v/\partial z| \gg |\partial v/\partial x|$. The neglect of the term $(\partial v/\partial x)(\partial A/\partial z)$ in Eq. (2) is therefore justified *a posteriori*.

II   The Non-Linear Problem

Consider the simplest non-linear dynamo problem in which, under the back reaction of the Lorentz force, the helicity is reduced according to

$$\alpha \to \alpha(1 - \xi \mathbf{B}^2), \tag{13}$$

(Steenbeck and Krause, 1969; Soward and Roberts, 1976; Ivanova and Ruzmaikin, 1977; Bräuer, 1979), where the parameter ξ is a measure of the back reaction.

In the solar convection zone $B_y \gg B_x$ and B_z, so that we may assume that $\mathbf{B}^2 \approx B_y^2 \equiv B^2$. We can, as in the linear problem, seek a solution of Eqs. (1) and (2) in two regions: $z < 1$ and $z > 1$ for A, and $z < 0$, $z > 0$ for B. In each of these regions the problem is reduced to solving a linear equation for one function (A or B) with the right-hand side determined by the other. Of course, this reduction is possible only for models with non-overlapping sources. In the helicity layer the non-linear term ($\sim B^3$), on the right-hand side of Eq. (1) for A results in frequencies which are multiples of the frequency of B (Bräuer, 1979). The A so created diffuses into the shear layer and gives rise to multiple frequencies in B etc. Thus, it is natural to seek a steady solution of the non-linear problem of the form

$$A = \sum_{n=-\infty}^{\infty} a_n(z) \exp[in(\Omega t - kx)],$$

$$B = \sum_{n=-\infty}^{\infty} b_n(z) \exp[in(\Omega t - kx)].$$

(14)

Here Ω is the basic frequency of the oscillation; also $a_n = a_n^*$ and $b_n = b_n^*$ since A and B are real. The values $n = \pm 1$ correspond to the linear problem. The expansion is a generalization of Bräuer's series (Bräuer, 1979). Eqs. (1) and (2) with α of the form (13) are reduced to the following system of non-linear ordinary differential equations:

$$(q_n^2 - d^2/dz^2)a_n(z) - \alpha b_n(z) = a\xi \sum_{n=l+p+r} b_l(z)b_p(z)b_r(z),$$

$$(q_n^2 - d^2/dz^2)b_n(z) + inkdvq_n(z) = 0$$

(15)

where

$$q_n^2 = in\Omega + n^2 k^2.$$

Equations (15) admit exact solutions in models that employ non-overlapping sources, because dynamo waves with multiple frequencies have the same form in both non-linear and linear cases.

Integrating Eqs. (15) for models (6) and (10), in the intervals $z < 0$, $0 < z < 1$ and $z > 1$, and imposing continuity conditions on both the fields and their derivatives at $z = 0$ and 1, we obtain for model 1

$$A = \sum_n \tfrac{1}{2} q_n^{-1} \left(b_n - \xi \sum_{n=l+p+r} b_l b_p b_r \right) e^{-q_n|z-1|} e^{in(\Omega t - kx)},$$

$$B = \sum_n b_n e^{q_n(1-|z|)} e^{in(\Omega t - kx)}, \tag{16}$$

and for model 2

$$A = \sum_n \frac{e^{in(\Omega t - kx)}}{4q_n^2}$$

$$\times \begin{cases} \left[\tilde{b}_n - 2\xi q_n \displaystyle\sum_{n=l+p+r} \frac{\tilde{b}_l \tilde{b}_p \tilde{b}_r}{(q_l + q_p + q_r + q_n)} \right] e^{q_n(z-1)}, & z < 1, \\[2ex] \tilde{b}_n \left[1 + 2q_n(z-1) - 2\xi q_n \displaystyle\sum_{n=l+p+r} \frac{\tilde{b}_l \tilde{b}_p \tilde{b}_r}{(q_l + q_p + q_r + q_n)} \right] e^{-q_n(z-1)} \end{cases}$$

$$+ 4\xi \sum_{n=l+p+r} \frac{q_n^2 \tilde{b}_l \tilde{b}_p \tilde{b}_r}{[(q_l + q_p + q_r)^2 - q_n^2]} e^{-(q_l+q_p+q_r)(z-1)}, \quad z > 1, \tag{17}$$

$$B = \sum_n \tilde{b}_n e^{q_n} e^{in(\Omega t - kx)} \begin{cases} (1 - 2q_n z) e^{q_n z}, & z < 0, \\ e^{-q_n z}, & z > 0. \end{cases}$$

The amplitudes $\tilde{b}_n$ of the dynamo waves and the frequency satisfy the conditions

$$\tilde{b}_n (1 + 4iq_n^2 e^{2q_n}/nkd) = \xi \sum_{n=l+p+r} \tilde{b}_l \tilde{b}_p \tilde{b}_r \tag{18a}$$

for model 1, and for model 2

$$\tilde{b}_n \left(1 + i \frac{16 q_n^2 e^{2q_n}}{nkd} \right) = \xi \sum_{n=l+p+r} \frac{q_n \tilde{b}_l \tilde{b}_p \tilde{b}_r}{(q_n + q_l + q_p + q_r)}. \tag{18b}$$

We now assume that d is close to d_{cr} and that ξ is small. Then the amplitudes of the first modes dominate the others, $b_1, b_{-1} \gg b_{|n|>1}$. Equations (18) then give for model 1

$$\tilde{b}_1 = [(1 - d^{-1} d_{cr})/3\xi]^{1/2}, \quad \Delta\Omega = 0, \tag{19}$$

and for model 2

$$1 - \frac{d_{cr}}{d} = 3\xi\tilde{b}_1^2\left(\frac{k^2 + 3K^2}{3k^2 + 5K^2}\right)\left\{1 - \frac{\omega_0^2[8^{1/2} + K^4/(k^4 + K^2)]}{(k^2 + 3K^2)(8^{1/2} + K^2)}\right\},$$

(20)

$$\Delta\Omega = -3\xi\tilde{b}_1^2(\omega_0 d/d_{cr})[8^{1/2}k^2 + (k^2 + K^6)^{1/2}(5K^2 - 3k^2)]^{-1}2^{1/2}K^4,$$

where $\Delta\Omega = \Omega - \omega_0$, $\omega_0 = \omega(d_{cr})$ and $K^2 = (k^4 + \omega_0^2)^{1/2}$. The amplitude of the stationary oscillations in both models is proportional to $(d - d_{cr})^{1/2}$. The change in frequency for the first model vanishes in this approximation, while for model 2 the frequency decreases slightly

$$\Delta\Omega/\omega_0 \sim 1 - (d/d_{cr}).$$

(21)

Mathematically, this difference is a consequence of the fact that the right-hand side of (18a) is real whereas the right-hand side of (18b) is complex. Physically, it reflects the spatial interaction of dynamo waves in model 2, in which the sources are non-localized.

In both models, if $\xi > 0$ subcritical $(d < d_{cr})$ generation (i.e. an inverse bifurcation) is impossible. This is evident for model 1. To prove the statement for model 2 it is sufficient to observe that the third factor on the right-hand side of expression (20) for $\tilde{b}_1$ is positive for all k and for all branches. (If, instead of (13), we assume that the non-linearity is proportional to B^{2l+1}, where $l > 1$, then an inverse bifurcation can occur for the second branch.)

In concluding this section we would like to summarize the basic properties of the steady non-linear solutions:

1. The frequency shift appears to result from the spatial interaction of dynamo waves. For the model with non-concentrated sources, the weak-field approximation gives $\Delta\Omega/\omega_0 \sim 1 - d/d_{cr}$. Apparently, the frequency of the non-linear solution is always less than the frequency of the associated linear solution with the same dynamo number. For sufficiently large $d - d_{cr}$ also, the frequency $\Omega(d)$ is less than $\omega(d_{cr})$, as numerical calculations with $\alpha \propto (1 + \xi_1 B^2)^{-1}$ have shown (Ivanova and Ruzmaikin, 1977). The model examined by Ivanova and Ruzmaikin is similar to model 1 with concentrated sources, and the basic period of their cycle is also unchanged for small $d - d_{cr}$.

2. Since the stationary solution is determined by $\tilde{b}_{\pm 1}$, non-linear corrections at subsequent orders appear, according to (18), only in

the $\tilde{b}_n$ of odd n. Indeed, $\tilde{b}_1 \sim (d - d_{cr})^{1/2}$ is small, so that Eq. (18) with a right-hand side of the form $\tilde{b}_{\pm 1} \tilde{b}_{\mp 1} \tilde{b}_{\pm 1}$ is satisfied only because d is near d_{cr}, i.e. the expression in brackets on the left-hand side is very small. If $n = \pm 2$ (or n has another even value) the right-hand side of Eq. (18) cannot be constructed from $\tilde{b}_{\pm 1}$ (or $\tilde{b}_{2n+1}$), so $\tilde{b}_{2n} \approx 0$. Hence we have an important property of the non-linear solution (Ivanova and Ruzmaikin, 1977; Bräuer, 1979)

$$B(t + \pi/\Omega) = -B(t).$$

3. Subcritical generation (an inverse bifurcation) is impossible in the present models.

III   Landau's Equation in Dynamo Theory

The transition to the solution established after bifurcation can be described by an equation which is analogous to the well-known Landau equation for the onset of hydrodynamic turbulence (Landau and Lifshitz, 1959). Let us emphasize though that the Landau equation only has a meaning in non-linear dynamo problems.

It will be recalled that in the hydrodynamic context, the objective is to study the stability of a steady flow, $u_0(r)$, as the Reynolds number is increased. In the linear approximation one obtains a basic frequency that has a negative imaginary part, with a corresponding growth in the amplitude, A, of the velocity disturbance

$$u_1(r, t) = A(t)f(r), \qquad A(t) = e^{\gamma t + i\omega t}.$$

The Landau equation describes the time evolution of $|A|^2$, averaged over times large compared with $2\pi/\omega$;

$$d\overline{|A|^2}/dt = 2\gamma\overline{|A|^2} - \delta(\overline{|A|^2})^2,$$

where the next (non-linear) terms have been taken into account. In the case of dynamo theory the situation is somewhat simpler because one can take $B = 0$ for the unperturbed field (in accordance with the sense of the dynamo problem). In the case of the axisymmetric $\alpha\omega$-dynamo, the Reynolds number above is replaced by the dynamo number. The growth rate γ is obtained by solving the linear problem, γ being negative when $d < d_{\mathrm{cr}}$ where $\gamma(d_{\mathrm{cr}}) = 0$, and being positive when $d > d_{\mathrm{cr}}$. The most difficult quantity to find is the factor δ. This requires us to solve the dynamo equations in the next (non-linear) approximation. The sign of δ is decisive. If $\delta > 0$, normal bifurcation to a new solution occurs at $d = d_{\mathrm{cr}}$. The stationary amplitude of this solution for small positive $d - d_{\mathrm{cr}}$ is $|B| = (2\gamma/\delta)^{1/2}$. On further increasing d this solution may become unstable, another bifurcation occurs, and so on. If $\delta < 0$, the transition to a new solution occurs at $d < d_{\mathrm{cr}}$ (an inverse bifurcation). In this case the solution rapidly becomes unstable and stochastic as d is increased, and possibly a strange attractor appears (Ruelle and Takens, 1971; Monin, 1978; Rabinovich, 1978). The constant δ can be obtained in our case just by knowing the amplitude and the rate of growth since $B^2 = 2\gamma/\delta$. For instance, in model 1

$$dB^2/dt = 2\gamma B^2 - 6[1 + (d_{\mathrm{cr}}/d)]\xi\gamma B^4. \tag{22}$$

A formal derivation of this equation is given in an appendix to the paper of Kleeorin and Ruzmaikin (1981).

Thus, in the models studied having $\xi > 0$, only normal bifurcation is possible. One can imagine a transition to stochastic behavior resulting from a sequence of successive bifurcations which occur as $d - d_{\mathrm{cr}}$ is increased. Ruelle and Takens (1971) believe that stochastic behavior sets in after three bifurcations. It should be borne in mind, however, that the solar dynamo number is apparently near d_{cr} (see for instance Ivanova and Ruzmaikin, 1977), so that the required number of bifurcations might not take place. Moreover, to explain phenomena like the Maunder minimum we need, not stochastic-turbulent behavior, but rather a strange attractor (Ruzmaikin, 1981; Zeldovich and Ruzmaikin, 1982). It is known from a study of the Lorenz convective system (see for instance Monin, 1978) that the appearance of a strange attractor is associated with the occurrence of an inverse bifurcation in the system ($\delta < 0$). In the dynamo models presented here this would require $\xi < 0$. Consequently, to find the sign of ξ is of prime importance. Instead of (13) we must construct an equation for $\partial\alpha/\partial t$ and, possibly, for $\partial\nabla v/\partial t$ also. Such additional equations must be solved in conjunction with Eqs. (1) and (2).

IV Mean Helicity Dynamics in a Magnetic Field

The phenomenological approach to the non-linear problem developed above [see (13)] results in a number of interesting and important results (the stabilization of field oscillations, the symmetry of the solution, the period dependence on amplitude etc.). As a next step it is important to compare these results with reality and to confront the approach with the fundamentals of the theory. The observational facts concerning stellar cycles will be discussed in the next chapter. Here we examine in more detail the physical principles governing the back reaction of the magnetic field on the helicity.

Let l be the length-scale at which the kinetic energy of the turbulence is generated. For simplicity we set the mean fluid velocity zero and divide the magnetic field into a large-scale mean field $\langle \mathbf{H} \rangle \equiv \mathbf{B}$ and a small-scale random component $\mathbf{h}$. The averaging is carried out over scales large compared with l and over times greatly exceeding the characteristic correlation time τ_0 of the turbulence. It is convenient to use for τ_0 the time of overturning of an eddy of scale l.

The change in the hydrodynamical helicity $\langle \mathbf{v} \cdot \nabla \times \mathbf{v} \rangle$ due to the Lorentz force has been examined in detail by Vainstein and Zeldovich (1972) and by Roberts and Soward (1975) [see also Vainstein et al. (1980)]. The main results are as follows. First, the hydrodynamical helicity has a short relaxation time (of the order of τ_0) so that it can keep pace with changes in the mean magnetic field, i.e. in the first approximation one can safely use a formula such as (13). Second, the magnetic field changes the helicity $\langle \mathbf{v} \cdot \nabla \times \mathbf{v} \rangle$ even when the Coriolis forces vanish. In a homogeneous isotropic turbulence the effect is proportional to B^4, while in a flow with a preferred direction it can be proportional to B^2, for $B \to 0$.

In reality, we are interested not so much in the change of the hydrodynamical helicity of the magnetic field as in the modification of Ohms law,

$$\langle \mathbf{v} \times \mathbf{h} \rangle = \alpha \mathbf{B} - \beta \nabla \times \mathbf{B}, \tag{23}$$

resulting from the Lorentz force. It appears that, in sufficiently weak magnetic fields, the law (23) preserves its form (Frisch et al., 1975; Vainshtein, 1980a). The expressions for α and β are, however, changed. In the first approximation we need only discuss modification to the helicity, leaving the diffusivity unchanged.

In the linear problem the helicity is completely determined by $\langle \mathbf{v} \cdot \nabla \times \mathbf{v} \rangle$. The back reaction of the magnetic field on the motion results not only in a change in its magnitude, as given for instance in (13), but also in the appearance of a qualitatively new contribution to α, the so-called mean magnetic helicity $\langle \mathbf{h} \cdot \nabla \times \mathbf{h} \rangle$ (Frisch et al., 1975). The origin of this effect may be easily understood from the following considerations.

The response of the velocity field to the action of the mean field $\mathbf{B}$ in the presence of a small-scale magnetic field $\mathbf{h}$ is

$$\mathbf{v} = \mathbf{v}_0 + (4\pi\rho)^{-1} \int_0^t (\nabla \times \mathbf{h}) \times \mathbf{B} \, dt.$$

We have omitted $O(h^2)$ terms because the third order $\langle hhh \rangle$ correlations, that arise after multiplying by h and averaging, can be reduced to second order correlations in the τ-relaxation approximation (following the lead of Orszag 1970). After averaging (in the isotropic case) a new contribution to the electric field arises in (23), namely an additional

$$\langle \mathbf{v} \times \mathbf{h} \rangle = (6\pi\rho)^{-1} \langle \mathbf{h} \cdot \nabla \times \mathbf{h} \rangle \mathbf{B}.$$

The presence of the magnetic helicity in (23) introduces a serious complication into the problem. Earlier we could express $\langle \mathbf{v} \times \mathbf{h} \rangle$ as a function of the mean magnetic field so as to obtain a closed equation for $\langle \mathbf{H} \rangle$. Now the small-scale field is involved in Ohms law. The problem is to find a connection between $\langle \mathbf{h} \cdot \nabla \times \mathbf{h} \rangle$ and the mean magnetic field.

A characteristic feature of the magnetic helicity is that, unlike its hydrodynamical counterpart, its relaxation time is, in general, not equal to $\tau_0 \sim l/v$. The magnetic helicity cannot keep pace with changes in the large-scale magnetic field. Its dependence on the mean magnetic field is therefore described by a differential equation (Vainshtein, 1980a; Kleeorin and Ruzmaikin, 1982). The reason can be found in the difference between the conservation laws of ideal magnetohydrodynamics and hydrodynamics. We will discuss this point in detail later, after deriving the equation governing the magnetic helicity.

The technique for deriving an equation that connects $\langle \mathbf{h} \cdot \nabla \times \mathbf{h} \rangle$ to the mean magnetic field is quite simple. Let us introduce a vector-potential $\mathbf{A}$ for the total magnetic field $\mathbf{H}$ and divide it into mean ($\langle\mathbf{A}\rangle$) and fluctuating ($\mathbf{A}'$) parts. We know that in a perfectly conducting fluid the topological invariant $\langle\mathbf{A} \cdot \mathbf{H}\rangle$, characterizing the linkages of the magnetic lines of force, is conserved, i.e.

$$\langle\mathbf{A}' \cdot \mathbf{h}\rangle + \langle\mathbf{A}\rangle \cdot \mathbf{B} = \text{constant.}$$

It follows that the mean of the product of the fluctuating components can be expressed immediately as the product of two mean fields. Provided $\langle\mathbf{h} \cdot \nabla \times \mathbf{h}\rangle$ can be easily related to $\langle\mathbf{A}' \cdot \mathbf{h}\rangle$, the problem is solved.

We will now give the derivation, taking into account also the finite conductivity of the fluid. Let us multiply the induction equation,

$$\partial\mathbf{H}/\partial t = \nabla \times (\mathbf{v} \times \mathbf{H} - \nu_m \nabla \times \mathbf{H}),$$

scalarly by $\mathbf{A}'$ and the corresponding equation for the vector potential,

$$\partial\mathbf{A}/\partial t = \mathbf{v} \times \mathbf{H} - \nu_m \nabla \times \mathbf{A},$$

scalarly by $\mathbf{h}$. After summing and averaging over turbulent fluctuations, we obtain an equation for the variable $\chi \equiv \langle\mathbf{A}' \cdot \mathbf{h}\rangle$, namely

$$\partial\chi/\partial t = -2\langle\mathbf{v} \times \mathbf{h}\rangle \cdot \mathbf{B} - 2\nu_m\langle\mathbf{h} \cdot \nabla \times \mathbf{h}\rangle - \langle\nabla \cdot [\mathbf{A} \times (\mathbf{v} \times \mathbf{h})]\rangle. \quad (24)$$

The term containing the divergence may be omitted in virtue of boundary conditions. The term $\langle\mathbf{v} \times \mathbf{h}\rangle$ may be expressed in terms of the mean field and the total mean helicity (or more precisely α), by using (23).

In relating χ to α, it is convenient to introduce a spectrum function $\chi(k, t)$ where

$$\chi(t) = \int \chi(k, t) \, dk.$$

Let us assume that this helicity is injected at the wave-number k_0. Like the energy, it will progressively "cascade" through the sequence of decreasing length scales of the inertial range (k_0, k_1), where k_1 is the upper wave-number bound of that range. For the sake of simplicity we shall also assume that the helicity has a power spectrum (k^{-q}), i.e.

$$\chi(k, t) = \chi(t)[(q - 1)/k_0][1 - (k_0/k_1)^{q-1}]^{-1}(k/k_0)^{-q}.$$

By choice of the normalizing factor, the mean helicity is $\chi(t)$. The exponent q is determined by the specific mechanism of non-linear interaction that causes the χ-cascade; we regard it as a phenomenological constant. For the purpose of making estimates, one can use $q = \frac{5}{3}$ (the Kolmogorov spectrum) or $q = \frac{3}{2}$ [an ensemble of interacting Alfvén waves (Kraichnan, 1965)]. The characteristic time of the helicity transfer for each scale coincides with the characteristic time for the energy transfer, and it is determined by the condition of constant flux in the inertial range

$$\tau^{-1}(k) \int_{k_0}^{k_1} \chi(k) \, dk = \text{constant},$$

whence it follows that

$$\tau(k) = \tau_0 (k/k_0)^{1-q}.$$

The magnetic helicity can now easily be calculated

$$\langle \mathbf{h} \cdot \nabla \times \mathbf{h} \rangle = \int_{k_0}^{k_1} k^2 \chi(k, t) \, dk$$

$$= \chi(t) k_0^2 \left(\frac{q-1}{3-q} \right) \frac{(k_1/k_0)^{3-q} - 1}{[1 - (k_0/k_1)^{q-1}]}.$$

Its value is decided principally by the largest scale, k_0^{-1}.

The contribution made by the magnetic helicity to α_m is given by the integral

$$\alpha_m = (12\pi\rho)^{-1} \int_{k_0}^{k_1} k^2 \tau(k) \chi(k, t) \, dk$$

$$\equiv \chi(t) I,$$

where

$$I = \frac{1}{18} \frac{1}{4\pi\rho\nu_T} \left(\frac{q-1}{2-q} \right) \frac{(k_1/k_0)^{4-2q} - 1}{[1 - (k_0/k_1)^{q-1}]},$$

and

$$\nu_T = \tau_0 \nu_0^2 / 3$$

is a turbulent diffusivity. Multiplying (24) by I we obtain the requisite equation

$$\partial \alpha_m / \partial t = (Q/4\pi\rho)[\mathbf{B} \cdot (\nabla \times \mathbf{B}) - (\alpha_m + \alpha_h)\nu_T^{-1} B^2] - (\alpha_m / T), \quad (25)$$

where

$$Q = \frac{1}{18}\left(\frac{q-1}{2-q}\right)\frac{(k_1/k_0)^{4-2q} - 1}{[1-(k_0/k_1)^{q-1}]},$$

$$T = \left(\frac{3-q}{q-1}\right)\frac{1}{2k_0^2\nu_m}\frac{[1-(k_0/k_1)^{q-1}]}{(k_1/k_0)^{3-q} - 1}.$$

In one limiting case k_1^{-1} is determined by the realizability condition

$$|\chi(k)| \leq 2k^{-1}M(k); \tag{26}$$

see for instance Moffatt (1978); k_1 is that k for which (26) holds with equality. Here $M(k)$ is the magnetic energy spectrum function so that

$$\langle H^2 \rangle / 8\pi = \int M(k)\,dk.$$

The physical meaning of condition (26) is that the presence of magnetic helicity on a given scale requires that magnetic energy also exists there.

When k_1 is close to k_0, the factors Q and T become universal constants, independent of the spectral index q,

$$Q \simeq \tfrac{1}{9}, \qquad T \simeq \tfrac{1}{2}k_0^{-2}\nu_m^{-1} = \tau_0 R_m, \tag{27}$$

where $R_m = v/k_0\nu_m$ is the magnetic Reynolds number. This gives a hint that, in this approximation, Eq. (25) also has a universal nature, i.e. its form does not depend on the details of its derivation. We draw particular attention to the fact that the characteristic time T of the α-relaxation in (27) is R_m times larger than the relaxation time τ_0 of the hydrodynamical helicity.

In the other limiting case k_1^{-1} is taken to be of the same order as the dissipation scale k_d^{-1}, which itself is determined by the obvious condition

$$\tau^{-1}(k_d) = \nu_m k_d^2,$$

implying

$$k_d = k_0 R_m^{1/(3-q)}.$$

Now

$$Q \approx \tfrac{1}{9}(3\delta/2+1)(1-3\delta)^{-1}R_m^{\frac{1}{2}(1-3\delta)/(1-3\delta/4)},$$
$$T \simeq \tau_0(1-3\delta/4)/(1+3\delta/2) \simeq \tau_0, \qquad \delta = q - \tfrac{5}{3}.$$

(28)

Only in this case is the relaxation time T of the order of τ_0.

Thus the characteristic relaxation time T crucially depends on the realizability bound, k_1. To obtain this one needs to know the magnetic energy spectrum function $M(k)$. It is clear, however, that at large magnetic Reynolds numbers (large k_d) T is large [see (27)].

In the purely hydrodynamical case we also have a realizability condition

$$|F(k)| \le 2kE(k),$$

where $E(k)$ is the energy spectrum function and $F(k)$ is the spectrum function of the helicity $\mathbf{v} \cdot (\nabla \times \mathbf{v})$. In a homogeneous isotropic turbulence these functions have an identical dependence on k (Moffatt, 1978). Hence, when the realizability condition holds on the scale k_0^{-1}, it must hold also on the dissipation scale k_d^{-1}. This is the reason for the difference between the hydrodynamical case and the magnetohydrodynamical one. A deeper insight shows that it arises for the following reason. The helicity $\langle \mathbf{v} \cdot \nabla \times \mathbf{v} \rangle$ is constant in time in ideal hydrodynamics. In contrast, the helicity $\langle \mathbf{H} \cdot \nabla \times \mathbf{H} \rangle$ is not constant in time in ideal magnetohydrodynamics; it is $\langle \mathbf{H} \cdot \mathbf{A} \rangle$ that is conserved. That is why the realizability conditions in hydrodynamics and magnetohydrodynamics differ by k^2. This also explains the contraction of the inertial range in magnetohydrodynamic turbulence [(k_0, k_1) as against (k_0, k_d)].

It must be noted that Eq. (25) describes the change in the magnetic helicity. The back reaction of the field on the hydrodynamical helicity $\langle \mathbf{v} \cdot \nabla \times \mathbf{v} \rangle$ can be taken into account in (23) directly by using forms like (13). On the other hand, the total helicity appears on the right-hand side of (25). Hence, in the stationary case ($\partial \alpha_m / \partial t = 0$), we have

$$\alpha > \alpha_0 + \frac{\tau_0 R_m^{2/3}(\mathbf{B} \cdot \nabla \times \mathbf{B})/4\pi\rho}{1+R_m^{2/3}B^2/4\pi\rho v^2} \cong \alpha_0(1-\xi B^2),$$

where $\xi \equiv R_m^{2/3}/8\pi\rho v^2$ and $\alpha_0 = \alpha_h(B=0)$. This form is, however, viable only when $k_1 \to k_d$, which is rarely the case in astrophysics.

V Numerical Simulation of Magnetohydrodynamic Turbulence

Hopes of progress in the non-linear turbulent dynamo problem rest on numerical modeling. This has two purposes: to test and quantify known physical effects, and to discover new ones. Because of the non-existence theorems (Chapter 5), only three-dimensional turbulence is of interest in the dynamo problem. The construction of numerical models of the mean field turbulent dynamo has been described in detail in a series of reviews and monographs. Some numerical non-linear models for the solar cycle are mentioned in the next chapter. Here we briefly dwell on the results of direct numerical simulations of non-linear magnetohydromagnetic dynamo processes performed by U. Frisch and his colleagues (Meneguzzi $et\ al.$, 1981; see also references in that paper).

The problem is to integrate the complete magnetohydrodynamic equations; with $\mathbf{H}$ scaled by $\sqrt{(4\pi\rho)}$, these are

$$(\partial/\partial t - \nu\nabla^2)\mathbf{v} = -(\mathbf{v}\cdot\nabla)\mathbf{v} + (\mathbf{H}\cdot\nabla)\mathbf{H} - \nabla P + \mathbf{f},$$

$$(\partial/\partial t - \nu_m\nabla^2)\mathbf{H} = -(\mathbf{v}\cdot\nabla)\mathbf{H} + (\mathbf{H}\cdot\nabla)\mathbf{v}.$$

Here $\mathbf{f}$ is a random force δ-correlated in time which drives the turbulent flow; P includes the kinetic and magnetic pressures. Incompressibility is assumed.

The turbulence was supposed to be homogeneous and isotropic. A 2π-periodicity in all three spatial directions was assumed so that the spectral technique (3-dimensional fast Fourier transforms) could be applied. Two codes were developed for the NCAR CRAY1 computer, for a 32^3 mesh that required 0.5 s per time step and for a 64^3 mesh that needed 16 s per time step. The maximum Reynolds and magnetic Reynolds numbers attained were 100; the magnetic Prandtl number ν/ν_m was unity. Flows with a mean helicity and reflectionally symmetric flows were considered separately.

(a) The dynamo in reflectionally invariant turbulence

Initially motion with an energy spectrum of the form $k^4 \exp(-k^2)$ was injected into a pure hydrodynamical flow with $\mathrm{Re}^{-1} = 1.2 \cdot 10^{-2}$. The Navier–Stokes equation was integrated until a statistically steady state was reached. This required a few turnover times (as measured for eddies of the injection scale, $k \sim 1$). At that moment, t_1, a weak magnetic field was introduced, the ratio of magnetic energy to kinetic energy being 0.02. The magnetohydrodynamic equations were then integrated for more than 10 turnover times. The ratio of energies rapidly grew to 0.1 and after that it appeared that a statistically steady state had been attained. The magnetic energy spectrum established had a typical bell-like shape with a maximum at quite large k (ten times more than the k of injection). There was a slight excess of magnetic energy at high wavenumbers ($>k_{max}$). Unfortunately, the calculations do not give a definitive form for the spectrum at low wavenumbers, and it is in this spectral range that the theoretical dynamo models unequivocally predict the spectrum $M(k) \sim k^4$ (see Chapter 8). This important question should be attacked in future numerical simulations.

An interesting result was the appearance of intermittency in the magnetic field in physical space. The calculations revealed that, while the magnetic energy in the stationary state was only 10% of the kinetic energy, the maximum value of **H** over space was comparable to the maximum of **v**. This suggests that the magnetic field is concentrated in small regions of physical space (Figure 10.4). It is a remarkable fact that the velocity distribution (but not that of the vorticity!) remained quite smooth. The intermittent magnetic structures were variable in time, having a lifetime of about one eddy turnover time.

It is natural to compare the appearance of intermittency with the findings of the analytical approach of Chapter 7 for the velocity distribution $v_l = c_{lk} x_k$. That revealed the existence (in the limit $R_m \to \infty$) of field concentrations in ropes and sheets. The magnetic Reynolds number of the integrations described here was not large (100) so that the ranges of intermittency had more complicated forms. It is more important that the numerical experiments should support the conclusions of Chapter 8 on the possibility of dynamo action in an, on average, reflectionally-symmetric turbulence. We recall, however, that, while the mean hydrodynamical helicity was absent in that case, strong fluctuations of the helicity existed.

(b) Turbulence with a Mean Helicity

In this case both kinetic energy and hydrodynamical helicity $\langle \mathbf{v} \cdot \nabla \times \mathbf{v} \rangle$ were injected, the latter being maximal [$F(k) = kE(k)$]. The energy injection spectrum was $\sim k^4 \exp(-0.08k^2)$, which peaks

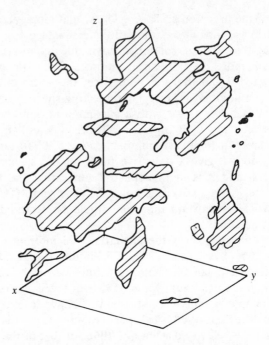

Figure 10.4 Spatial intermittency of the magnetic field in a nonlinear dynamo. The mean helicity is zero, $R_e = R_m = 100$, $t = 23$. The shaded regions have h within 95% or more of its maximum value (after Meneguzzi *et al.*, 1981).

at $k = 5$. Then, after a state had been achieved, a seed of 2% magnetic energy was introduced and the integration of the magnetohydrodynamic equation was continual for about 130 further turnover times. The magnetic Prandtl number was taken to be 5.

Figure 10.5 shows the evolution of the magnetic energy spectrum. Initially magnetic energy built up near the injection scale. Then the peak of the spectrum drifted to larger scales until it reached the maximal scale $k_{min}^{-1} = 1$. It then continued to grow in amplitude on the large scales. At the end of the integration ($t = 45$ in Figure 10.5) $M(k)/E(k) = 50$ at $k = 1$ and the relative magnetic helicity was $k\chi(k)/M(k) = -0.96$ at $k = 1$, i.e. it was negative and nearly maximal. Hence the large-scale magnetic field was almost force-free and produced practically no back reaction on the large-scale motions.

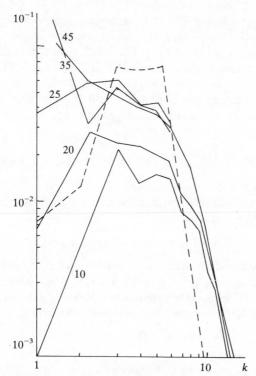

Figure 10.5 The nonlinear dynamo with mean helicity. Magnetic energy spectra are shown at different times, and the kinetic energy spectrum at $t = 45$ (after Meneguzzi *et al.*, 1981).

The force-free field appears here in a periodic flow. In a real situation, where the magnetic field vanishes like $O(r^{-3})$ as $r \to \infty$, a totally force-free solution is not possible (see, for instance, Moffatt, 1978).

In the case considered the back reaction on the motion was due entirely to the small-scale magnetic fields. According to Figure 10.5 the small-scale ($k \sim 5$) field was an order of magnitude less than the large-scale ($k \sim 1$) field at the end of the integration period ($t = 45$). The energy of this field was (after $t \approx 20-25$) in approximate equipartition with the kinetic energy. At small-scales ($k \sim 5$), magnetohydrodynamic turbulence can apparently be pictured statistically as a sea of interacting Alfvén waves riding on the large-scale ($k \sim 1$) magnetic field. It is interesting that the large-scale magnetic field can gain energy from the small-scale turbulence over a long time (Kraichnan, 1979; Meneguzzi et al., 1981). The effect is due to a mean helicity created by the small-scale turbulence. The magnitude of this helicity can be estimated as follows (Moffatt, 1978; Kraichnan, 1979).

Suppose first that the large-scale field **B** is uniform and directed along the z-axis. The helical standing waves are

$$\mathbf{v}(r, t) = v_0 \cos (V_A kt)[\sin kz, \cos kz, 0],$$

$$\mathbf{h}(r, t) = (4\pi\rho)^{1/2}v_0 \sin (V_A kt)[\cos kz, -\sin kz, 0],$$

where $V_A = B/(4\pi\rho)^{1/2}$ is the Alfvén velocity, ρ is the fluid density, **v** and **h** are the velocity and small-scale field, and produce an electric field directed along the z-axis

$$(\mathbf{v} \times \mathbf{h})_z = -\tfrac{1}{2}v_0^2 V_A^{-1} \sin (2V_A kt)B.$$

Now let a statistically stationary excitation of such waves be maintained by statistical forcing (with a white noise spectrum) of the velocity, there being an effective wave-damping rate γ (due to resistivity or non-linear effects). Then the mean helicity is

$$\alpha = \langle \mathbf{v} \times \mathbf{h} \rangle_z / B$$

$$= \tfrac{1}{2}B^{-1} \int_0^\infty \sin (2V_A kt) \, e^{-\gamma t} \, dt$$

$$= -\gamma k \langle v^2 \rangle / (\gamma^2 + 2k^2 V_A^2),$$

where

$$\langle v^2 \rangle = v_0^2 \int_0^\infty \cos^2 (V_A kt) \, e^{-\gamma t} \, dt.$$

It is clear that in the strong-field regime $\alpha \sim B^{-2}$.

By a similar calculation for the case of a large-scale magnetic field with constant gradient (Moffatt, 1978) we may obtain an expression for the turbulent diffusivity which is also proportional to B^{-2} in the strong-field regime. Therefore, by Eq. (8.9), linear growth (or decay) of the large-scale magnetic field is possible when we neglect the ohmic dissipation. In the numerical calculations of Meneguzzi et al. (1981), $\nu_m = 2.5 \cdot 10^{-3}$ in dimensionless units. Thus, the ohmic decay, which will stabilize the large-scale force-free field, will start to act only after a characteristic time of $(\nu_m k^2)^{-1} \sim 400$. We recall that the calculations were terminated when $t = 45$. The linear growth of the large-scale magnetic energy was therefore temporary and was due to the gain of energy from the small-scale magnetohydrodynamic turbulence. It is emphasized once more that this growing force-free field could arise only because periodic boundary conditions were postulated.

CHAPTER 11

Stellar Cycles

I Problems of Stellar Taxonomy

Taxonomy, the science of classification, was long ago applied with conspicuous success to plants and animals and, as is well known, this became the basis on which theories of natural evolution rest. The classification and identification of stellar spectra are likewise the empirical basis of the theory of stellar evolution, and for astronomers the names Hertzsprung, Russell and Eddington are as evocative as Linnaeus and Darwin are for naturalists. In classical astronomy stars are plotted as points on the two-dimensional Hertzsprung–Russell diagram (Figure 11.1), where one of the axes labels the spectral class, and the other the stellar luminosity. The spectral class is determined basically by the degree of ionization of atoms in the stellar atmosphere, and this itself is given by the effective temperature and density of the stellar surface. The temperature of the hottest O-type stars is $(2.5–3.0) \cdot 10^4$ K. The Sun belongs to spectral type G ($\sim 0.6 \cdot 10^4$ K), while the coolest stars with temperatures of about $0.2 \cdot 10^4$ K are referred to as of M-type. Stellar luminosity proves to depend non-uniquely on the spectral type (at the same degree of ionization the temperatures of stars with dense and extended atmospheres are different). We therefore see several curves on the H–R diagram. The central one is called the main sequence; above it (in luminosity) are the curves of subgiants, giants, bright giants and supergiants; below are the curves of subdwarfs and white dwarfs. This non-uniqueness is taken into account by adding to the spectral type of each star the letter d (dwarf), g (giant) or s (supergiant); for example, the Sun is dG. Since the effective temperature changes even within one spectral type, further subclasses are defined and are distinguished by the addition of a digit to the two letters; in this fashion the Sun is a dG2 star. As

194

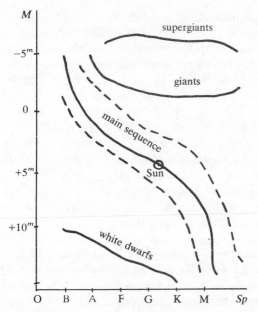

Figure 11.1 Hertzsprung–Russell diagram: *M* is the absolute magnitude, *Sp* is the spectral type (class). Subgiants and subdwarfs are situated on the dashed lines above and below the main sequence.

observational astronomy developed and new features of stellar spectra were identified (for the same temperature and luminosity, the spectral lines may be narrow or broad, polarized or unpolarized, etc.) additional letters were added to these spectral designations, and the two-dimensional classification lost its transparency. Now astrophysicists are keenly aware of the necessity for progress in stellar taxonomy (Cram, 1980).

The classical two-dimensional classification was in good agreement with models of stellar atmospheres, in which thermal properties were fully determined by their hydrostatic and their radiative or convective equilibrium. Those simple models are however, incapable of explaining some important observational facts established during the rapid progress in astrophysical observations at all wavelengths, from the IR to the X-ray and γ-ranges. Particularly deep impressions were made by the discoveries of the quintuply ionized oxygen O(VI) lines (ionization potential 114 eV) in the UV emitted from stellar atmospheres with temperatures of no more than $5 \cdot 10^4$ K, of the resonant ionized calcium Ca(II) lines in the visible band (a direct indication of the presence of a chromosphere—a thin nonthermal envelope—in many stars), and of stellar coronas with temperatures of about 10^6 K. Prolonged observations of the Ca(II) lines revealed a cyclic variability in a number of stars (Wilson, 1978, see Section II below).

It is interesting that all these phenomena can be observed on the Sun. Due to the closeness of this star we understand their physical origins. The most important of these, along with rotation and turbulence, is the presence of a magnetic field. Both the solar cycle, and the heating of the solar chromosphere and corona, owe their existence to this field. It appears to be the major factor in stellar activity. The discovery and study of such activity shook classical taxonomy to its very roots where lies buried the idea of a quiet star, in approximate mechanical and thermal equilibrium.

The activity of the Sun is not great; it is known that during periods of maximum activity sunspots cover no more than 2% of the solar surface, and the X-ray luminosity of the solar corona is only a millionth of the full luminosity of the Sun. Solar activity can be considered, therefore, as a small departure from equilibrium. Observations show, however, that the activity of some stars is so high that such a picture would be untenable. For example, the spots of the star BY-Draconis in the Draco constellation located between the Ursa Major and Ursa Minor cover about 20% of the star's surface! The X-ray luminosity of the coronas of some dwarfs is as much as 10% of their total luminosity.

II   Evidence of Stellar Activity

Astronomers are fond of saying that the Sun is a typical star. So it was obvious long ago that its activity, and apparently its periodicity, must be typical of many stars. It was only recently that observational evidence of stellar activity in some other stars became available. The Sun is unique in its closeness and, because of this, its relatively weak activity can be studied in detail. It is found (see above) that sunspots never cover more than 2% of the solar surface; the X-ray luminosity of the solar corona is 10^6 times weaker than the optical luminosity of the Sun. Evidently such weak phenomena would be very difficult to detect for remote stars. Progress in observation was due, on one hand, to the discovery of stars more active than the Sun and, on the other, to a substantial increase in observational accuracy as well as to the broadening of the spectral range studied. Chromospheres, coronas, winds, spots and bursts bear witness to stellar activity. Such hallmarks of activity are seen on the Sun, and from their direct and detailed observations comes an understanding of the basic reasons for all stellar activity: motion and magnetic fields.

A chromosphere is an optically thin layer of a stellar atmosphere, transparent to continuous radiation and emitting lines of hydrogen, helium, calcium and other elements. The solar chromosphere extends up to a height of about $10^{-2}R_{\odot}$, where it gradually merges into the corona. An important feature of the chromosphere is the increase of its temperature with height. Above the Sun the temperature increases from about 4500 K at the base of the chromosphere to 30 000 K on its upper levels. This positive outward temperature gradient requires a thermal energy input to the chromosphere. The rôle of this heat source becomes increasingly significant with height since the density decreases and the radiation from a unit volume therefore diminishes. Numerous investigations (see for instance, Pikelner, 1966) have shown that solar chromospheric heating can be accomplished by magnetohydrodynamic waves. These waves are generated by convective motions in the solar magnetic field, and dissipate their energy due to the growth in the amplitude and the steepening of such waves when they propagate into a medium of decreasing density. Observations show that the structure of the solar chromosphere is closely associated with magnetic field. A chromospheric grid can be seen distinctly and resembles a honeycomb with a diameter of about $3 \cdot 10^{4}$ km, so mimicking the structure of the on average stationary solar convection in lower layers, the so-called supergranulation. Gas in visible cells flows from their centers to their peripheries with a velocity of about 0.5 km/s and concentrates the field at the cell boundaries. The existence of the magnetic field and its significance for the chromosphere are proved by the bursts that are observed in many frequency bands, as well as by other active phenomena (for example, thin vertical jets—spicules and prominences). Comprehensive evidence for the existence of chromospheres on other stars was gathered by optical astronomers, who detected the lines Ca(II), Mg(II), the hydrogen lines L_{α}, H_{α} amongst others, and the helium lines He(I), He(II) (Linsky, 1977). Chromospheres were discovered in completely evolved F- to M-type stars (the quiet Sun is of G2-type), young stars of the same spectral types (similar to the active Sun), the G- to M-type giants and supergiants, long periodic variables (O Ceti-type), spectroscopic binaries, novae, and protostars, amongst others.

The K and H lines of ionized calcium Ca(II), (ionization potential 11.87 eV) with $\lambda = 3933$ Å and 3968 Å respectively, proved to be the most convenient "tool" for the study of radiating chromospheres. These lines are excited mainly by collisions of particles in the chromosphere itself (not by the radiation coming from the photosphere) and have a considerable optical depth. By observing these lines for more than 10 years, O. Wilson (1978) established the existence of periodicity in stellar activity. Using the Mount Wilson 60-inch telescope he has been measuring, since 1967, the flux in 1 Å bands centered on the line frequencies relative to that in two continuous bands displaced to either side of these lines by 250 Å. Observations were carried out for 91 stars of the main sequence of F- to M-type; 18 of these were used as standards, their fluxes were minimal and remained practically constant in time. Most of the other stars revealed flux changes in the K–H lines and about 10 stars showed a variation in the time derivative of their flux, i.e. they completed full cycles.

An example is the star HD 81809, which is of spectral class G2 and which exhibited a complete 10-year cycle. In its spectral type and activity period, this star is similar to the Sun but the amplitude of its activity proved to be four times larger than that of the Sun during its solar cycle. Wilson noticed that the amplitudes of the variation were larger for stars of the later spectral types. This is important for the theory since the convection zones of stars increase in depth with the lateness of their spectral class and so the intensity of the convective motions capable of generating a magnetic field also increases.

Much stronger variations in the visible luminosity (of up to 30%) had previously been seen in spectroscopic binary dwarfs (Chugainov, 1966). An example is the star BY Dra (HD 234677, K7Ve) from the Draco constellation located between Ursa Major and Ursa Minor (Figure 11.2). The period of its luminosity variations (3.826 days) proved to be different from its orbital period (5.981 days). Periodic variations can be observed over a period of a few weeks after which

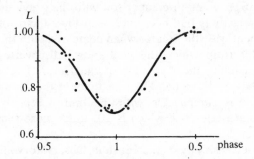

Figure 11.2   Observed brightness variations of *BY* Dra in 1965 (after Chugainov, 1966).

they disappear, only to reappear with another phase but the same period (Bopp and Evans, 1973). Such luminosity behavior can be explained only by supposing that one of the stars in the binary system has a spot (or spots) with a temperature several hundred degrees below that of the rest of its surface ($T_{Sur} = 3500$ K, $T_{Spot} = 2000$ K), and that the maximum coverage of the surface by the spot (or spots) is about 20%. The 3.826-day period is identified with the rotation period of the star. The discovery of a corona and the identification of bursts from this star corroborated this interpretation. Burst activity associated with surface spots seems to be incidental to a broad class of the K- to M-type dwarfs as well as to young evolving stars (Gershberg, 1975). And there are many stars whose activity is much higher than the Sun's and which also sometimes exhibit cyclic variations. In particular, astronomers of the Harvard Smithsonian Astrophysical Center proved that BY Dra type stars have cycles with periods of about ten years.

Essential evidence of stellar activity was furnished by the discovery of stellar coronas, made with the aid of the 58-cm X-ray telescope mounted on board the High-Energy Astronomical Observatory (HEAO-2) satellite named after Einstein (Giacconi, 1980). It was launched in November, 1978 and operated until the spring of 1981. The sensitivity of the X-ray telescope which functioned in the range 0.1 to 4 Kev was outstanding. According to Giacconi, the improvement in its sensitivity, when compared with the first X-ray telescopes of 1962, was as great as the 10 million-fold increase in sensitivity of the famous Palomar 200-inch optical telescope over that of Galileo's telescope of 1610. Even in comparison with the previous HEAO-1 satellite, the sensitivity had been increased (by a factor of 10^3), and the improvement was still greater when compared with the UHURU satellite (1970) from which the well-known discoveries in X-ray astronomy (including the discovery of the Her X-I, Cen X-3 and Cyg X-I sources) were made. The sensitivity of the X-ray telescope on HEAO-2 was comparable with the optical sensitivity of the Palomar 200-inch telescope which is 10^6 times more sensitive than the human eye.

This extreme sensitivity of the Einstein Observatory allowed it to detect an X-ray luminosity of 10^{28} erg/s from a star 100 pc. distant from us. It became possible therefore to discern coronas on stars of practically all spectral classes, with effective surface temperatures ranging from $3 \cdot 10^3$ K to $4 \cdot 10^4$ K and luminosities 10^4 times weaker or stronger than that of the Sun (Vaiana *et al.*, 1981). What proved to be unexpected was that the X-ray radiation came from young stars (O, B) and stars of late spectral types (K, M). The former are distinguished by their fast rotation (about 200 km/s) but, according to theoretical speculations, have no convective envelopes. The latter, in contrast, have deep convective envelopes but rotate slowly (with rotational surface velocities of less than 5 km/s). The existence and structure of stellar coronas are explained by the active dissipation of loops of magnetic flux that have emerged from the stellar surfaces.

These ideas were first developed on the basis of X-ray observations of the solar corona made during the Skylab missions (Vaiana and Rosner, 1978). The Skylab observations showed that the X-ray radiation from coronas was not emitted in a regular way, but came from loops of flux of the photospheric and chromospheric magnetic fields. The topology of the coronal magnetic field is essential here. The open configurations scarcely radiate X-rays at all; they are coronal holes that emit the bulk of the solar wind. The magnetic energy in the closed configurations is released by the reconnection of the magnetic field lines at the rate

$$\varepsilon_H \sim 10^{-2} v_A H^2 / 4\pi r_l \quad (\text{erg/cm}^3 \text{ s}),$$

where r_l is the loop radius, v_A is the Alfvén velocity, and H is the magnetic field. Reconnections of the field lines and the associated solar flares have long been studied as an energy source for the corona

(Syrovatsky and Somov, 1980). Such discussions are important to the theory of the large-scale heating of a stellar corona.

The magnetic energy released from the loops due to the magnetic field line reconnections is removed by thermal conduction and by X-radiation at the rate (Galeev *et al.*, 1979)

$$\varepsilon_R \sim 10^5 (nkT_e)^2 T_e^{-3/2} \quad (\text{erg/cm}^3 \text{ s}).$$

The most probable cause for the reconnections and dissipation of the nonpotential components of the field is assumed to be loop twisting due, for example, to differential motions at the bases of the emerging loops. It is interesting that individual loops in the corona do not seem to be in hydrostatic equilibrium with the ambient medium. Equilibrium may however be implemented dynamically through the involvement of plasma motion in the loops (Vaiana and Rosner, 1978).

We shall start the next section by considering solar activity.

III Mechanisms of Solar Activity

The quiet Sun can be represented by a gas sphere whose equilibrium is determined by a balance between the gravitational force and the pressure gradient. The thermonuclear fusion taking place at its center is the energy source. The outer envelope of the Sun is in a state of stationary convection, the cells of which are assumed to be the granules and supergranules visible on the surface. This convection is independent of latitude and longitude. The Sun rotates ($\omega \simeq 3 \cdot 10^{-6} \text{ s}^{-1}$ at its equator, i.e. its period is about a month) and possesses a magnetic field, but the rotation and field are so weak that they do not affect the equilibrium of the quiet Sun.

In the active Sun the rôle of motions and magnetic fields becomes dominant. All the manifestations of solar activity (plages, spots, flares, prominences, the coronal configuration, etc.) are associated with magnetic fields and it is widely acknowledged that the magnetic field is the thread binding these phenomena together. It is clear then that magnetohydrodynamics and plasma dynamics must be the bases on which the nature of solar activity and predictions of its future behavior are founded.

The most striking feature of solar activity is its recurrence or periodicity. The 11-year cycle is widely known (in the scientific community as well as among the general public). It was discovered in 1893 by an amateur astronomer, Heinrich Schwabe from Dessau, who spent all his spare-time hunting for sunspots. The number of spots, their total area on the solar disk, and some other features of the activity are now known to be associated with the variation in the magnetic field strength over an 11-year periodicity. Recognizing the change in sign associated with its reversal, the large-scale magnetic field of the Sun, whose azimuthal component is closely connected with the spots and whose poloidal component resembles that of a dipole, has in reality a 22-year periodicity.

This periodicity has been explained by magnetic dynamo theory (Parker, 1955; Babcock, 1961; Leighton, 1969; Steenbeck and Krause, 1969; Yoshimura, 1975a; Stix, 1976; Ivanova and Ruzmaikin, 1977). The mechanism operating the solar dynamo is shown schematically in the flow chart. The thermonuclear fusion at the Sun's center maintains the stationary convection in the surface envelope (a convective zone) and provides the energy source for the dynamo-machine. The oscillating system is the large-scale magnetic field averaged over scales and times exceeding the characteristic length and time scales of the turbulence. Mechanisms that drive the oscillating system (or, in other words, pump energy in and out of it) operate through the differential rotation of the Sun, the mean helicity of its turbulent

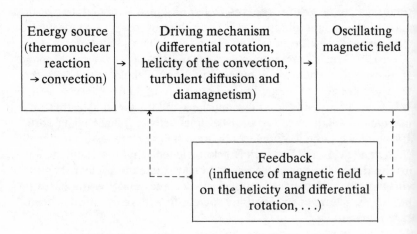

convection, and the turbulent diffusion of the magnetic field lines. An important rôle is played by some other factors, specifically, by the diamagnetic pumping of the field from the convection zone (Zeldovich, 1956; Rädler, 1968; Vainshtein and Zeldovich, 1972; Ivanova and Ruzmaikin, 1976), by magnetic buoyancy (Parker, 1979) and probably by topological pumping (Drobyshevsky and Yuferev, 1974).

The dependence of the solar rotation on latitude can be observed directly (see the review by Howard, 1978). The equator rotates about 20% faster than the poles. Analysis of the observational data strongly indicates that the rotation is also differential in depth (Stenflo, 1976; Deubner et al., 1979). Although the theory of the differential rotation of the Sun [see, for example, reviews by Durney (1976) and Vandakurov (1976)] has not yet been settled, being hampered by a lack of knowledge of the specific dependence of the angular velocity on the radius and heliolatitude, the fact that the rotation is non-uniform is beyond doubt. The fact that the radial direction in the solar convection zone is preferred decides the character of the differential rotation; most of the convective elements rise and sink in this direction, and a strong radial density gradient exists. The dispersion of the radial velocities is therefore larger than that of the tangential velocities. This results in an angular momentum transport to compensate for the smoothing action of the turbulent viscosity (Rüdiger, 1977). This was first noticed by Kippenhahn (1963), who considered it to be a consequence of an anisotropic viscosity. Actually the effect is proportional to the velocity (not its gradient, as for viscous stresses), and is due to the action of Coriolis forces on the convective elements (Ruzmaikin and Vainshtein, 1978). A similar but somewhat different point of view as to the origin of the differential rotation invokes Coriolis forces acting on giant convective cells that possibly exist in the depths of the solar convection zone (Gilman, 1976; Monin, 1980). It is interesting that the differential character of the rotation and its intensity follow simply from similarity considerations (Golitsyn, 1974).

The helicity of turbulent convection means that vortices which twist in one direction dominate over those with the opposite twist (say, vortices with a left-hand screw dominate over vortices with a right-hand screw), i.e. solar convection is not mirror-symmetric. Such an effect occurs naturally in an inhomogeneous convection zone due to

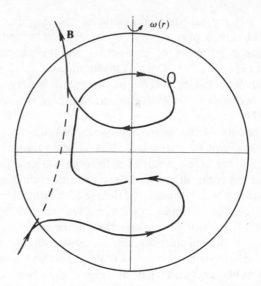

Figure 11.3 The effect of differential rotation on magnetic field. The axisymmetric dipole-type field creates two oppositely directed rings of azimuthal field.

the additional twisting of the rising and sinking convective elements by the Coriolis forces. It is usually described by the mean helicity density (Chapter 8)

$$\alpha = -\tfrac{1}{3}\tau\langle \mathbf{v} \cdot \mathbf{\nabla} \times \mathbf{v}\rangle \approx \rho^{-1} l^{2}(\boldsymbol{\omega} \cdot \mathbf{\nabla}\rho),$$

where ρ is the density, l the correlation length, and τ the correlation time of the turbulent convection.

A simplified representation of the action of the solar dynamo-mechanism may be given as follows. The differential rotation of the conducting fluid in the convection zone leads to the zonal deflection of the poloidal field lines, i.e. to the creation of an azimuthal field (Figure 11.3). Under the action of convection, which is helical on average, the azimuthal field loops rise, turn, tear and merge, creating a poloidal field out of the azimuthal field (Figure 11.4). At appropriate values of $\mathbf{\nabla}\omega$ and α such wave-like processes become self-generating. In this simple case the condition for self-generation is determined by a dimensionless dynamo number, D, defined by (4) below.

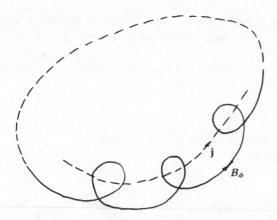

Figure 11.4 The action of spiral vortices on the magnetic field. A set of "small" vortices for which $\mathbf{v} \cdot \nabla \times \mathbf{v} \neq 0$ can generate a large-scale magnetic field of poloidal type (through the current $\mathbf{j}$) from the azimuthal (B_ϕ) field.

Dynamo theory has been a great success even in its simplest form, i.e. the form linear in the field B (when the flow chart has no feedback loop). It explains the magnetic field periodicity (the polarity reversals and the changes in sign of the sunspot fields), the migration of field maxima to lower heliolatitudes (Spörer's law, Maunder's butterfly diagrams), the observed phase relation between the radial and azimuthal field components, and other features of the solar cycle.

The linear kinematic problem of the solar dynamo is rather simple. The forms of the differential rotation, the helicity, the turbulent viscosity and the turbulent magnetic permeability are taken from semi-empirical or theoretical models. Qualitatively the problem involves only the dimensionless dynamo-number. An oscillating solution with constant amplitude is obtained for a certain value D_0 of D; this is found by solving the dynamo equations and the associated boundary conditions. The solutions have the form of waves propagating over certain surfaces [in the simplest case, over the surfaces $\omega = $ constant; Yoshimura (1975b)]. The wave period is identified with the observed period of cycle. It is important that the period and direction of the wave propagation have only a weak dependence on $D - D_0$ since then the linear theory is a reliable and valuable indicator of these features in the more realistic but more difficult nonlinear problem.

As linear solar dynamo theory has been described in detail in a number of reviews and monographs, we shall content ourselves here with a brief summary. We shall discuss only those features which somehow survive in the full nonlinear problem and can be used for comparison with observations.

IV The Capabilities of Kinematic Dynamo Theory

Although there is observational evidence that large scale asymmetric fields exist (Altschuler et al, 1974), we shall consider only axisymmetric mean magnetic fields. Then instead of the three components of the vector **B**, it is convenient to consider two quantities: the azimuthal field $B \equiv B_\phi(t, r, \theta)$ and the azimuthal component of the vector potential $A = A_\phi(t, r, \theta)$ of the poloidal field

$$B_r = \frac{1}{r \sin \theta} \frac{\partial}{\partial \theta} (\sin \theta A), \qquad B_\theta = -\frac{1}{r} \frac{\partial}{\partial r} (rA).$$

First we assume that the turbulent magnetic permeability is constant in the convection zone. Then the actions of the parametrized helical turbulence and the differential rotation on the magnetic field are described by equations that follow from (8.9)–(8.10) of Chapter 8:

$$\partial A / \partial t = \alpha B + \nu_T (\nabla^2 - \mathrm{cosec}^2 \, \theta) A, \tag{1}$$

$$\partial B / \partial t = (\nabla \omega \times \nabla)_\phi r \sin \theta \, A + \nu_T (\nabla^2 - \mathrm{cosec}^2 \, \theta) B. \tag{2}$$

The most important property of these equations is that they admit wave-like solutions

$$A \sim \exp [i(\Omega t - \mathbf{k} \cdot \mathbf{r})], \qquad B \sim \exp [i(\Omega t - \mathbf{k} \cdot \mathbf{r} + \Phi)]. \tag{3}$$

Parker (1955), who first discovered these solutions, called them "dynamo waves". In the linear regime, dynamo waves are unstable, their amplitudes exponentially decay at short wavelengths and grow at long ones. There is only one value $k = k_{cr}$, for which the dynamo wave propagates with constant amplitude.

To estimate qualitatively from Eqs. (1) and (2) the frequency of the wave Ω, the phase shift Φ between the field components, and the critical wavenumber k_{cr}, one may ignore the effects of curvature and assume that α and $\nabla\omega$ are constant. The substitution of (3) into Eqs. (1)–(2) then yields

$$\Omega^2 = -\tfrac{1}{2}\alpha R (\nabla\omega \times \mathbf{k})_\phi, \qquad k_{cr}^3 = R\nabla_r\omega\alpha/2\nu_T^2, \qquad \Phi = \pi/4,$$

where R is the stellar radius. Thus $\mathbf{k}$ is perpendicular to $\nabla\omega$, i.e. the dynamo waves propagate along the surfaces $\omega = $ constant (Yoshimura, 1975b). In particular, when ω depends only on the radius r the waves propagate along the meridians $\theta = $ constant. This property allows us to explain a remarkable phenomenon observed during the solar cycle: the migration of sunspot activity over the solar surface from high latitudes to the equator. For this one needs to identify the occurrence of a spot with the crest of the dynamo wave. The correct direction of propagation is obtained only when $\alpha\nabla_r\omega < 0$ in the northern hemisphere ($k_\theta > 0$) and $\alpha\nabla_r\omega > 0$ in the southern hemisphere ($k_\theta < 0$).

For the sake of simplicity it has been assumed above that the product $\alpha\nabla_r\omega$ as a function of position is non-zero, but this is not necessary. Sources of differential rotation and helicity can be localized at different places in either of which the product $\alpha\nabla_r\omega$ vanishes. Even in this case dynamo waves can be generated due to the turbulent diffusion of the mean field from one source to the other (Steenbeck and Krause, 1969; Ivanova and Ruzmaikin, 1976; 1977, 1980), their frequency again being determined by $\alpha\nabla_r\omega$, understood now as the product of the source amplitudes in different locations.

Another important property of dynamo waves is the phase shift between A and B. The observable quantities are the azimuthal field B (the field of the spots) and the radial field component B_r (more precisely, the field directed along the line of sight is measured). One can see that B_r will be displaced relative to B by $3\pi/4$. For fixed phase shift, the relationship between B and A (or B_r), as can easily be seen from Eq. (2), depends only on the sign of the angular velocity gradient. Hence the observed relationship between the B and B_r phases yields information about the sign of $\nabla\omega$ (Stix, 1976). The analysis of the observations of the Sun's field during the period 1965–1975 carried out by Stix showed that the angular velocity must increase with depth in regions where the azimuthal field is generated ($\partial\omega/\partial r < 0$). From this it follows that $\alpha > 0$ in the northern hemisphere near the surface of the Sun. This agrees with the theoretical argument that the sign of the mean helicity should be proportional to $-(\boldsymbol{\omega} \cdot \nabla\rho/\rho)$.

Instead of k_{cr} it is preferable to employ the dimensionless wavenumber $k_{cr}R$. The linear problem is usually characterized by the dimensionless dynamo number

$$D = \alpha_0 \omega R^3 / \nu_T^2. \tag{4}$$

Its critical value, corresponding to a dynamo wave of constant amplitude, is $D_0 \sim (k_{cr}R)^3$. By substituting estimates of the parameters for the solar convection zone into the expression for k_{cr}, we obtain $D_0 \gtrsim 10^4$.

One can estimate the period of the cycle to be

$$P = 2\pi/\Omega \simeq 2\pi (\alpha \nabla \omega k_{cr}R)^{-1/2}.$$

We take $\alpha \approx \frac{1}{3}l\omega_0$, $\nabla\omega \approx \omega_0/0.3R$, $\nu_T = \frac{1}{3}lv$, where l is the mixing length in the convection zone. By substituting $\omega_0 = 3 \cdot 10^{-6} \, \text{s}^{-1}$, $R = 7 \cdot 10^{10}$ cm, $l \simeq 10^{-2}R$ and $v \simeq 5 \cdot 10^4$ cm/s, we find that the period P is less than one year, despite the fact that the observed solar cycle period is known to be 22 years. Why does this difference arise? Admittedly, the estimate of P was somewhat crude. More refined calculations of the dynamo-wave period have, however, been carried out by a number of authors who, finding the same discrepancy, argue that the mean helicity density used is not reliable. But to bring the theoretical value of P into agreement with observation, α must be decreased by more than two orders of magnitude, and this cannot be justified from a theoretical point of view.

Agreement can be achieved, however, by restoring turbulent diamagnetism (Ivanova and Ruzmaikin, 1976). The existence of this phenomenon is inevitable, since hydrodynamic models of convection zones (Baker and Tamesvary, 1966) have shown that the intensity of turbulent motions must vary with depth.

In order to assess the rôles of diamagnetic and other effects one must know the distribution of the kinematic parameters in stellar convection zones. Unfortunately, the hydrodynamic problem of the convection in the external envelope of the Sun and other stars has not yet been properly solved (see Durney, 1976; Vandakurov, 1976). The main difficulty is associated with the stratification of the convection zone. From the base of this zone to its surface ($0.7R$ to $1R$, for the Sun), the density changes by several orders of magnitude. The change is most abrupt near the surface, it being smoother near the base. For a motion whose scale exceeds that of the density change (the scale height) one must take into account compressibility and thus go beyond the limits of the so-called Boussinesq approximation, in which the compressibility appears only in the buoyancy force. There is a nice discussion of these problems in a paper by Simon and Weiss (1968) where they proposed a cellular convection model with three types of cells: (1) *giant cells* with a horizontal dimension comparable with the zone depth, extending vertically to a depth of about 150 000 km (measured from the surface), the characteristic velocity in these cells being about 0.05 km/s, and lasting for about a month; (2) *supergranules* with a characteristic radius $l \simeq 15\,000$ km and velocity $v \simeq 0.5$ km/s, i.e. with a characteristic time of no more than a day; (3) *granules* with a characteristic scale corresponding to the scale height at the solar surface, $l \simeq 10^3$ km and $v \simeq 1$ km/s; it is assumed that the granulation is produced by the supergranules rising from below. The granules and supergranules can be observed directly on the solar surface, but flows corresponding to the giant cells have not yet been reliably detected. Although indirect arguments based on observations of large-scale phenomena (Bumba and Howard, 1965) indicate that these cells are present, these still cannot be regarded as proof that such large-scale convection occurs.

We shall therefore restrict ourselves to crude qualitative estimates of the kinematic parameters, taking as a basis calculations of the structure of convection zones that employed the mixing length approximation (Baker and Tamesvary, 1966). The characteristic scale here is the mixing length, proportional to the scale height and increasing from $2 \cdot 10^2$ km at a conventionally chosen solar surface (where the optical depth is of order unity) to $8 \cdot 10^4$ km at a depth $0.3R$, where the base of the convection zone is located. By contrast, the characteristic velocity of convective motion on the scale l decreases towards the base of the zone (from $2 \cdot 10^5$ to $2 \cdot 10^3$ km/s). Hence the turbulent viscosity of the solar convection zone, $\nu_T = \frac{1}{3}lv$, has a maximum, of about 10^{13} cm^2/s, within the convection zone (Figure 11.5).

In order to estimate the variation of α with depth one should recognize that mean helicity arises in a stratified convection zone due to the influence of Coriolis forces. The characteristic time over which the Coriolis force acts on a convective element depends on the relationship between the angular velocity, ω, of the star, and the time of convective overturning, $\tau \sim l/v$. In the solar convection zone this

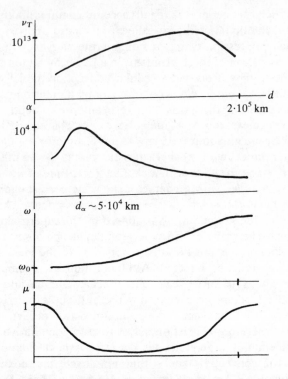

Figure 11.5 Qualitative representation of kinematic parameters deep within the solar convection zone where field generation takes place.

relationship changes with depth. Near the surface $\omega\tau < 1$, i.e. the shortness of τ is the determining factor. At the base of the zone, the opposite occurs, $\omega\tau > 1$, i.e. the convective element has time for many rotations during the period ω^{-1} before it ceases to exist. Therefore, when estimating the helicity as the time average

$$\alpha \approx \int \mathbf{v} \cdot \nabla \times \mathbf{v} \, dt,$$

where $\mathbf{v} \cdot \nabla \times \mathbf{v}$ is created by the Coriolis forces due to the general rotation (crudely $\mathbf{v} \cdot \nabla \times \mathbf{v} \sim v\omega$), one should set the integration interval as τ in the upper half of the zone and as ω^{-1} in the lower half. As a result we have

$$\alpha \sim \begin{cases} l\omega, & d < d_\alpha, \\ v, & d > d_\alpha. \end{cases}$$

Since $l\omega$ grows and v decreases with the depth d in the zone, α has a maximum at approximately $d = d_\alpha \sim 5 \cdot 10^4$ km (Figure 11.5). For the numerical estimates it can be supposed that $\alpha \simeq 10^3 - 10^4$ cm/s. The latitudinal variation in the mean helicity can be estimated from symmetry considerations. Indeed, the pseudoscalar function $\alpha \sim \boldsymbol{\omega} \cdot \boldsymbol{\nabla}\rho \sim \cos\theta$, i.e. it must have opposite signs in opposite hemispheres.

Although the mean helicity of the solar convection is by itself sufficient to enable the dynamo to act, differential rotation is also present, and plays a very important rôle. The dimensionless number characterizing its action in the convective zone is $R_\omega = R^2 \boldsymbol{\nabla}\omega / \nu_T \sim 10^3$ (compare this with $R_\alpha = R\alpha / \nu_T \sim 10$). Its importance is clearly seen through the average axisymmetric azimuthal magnetic field, which relates the emerging spots to one another, and which is the strongest component of field observed. From symmetry considerations, this axisymmetric component is, in the pure helical α^2-dynamo, not dominant in size. On the other hand, it has been shown in Chapter 4, in the simplest example of a homopolar generator, that a differential rotation naturally generates a strong axisymmetric magnetic field. This fact was first noted by Elsasser, and the importance of this mechanism under cosmic conditions was stressed by Parker (1955), who included it in his first theoretical model of the solar dynamo.

Nonuniformity of the solar rotation with latitude can be directly observed through the motion of sunspots (Newton and Nunn, 1951) and the Doppler line shift in the solar atmosphere (Howard and Harvey, 1970): it is

$$\omega = \omega_0(1 + a \cos^2\theta + b \cos^4\theta),$$

where $\omega_0 \simeq 2.90 \cdot 10^{-6}\,\text{s}^{-1}$, $a = -0.18$, $b \simeq 0$ from sunspot observations and $\omega_0 \simeq 2.78 \cdot 10^{-6}\,\text{s}^{-1}$, $a = -0.13$, $b = -0.16$ from Doppler shift measurements. Stenflo (1976) noted that measurements of the angular velocities of the magnetic structures emerging at the surface of the Sun indicate that ω increases with depth. An analysis of the spectral properties of the so-called 5-min oscillations suggests the same conclusion (Deubner et al., 1979). These oscillations represent standing waves in the convection zone considered as a resonator. To distinguish the effects of rotation, they are decomposed into waves propagating in the direction of rotation (increasing longitude ϕ) and in the opposite direction. Then by analyzing different radial harmonics, i.e. waves penetrating to different depths, it can be shown that the rotation varies with radius, r. Comparison with observations gives $\partial\omega/\partial r < 0$ up to a depth of about $1.5 \cdot 10^9$ cm. A deeper examination of this method is rather difficult, but it is imperative because the depth reached is only of the order of the supergranulation scale. (Recall that the dynamo theory of mean fields employs an angular velocity averaged over many cells.) Stronger evidence that $\partial\omega/\partial r < 0$ in the solar interior has recently been provided by Gough (1982), Hill et al (1982) and Dicke (1983) through analysis of the low l-modes in global solar oscillations.

A diamagnetic effect is exhibited in the pumping of magnetic field from areas of low turbulent intensity to places where it is higher. Therefore its strongest manifestations occur at the bottom and at the surface of the convection zone. Figure 11.5 sketches the effective magnetic permeability. The velocity v_μ of pumping is proportional to $-\beta\nabla\mu^{-1}$ (Ruzmaikin and Vainshtein, 1978).

After assigning the kinematic parameters (Figure 11.5), the mean magnetic field can be obtained by solving the generation equation (see Chapter 8)

$$\partial\mathbf{B}/\partial t = \nabla \times \{[(\boldsymbol{\omega} \times \mathbf{r}) \times \mathbf{B}] + \alpha\mathbf{B} - \beta\nabla \times (\mu^{-1}\mathbf{B})\} \qquad (5)$$

in the sphere $r \leq R$ and matching the solutions to those for the region outside the Sun. In the simplest case, when this region is supposed to be a vacuum, $\nabla \cdot \mathbf{B} = 0$ and $\nabla \times \mathbf{B} = 0$ for $r > R$. The solution of this system is described in detail in a number of monographs (Moffatt, 1978; Parker, 1979; Krause and Rädler, 1980; Vainshtein et al., 1980). It largely confirms the qualitative models discussed at the beginning of this section. The field propagates as a dynamo wave along the surfaces $\omega = $ constant. The wave period is determined by the intensity of the helical motion and the differential rotation; it also

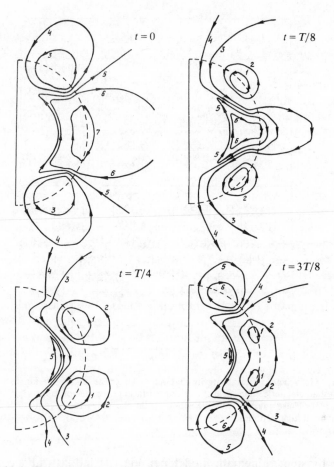

Figure 11.6 Four "snapshots" of the poloidal field from the dynamo cycle operating in the solar convection zone. The numbers on the field lines mark (in dimensionless units) the values of the vector-potential, which is constant on each line.

depends on the turbulent magnetic permeability. The poloidal components of the large-scale field generated at four consecutive epochs and the theoretical butterfly-diagrams for $D \simeq D_0$ obtained from the model by Ivanova and Ruzmaikin (1977) are shown in Figures 11.6–11.7.

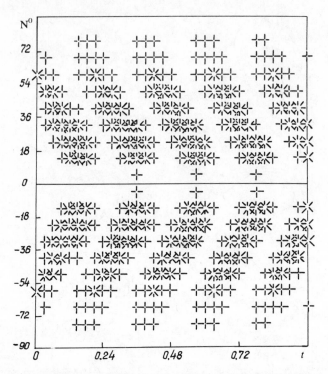

Figure 11.7 Theoretical butterfly-diagram. The crests mark the maxima of the azimuthal magnetic field near the solar surface (+ corresponds to $B_{max}/3$, × to $2B_{max}/3$). These maxima migrate slowly from the poles to the equator and then reappear at high latitudes.

In the linear kinematic model, periodic oscillations with a constant amplitude occur only for $D = D_0$. A departure of D from this value results in a growth or decay of the oscillation amplitudes. On the other hand, the observed 11-year (or 22-year) period is not purely harmonic, nonlinear distortions (Figure 11.8) and amplitude modulation being evident. These peculiarities of the solar cycle could be explained only in the framework of a nonlinear theory, but no such theory has yet been developed, although there have been several ideas and attempts to create one. The remaining sections of this chapter discuss fragments of the future theory, with the observed phenomena borne constantly in mind.

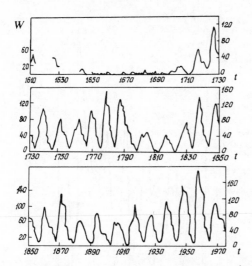

Figure 11.8 Annual mean sunspot number, *W*, from 1610 to 1970.

V Portrayal of the 11-Year Activity

We shall discuss first the observed picture. The basic 11 year (or 22-year) periodicity in solar activity has being studied for a long time and very thoroughly [see the monograph by Waldmeier (1941)]. Systematic observations of sunspots have been made since 1821. Wolf determined smoothed averages of the monthly numbers of sunspots, $W(t)$, from data collected by telescopic observations since 1749. He also estimated the epochs of extrema of the 11-year cycle with an accuracy of 0.1 year from Galileo's time (1610). A segment of the record of the average annual dependence is shown in Figure 11.8. From this it is clear that the lengths of the cycles are not constant, and their magnitudes change significantly. The heights of the maxima vary from 46 to about 190. An isolated peak does not have a harmonic form either. The record has a quasi-timbral character and, as Waldmeier has noted, the higher the maximum the briefer the time, τ_r, for the rising branch between a minimum and maximum, while the duration of the falling branch, τ_d, is practically independent of this.

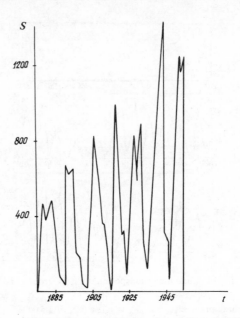

Figure 11.9   Observed total area of sunspots in the northern hemisphere for 1879–1954 (after Vitinsky, 1973).

It immediately follows that the curve $W(t)$ cannot be described by one parameter. Moreover, quite apart from the extremals, points of inflection can also be seen which presage slowing down of the next bursts of activity. Forked maxima occur that resemble the back of a two-humped (Bactrian) camel. This last effect is especially visible on the curve of total sunspot area constructed separately for the northern and southern hemispheres of the Sun (Figure 11.9). Looking at these curves (Vitinsky, 1973), one notices the alternation of one and two peaked maxima (transformations of an Arabian camel into a Bactrian camel and vice versa!). Also worthy of note is the fluctuating background of the average monthly curve $W(t)$. This shows relative amplitudes of 0.01–0.6 and characteristic times of 2–4 months, with fluctuations of short duration on the growing branch dominating.

In the paper by Gudzenko and Chertoprud (1964) a phase portrayal of the 11-year activity (Figure 11.10) was constructed through a statistical analysis of the average annual observational data from the zeroth up to nineteenth cycle (the zeroth cycle is counted as starting

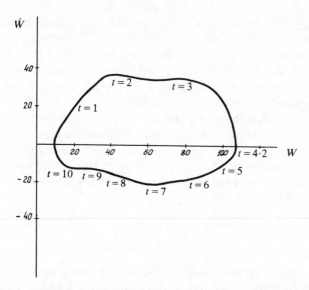

Figure 11.10 Limit cycle constructed by statistical analysis of the observed sunspot numbers in the phase plane $(W, \dot{W})$: $(W, \dot{W})_{min} = (62, -22.5)$, $(W, \dot{W})_{max} = (108, 36)$, $\tau_r = 4.2$ years, $\tau_d = 7.0$ years, $T = \tau_r + \tau_d = 11.2$ years (after Gudzenko and Chertoprud, 1964).

in 1755). All the trajectories tend asymptotically to a mean statistical closed curve (an analog of Poincaré's limit cycle) which is nearly symmetrical with respect to the phase minimum $t = 0$. Crudely one may represent the limit cycle by an ellipse, i.e. a two-parameter curve. The representative point describes the upper part of the limit cycle faster (in about 4.2 years) than the lower part (7.0 years). The total time for one circuit, i.e. the period of cycle, is $T = 11.2$ years. The authors also noted a peculiar property of the "rigidity" of the cycle—the quantity describing the deviation along the normal of the representation point from the limit cycle. Near a maximum the rigidity suffers a sharp burst, i.e. the system tries to wipe out the memory of the previous 11-year period.

The important revelation of this paper (Gudzenko and Chertoprud, 1964) was the 11-year cycle as a real Poincaré limit cycle in the phase plane $(W, \dot{W})$. As in the theory of nonlinear oscillations (Andronov, 1929), the term cycle took on a concrete physical and mathematical meaning. The absence of a mean statistical dependence of the amplitude on the cycle period became clear (Yoshimura, 1979). An individual trajectory in the phase plane can lie outside or inside the limit cycle and have a correspondingly larger or smaller period. The mean period, however, will be $T = 11$ years since all trajectories asymptotically approach the limit cycle.

A steady solution corresponding to the 22-year limit cycle can be obtained from magnetic dynamo theory by invoking the back reaction of the magnetic field on the helicity. Consider the simplest model of the solar dynamo in which the convection zone is approximated by a thin flat slab and the generation sources are concentrated at two different levels (see Chapter 10)

$$\partial v / \partial z = -v_0 \delta(z),$$

$$\alpha = \alpha_0 (1 - \xi B^2) \delta(z - 1).$$

The z-axis is perpendicular to the slab, z being equivalent to the radial coordinate of the star, and x and y to the θ and ϕ coordinates, respectively. The parameter ξ measures the strength of the nonlinear action of the field upon the helicity; $B^2 \simeq B_y^2$; the velocity shear $(v = v_y)$ describes the differential rotation of the star. In this model the dynamo problem admits an analytical solution (Chapter 10). When $D = D_0$ in the linear approximation ($\xi = 0$), where $D = v_0 \alpha_0 h^2 / v_T^2$ is the dynamo number, h the slab thickness, and v_T the turbulent magnetic viscosity, the field performs undamped oscillations, with an amplitude determined by the initial conditions. When $D > D_0$ the amplitude of the oscillation increases exponentially. Recognition of the nonlinearity allows the oscillations to stabilize. For small deviations $D - D_0 > 0$, undamped oscillations are set up with a frequency close to that of the linear solutions and an amplitude proportional to $(D - D_0)^{1/2}$. By using the solutions given by Kleeorin and Ruzmaikin (1981, also see Chapter 10), it can be shown that in the phase plane (B_x, B_y) the limit cycle is the ellipse

$$B_x^2 / a^2 + (B_y - 2 e^{-2q_1} \cos q_2 B_x)^2 / b^2 = 1,$$

where

$$b = b_1 \exp[-q_1(z + 1)], \qquad a = \tfrac{1}{2} b_1 (1 - \xi b_1^2) \exp[-q_1(z - 1)],$$

$$b_1 = \{\tfrac{1}{3} \xi^{-1}[1 - (D_0 / D)]\}^{1/2},$$

and $q = q_1 + iq_2$ is determined by solving

$$q^2 = \tfrac{1}{4}ikD \; e^{-2q},$$

(k is the wave number in the x-direction).

The process of establishing undamped oscillation in the solar dynamo and the stationary regime of oscillation, have been studied by Ivanova and Ruzmaikin (1977) for a complex but more realistic dynamo model. Their results agree with those of the simple model above. In this connection we also mention the early papers by Stix (1972), Jepps (1975) and Yoshimura (1975a). For one specific model, Bräuer (1979) noted the possibility of an inverse bifurcation and the associated instability of the nonlinear solution for small deviations of D from D_0 [in this context see Kleeorin and Ruzmaikin (1981) and also Chapter 10 above].

An important feature of the dynamo model under consideration should be noted. The nonlinearity leads to a change in the oscillation amplitude $b \sim (D - D_0)^{1/2}$ but, in the first approximation, it only weakly affects the oscillation period. In addition, the nonlinear solution has a peculiar symmetry (Bräuer, 1979; Ivanova and Ruzmaikin, 1977; Kleeorin and Ruzmaikin, 1981):

$$\mathbf{B}(t + \tfrac{1}{2}T) = -\mathbf{B}(t).$$

Such a symmetry seems to be inherent to models of $\alpha\,(\mathbf{B})$ that involve $\mathbf{B}$ quadratically, for then only odd harmonics appear in the solution. It is interesting that, in the first approximation in which only the first and (weaker) third harmonics are included, a forked maxima appears in their sum. This may be relevant to the observed two-humped maxima in solar activity (see the beginning of this section).

VI   Secular Modulation

Even the first investigators of the solar cycle noticed that, superimposed on the 11-year activity, a longer secular modulation occurs. It was supposed that the main component of the modulation has a quasi-harmonic form with a period of about 80 years (Gleisberg, 1958; Waldmeier, 1949). This secular cycle seems to have sufficient statistical justification but its quasi-period, T_s, so far does not have an exactly determined value (Vitinsky, 1973). Different authors have obtained different values for T_s, ranging from 55 to 90 years. An exact knowledge of T_s is, however, not crucial for an understanding of the qualitative aspects of the modulation. The following two features of the secular modulation are really significant.

In every 11-year cycle the solar activity is approximately the same in the northern and southern hemispheres, but during the longer T_s-interval a noticeable "north–south" asymmetry can be discerned. This manifests itself, for example, in the different forms of the 11-year curves of relative sunspot numbers, in the excess of the total area and numbers of sunspot groups in one of the hemispheres, as well as in a phase difference in the extrema of the 11-year cycles (Vitinsky, 1973).

Another peculiarity of the secular modulation was discovered by Yoshimura (1979). Having studied the dependence of the 11-year periodicity on the maximal amplitude from 1698 to 1977 (25 periods of the main cycle), he noted that the dependence of $T(W_{max})$, was almost periodic with $T_s \simeq 57$ years, see Figure. 11.11. Of the five T_s-periods examined by Yoshimura, four demonstrated a new kind of limit cycle in which the system is represented by a point (in a three dimensional phase space) which approaches a limit surface having the form of a torus (a two-frequency limit cycle). Winding round the small circle of the torus corresponds to the basic 11 year (or 22-year) cycle; that over the large circle corresponds to the modulation, with period T_s. One of the five period (1867–1924) showed that this torus is unstable.

Ivanova and Ruzmaikin (1976) gave a possible explanation for this quasi-periodic modulation of the basic solar cycle. In their solar dynamo model it was shown that, along with the basic dipole mode oscillating with the period T, a quadrupole mode can be excited with the period $T_Q \neq T$ (where $|T_Q - T| \ll T$) at close, but somewhat larger, dynamo numbers. The proximity of the periods of these two modes indicates that the system can exhibit beats with period

$$T_s = TT_Q/(T_Q - T).$$

In the model by Ivanova and Ruzmaikin, $T \simeq 0.50$, $T_Q \simeq 0.54$ (in dimensionless units), so that $T_s \simeq 13T$, which is of the right order of magnitude. Moreover, the sum of the dipole and quadrupole fields

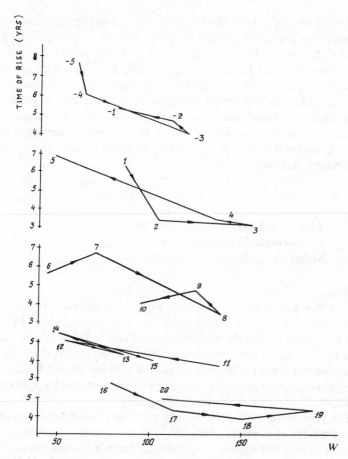

Figure 11.11 Quasiperiodic modulation of the 11-year cycle (after Yoshimura, 1979).
The starting point of the −5 cycle corresponds to 1698, the end point of the 20 cycle
to 1977.

is asymmetric with respect to a reflection through the equatorial plane.
It is just this north–south asymmetry that is associated with the secular
T_s-modulation (Waldmeier, 1941; Vitinsky, 1973; Yoshimura, 1978;
Gleisberg and Damboldt, 1979). The problem is difficult because we
are dealing here with the summation of oscillations of closely similar
frequencies in a nonlinear regime. In fact at $D = D_{0Q}$ (when the
quadrupole mode is excited) the dipole mode with $D_0 < D_{0Q}$ has, in

linear theory, an exponentially increasing amplitude. In the nonlinear regime, not this quasi-periodic regime, but a synchronization of the oscillations may take place for a wide range of parameters, i.e. transition to a one-frequency regime with $T = T(D)$ may occur. In fact the first numerical calculations, carried out for a nonlinearity of type (10.13), exhibited such one-frequency solutions. To isolate the regions of parameter space (or the forms of nonlinearity) where the quasi-periodic regime (beats) occurs, one can use a method similar to that applied in the theory of nonlinear oscillations [see, for example, Andronov and Witt (1930)].

VII The Stochastic Nature of the Prolonged Weakening of Solar Activity

A new challenge has recently been made to investigators of the solar cycle. In *The Case of the Missing Sunspots* by J. Eddy (1977), a brilliant result of a hobby and a profession combined (a love of history and astronomy), convincing arguments are assembled showing that solar activity was substantially weaker during the period 1645–1715 This phenomenon was first noted in the last century by Spörer and Maunder.†

The period was called the Maunder minimum in recognition of the unsuccessful attempts of Maunder to convince his conservative contemporaries of the reality of the effect. One may believe that Maunder's failure was due as well to the distrust in which his contemporaries held historical data. Eddy collected a variety of evidence for the weakened solar activity: the fewer number of sunspots, data on aurorae and on climatic variations, and particularly convincing data on the history of the ^{14}C isotope abundance. This isotope forms in the Earth's atmosphere through bombardment by cosmic rays of Galactic and solar origin, and during photosynthesis it is assimilated

---

† Connoisseurs (J. and N. O. Weiss, 1979) assert that the first reference in English to the disappearance of sunspots in the seventeenth century was in A. Marvell's satirical poem "The Last Instructions to a Painter" written in 1667 and published in Great Britain in 1689.

as CO_2 into the annual growth rings of trees. There live today trees such as bristle-cone pines (Pinus aristata) that are 3000 years old. The study of such trees, through a radiocarbon analysis of their rings, represents a symbiosis of botany and nuclear physics.

Solar activity or, more precisely, the magnetic field of the Sun regulates the intensity of the bombarding cosmic rays. In active years the cosmic flux is weak, while in quiet years it is strong (Figure 11.12). Three extremal periods are visible on this curve: the Medieval minimum (1280–1350), the Spörer minimum (1460–1550) and the Maunder minimum (1645–1715) (Eddy, 1976; Clark and Stephenson, 1978). Associating extremal points with the years 1310, 1500 and 1690, an enveloping curve can be drawn with an approximate period of 190 years. This curve cannot, however, be continued into the future (there was no marked extremum in 1890) nor, apparently, into the past. Hence a periodic or quasi-periodic modulation is out of the question. We conclude that such extreme weakenings of cyclic solar activity occur at random times, i.e. stochastically.

Early discussions of the stochastic nature of the solar cycle were confined to small-scale fluctuations. All global variations were considered as long period modulations of the main 11-year cycle. The discovery of the prolonged minima of solar activity now provides, however, evidence that the solar cycle is stochastic on the large scale.

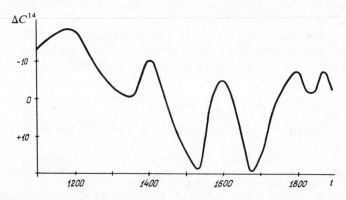

Figure 11.12 Carbon-14 concentration (increasing downwards!) from a tree-ring analysis for the period 1050–1900 (after Eddy, 1976). Three main anomalies are visible: the Medieval minimum 1280–1350, the Spörer minimum 1460–1550 and the Maunder minimum 1645–1715.

Can dynamo theory based on the equations of magnetohydro-dynamics explain such stochastic weakenings of solar activity? At first sight, one might think that random forcing should be added to those equations. In fact, the answer should be sought from the determining MHD-equations themselves. Recent progress in the qualitative theory of differential equations gives hope of success.

It is now known that the phase space trajectories of dynamical systems described by ordinary differential equations do not generally close. Exceptions are the singular points: the center, the focus and the node. An attracting point at the origin corresponds to a state of rest.† Steady nonlinear oscillations of a system correspond to Poincaré limit cycles. No other attractors exist for autonomous dynamical systems with less than three degrees of freedom, N. When $N \geq 3$, however, new attracting sets may appear in phase space. Ruelle and Takens (1971) called them "strange attractors", while Ya. Sinai proposed the term "stochastic attractors".

One can imagine a strange attractor in phase space as follows (Ruelle and Takens, 1971). Consider successive points of intersection of a trajectory in phase space with an arbitrary plane. It is clear that the evolution of the system can be described by $x_{n+1} = \hat{P}x_n$, the Poincaré mapping. If the mapping $\hat{P}$ is such that, by applying it to a continuous torus, we fold it into a figure "eight", and then double it by folding it onto itself (Figure 11.13), then the set formed by the points of intersection of the phase trajectory with the plane will describe a one-dimensional manifold in one direction and a Cantor set in the other.‡ This set portrays the strange attractor in the given plane. An example of a strange attractor is provided by the system of three equations first studied by Lorenz (1963) in a convection problem:

$$\dot{x} = -\sigma(x - y), \qquad \dot{y} = -y + rx - xz, \qquad \dot{z} = -bz + xy. \qquad (6)$$

For a certain range of parameters, in particular for $b = \frac{8}{3}$, $\sigma = 10$,

---

† All these peculiarities of phase space can easily be demonstrated by a pendulum. The resting pendulum corresponds to a center. Frictionless oscillations are represented by circles, oscillations with friction by spirals winding in towards the origin, i.e. a focus. A node corresponds to heavy friction, a saddle point to the inverted unstable state of the pendulum.

‡ Let us recall that a Cantor set is formed when a unit segment is divided into three parts, the middle being extracted and the remaining parts being similarly treated, the whole process being continued indefinitely.

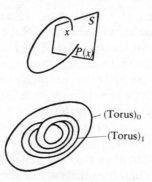

Figure 11.13 Illustrating the mapping, $\hat{P}$, associated with a strange attractor in phase space, and how it results in a folding of tori (after Ruelle and Takens, 1971).

$28 < r < 200$, a strange attractor occurs in the phase space of this system [see, for example, reviews by Monin (1978), Rabinovich (1978); Gaponov-Grekhov and Rabinovich (1979)]. In this case the phase trajectories behave in a pseudo-random manner. The amplitudes x and y, having performed a number of oscillations about one of the symmetric points $S_\pm = (\pm[b(r-1)]^{1/2}, \pm[b(r-1)]^{1/2}, r-1)$, make at a random time a transition to the other symmetric point, and start to oscillate in its vicinity. The system also has a stationary point $(0, 0, 0)$ to which the trajectory sometimes (rarely) descends via a two-dimensional manifold. After lingering in its neighborhood, the orbit leaves along one of two symmetric one-dimensional "feelers".

Another example of a system with stochastic behavior is the two-disk dynamo (Rikitake, 1958; Allan, 1962; Cook and Roberts, 1970) described by the (dimensionless) equations

$$\dot{I}_1 = -\mu I_1 + \Omega_1 I_2, \qquad \dot{I}_2 = -\mu I_2 + \Omega_2 I_1, \qquad (7)$$

$$\dot{\Omega}_1 = \dot{\Omega}_2 = 1 - I_1 I_2,$$

where Ω_1, Ω_2, I_1 and I_2 are the angular velocities and currents in the connecting wires, and the parameter μ is given by the ratio between the stored mechanical and electromagnetic energies. This system is simpler than the Lorenz one. There are only two parameters: μ and the integral of motion $A \equiv \Omega_1 - \Omega_2$ and, correspondingly, only two singular points $C_\pm = (\pm k, \pm k^{-1}, \mu k^2)$ where $A = \mu(k^2 - k^{-2})$ in the phase space (I_1, I_2, Ω_1). The remarkable phenomenon of the stochastic reversal of the sign of the Earth's magnetic field was explained qualitatively with the aid of this system. Robbins (1977) showed that, by adding a shunt to the two-disk dynamo, a Lorenz-type system and its concomitant strange attractor could be recovered.

An analysis of these and other examples (Rabinovich, 1978) leads to a picture of the strange attractor as a "sack" of zero volume in phase space; trajectories can enter it but they have no possibility (except, perhaps, for a set of trajectories of zero measure) of leaving. The other important feature of the attractor is the instability (scatter) experienced by trajectories that enter it. This is the cause of the stochastic behavior; figuratively speaking, the unstable trajectories in the closed sack have no way out and quickly begin to tangle. Certainly, this cannot happen in a plane since trajectories cannot cross, and such behavior can therefore be observed in autonomous dynamic systems only when the number, N, of their degrees of freedom is 3 or greater.

We conclude that there are observational and general theoretical reasons why the phase pattern of solar activity should exhibit a strange attractor that imparts a global stochastic behavior to the system. The interesting challenge is to construct a nonlinear model that performs appropriately.

Parker (1976) supposed that another type of zonal circulation, different from the one we observed today, could arise in the Sun. He imagined that Maunder type circulation is not effective enough to generate magnetic field but can persist for some time (say, 70 years). Dogel and Syrovatsky (1979) implemented this idea in a real model. They considered Boussinesq convection in a thin spherical layer rotating with angular velocity

$$\omega = \omega_0(1 - p \cos^2 \theta).$$

They showed that, for small latitudinal gradients $(p < p_{cr})$, a convection mode is excited with large-scale, banana-like cells stretched along the meridians ($m = l$ for $P_l^m(\theta)\, e^{im\phi}$ perturbations). When $p > p_{cr}$, a mode with $m = 0$ is excited, corresponding to axisymmetric ring-shaped convective cells stretched along the lines of latitude. It is known that the first structure ($m = l$) can generate angular velocity gradients (Busse, 1972; Durney, 1976), while the second one cannot maintain $\nabla \omega$. It is also assumed that a stationary state of rotation occurs in the solar convection zone when $p_s > p_{cr}$, in which the redistribution of angular momentum by the banana-like convective cells is balanced by the viscosity. A self-oscillating transfer can occur in which the "bananas" are transformed into ring-shaped cells and vice versa. Indeed, let us consider a solid-body or a nearly solid-body rotation (p small). Then the convection cells will have the banana shape and the angular velocity gradient will begin to grow. When $p = p_{cr}$ a bifurcation takes place: the cells acquire the ring shape, the gradient of ω decreases, after which "bananas" can again form, and so on. It is assumed that, in accordance with general principles (the non-existence theorem), the solar dynamo is switched off in the axisymmetric convection regime ("ring-shaped cells"), and that this explains the Maunder-type behavior. In confirmation of this idea Dogel and Syrovatsky pointed to the fact that during 1642–4, just prior the Maunder minimum, an anomalously large (one that was three times greater than usual) latitude gradient of the angular velocity had been inferred (Eddy *et al.*, 1977). Wöhl from Göttingen University, however, showed that an analysis of the same observational data (Hevelius's drawings) did not reveal any change in angular velocity during that period. Also, the rôle of small-scale convection (granulation and supergranulation) is unclear in this picture. Observations show that it is dominant, while even the existence of giant convection cells in the Sun is open to doubt. So far no reliable observational data on the large-scale velocity field exists, and the circumstantial evidence (active longitudes, large-scale magnetic structures, coronal holes, etc.) may have other explanations that do not involve giant convective cells. Nevertheless, the idea of Dogel and Syrovatsky is interesting and merits further investigation. In order to explain the Maunder minimum it seems to be more important to demonstrate that stochastic behavior (a strange attractor) can arise naturally from existing models of the solar convection zone.

In the case just considered, the magnetic field played a passive rôle; it "blindly" followed the auto-oscillations of the hydrodynamic system. But the solar dynamo is, in fact, a nonlinear system in which the magnetic field affects the motion. Exposing stochastic properties of this system might also solve the problem of the random prolonged weakenings of solar activity.

Zeldovich and Ruzmaikin (1982) exhibited an axisymmetric model of the solar dynamo in which the effect of the magnetic field on the helicity could be reduced to a Lorenz system of three variables: the azimuthal field $B \equiv B_\phi$, the vector potential $A \equiv A_\phi$, and the deviation of the helicity density from its linear value, $\alpha - \alpha_l$. Their model was

$$\dot{A} = -A + \sigma DB - CB, \qquad \dot{B} = -\sigma B + \sigma A, \qquad \dot{C} = -\nu C + AB, \qquad (8)$$

where $\sigma \simeq (\Delta B/B)(\Delta A/A)^{-1}$ determines the ratio of the characteristic diffusion times of A and B. We have taken the characteristic diffusion time of A to be the unit of time and have introduced the dynamo-numbers

$$D = \alpha_l \omega h^3 \nu_T^{-2}, \qquad C = \sigma \omega h^3 \nu_T^{-2}(\alpha_l - \alpha).$$

As noted above, in certain ranges of the parameters σ, D and ν the trajectories of system (8) behave stochastically. They make some revolutions in the neighborhood of one of the singular points $S_\pm = (\pm[\nu(D-1)]^{1/2}, \pm[\nu(D-1)]^{1/2}, D-1)$ and then pass to the other one, i.e. transitions take place in which the signs of A and B reverse. The transition from S_+ to S_-, and vice versa, can be identified with the basic limit cycle described in Section 2, and the oscillations around these points with the small-scale nonlinear distortions of the basic cycle. The most interesting feature is the singular point $(0, 0, 0)$. In the Lorenz-type system the trajectory now and then (though rarely) finds itself near this point and then remains in its neighborhood for a long time (compared with the oscillation period). This resembles rare random weakenings of the solar cycle. In this connection it is very important to calculate the characteristic oscillation times and "weakening" periods. Unfortunately, this problem is difficult even for the thoroughly investigated Lorenz system (Lücke, 1976). Let us present some crude estimates. The frequency of oscillations can be estimated in the linear regime $D = D_0$ from $\Omega^2 = 2\nu\sigma(\sigma + 1) \times (\sigma - 1 - \nu)^{-1}$ McLaughlin and Martin, 1975), the period of the basic cycle is $T \sim (\sigma D)^{1/2}$, while the time the system remains in the neighborhood of the singular point essentially depends on the parameter ν, where

$$\tau_M \sim \nu^{-1}(D - D_0)^{1/2}.$$

This can be estimated as the time taken by the representative point to describe a trajectory close to the separatrix crossing $(0, 0, 0)$. For a small deviation of D from the critical D_0,

$$\tau_M \sim \nu^{-1} \ln |D - D_0|^{-1}.$$

According to observation, τ_M^{-1} is small, i.e. the parameter ν must also be small. When D is large but $\nu(D - D_0)$ is small enough, the regime $\Omega^{-1} \ll T \ll \tau_M$ can occur.

Of course, the model just described is greatly simplified and crude. By a more careful treatment of the dynamo equations together with an additional equation for $\partial\alpha/\partial t$ (see Chapter 10), we obtain a system of more than three differential equations. Preliminary calculations indicate that this system also can exhibit global stochastic behaviors, with the singular point $(0, 0, 0)$ playing a major rôle.

The two following properties of the Lorenz approximation should be noted. First, a trajectory entering the neighborhood of the point $(0, 0, 0)$, i.e. the global minimum, "forgets" the phase of its basic oscillation. It is unclear, however, whether this feature is necessarily preserved in the more general model (only the amplitude-period relationship of the basic oscillation can be subject to a random modulation). The question of preserving the phase after a minimum has been enthusiastically discussed (Gleisberg and Damboldt, 1979; Vitinsky, 1979). Second, the approach to the global minimum is smooth, over say a few basic periods. This point is important in relation to observation (Vasiliev and Dergachev, 1980).

In the model described above only the effect of the magnetic field on the helicity was invoked. But in general the back reaction of the field on the differential rotation must also be taken into account. Certainly this makes the problem more complex, but it allows a bridge to be built between our model and that of Dogel and Syrovatsky. The development of such a complete nonlinear model of the solar dynamo, not in a simplified form but using the full partial differential equations, is an important task for theorists investigating the nature of solar activity.

It may seem that explaining the prolonged weaknesses of solar activity like the Maunder minimum is a purely theoretical matter. This is not the case. First, a theory capable of explaining the Maunder minimum along with other features of the solar cycle would provide a concrete base on which explanations of other observed phenomena could be built. Second, an answer should be given to the very practical question: is it possible that solar activity will be greatly reduced in the future? It is interesting that Gleisberg and Damboldt (1979) believe that a new grand minimum is not to be expected in the immediate future. They drew a curve of the time dependence of the asymmetry of the basic 11-year cycle $A_{sym} = (\tau_r - \tau_d)/(\tau_r + \tau_d)$, where τ_r is the time of rise of the 11-year curve from minimum to maximum and τ_d is the time of its descent from maximum to minimum (Figure 11.14). During the Maunder minimum this was rather small and negative. After the Maunder minimum it became positive and kept growing, experiencing strong variations but remaining positive. Today A_{sym} is still far from zero, we may therefore hope that sunspots will not disappear over long periods during our lifetime.

The versatile Robert Hooke, who lived by the way at the time of the Maunder minimum (he was born on July 18, 1635 and died on March 3, 1703), noted that "the Sun's colour resembles that of a flame". Hence, Hooke believed that the Sun could emit smoke and soot, and that this could explain sunspots and flares (Woller, 1705).

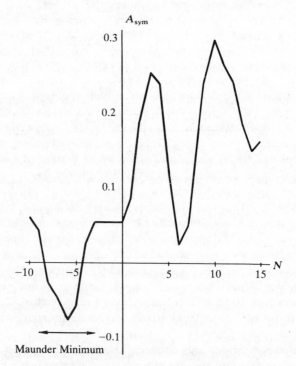

Figure 11.14 The dependence of the average asymmetry of field on the number of the 11-year cycle (after Gleisberg and Damboldt, 1979).

Today we are privileged to know much more than the famous Hooke about the nature and behavior of sunspots.

But are we much closer to the truth?

VIII Peculiar Magnetic Stars

Although Hale measured the Zeeman splitting (see Chapter 2) of lines in the spectra of sunspots as early as 1908 (finding fields of from 1 to 4 kG), the magnetic fields of other stars were discovered only in the fifties. The main difficulty is associated, of course, with the weakness of the light received from stars. H. Babcock (1947) measured a mean field weighted over the stellar surface by using a magnetograph, invented by him, and a special measurement technique. He discovered the magnetic fields of nearly 90 stars. Subsequent investigations showed that stars with strong fields (from 1 to 34 kG) are peculiar. They lie on the main sequence in a spectral interval near class A (from B5 to F0, with surface temperatures from $1.8 \cdot 10^4$ to $7.6 \cdot 10^3$ K) and have sharp anomalies in their chemical composition, which is why they are called A peculiar (Ap) stars (Ledoux and Renson, 1966; Pikelner and Khokhlova, 1972). "Magnetic" stars is another name for them, though it is less attractive because magnetic fields have now been discovered for a great variety of other stars (see, for example, Chapter 2 and Section II above).

The intensities of the lines in the spectra of Ap stars are anomalous in comparison with those of solar type, and testify that these stars have peculiar chemical compositions, and non-uniform distributions of chemical elements over their surfaces. We note some of the anomalous features: (a) a ten-fold deficit of helium; (b) an excess of light elements Al, Si, P, S, Cl, sometimes by 2–3 orders of magnitude, and elements of the iron group (Sc, Ti, Gr, Mn), sometimes by a factor exceeding 10 to 100; (c) an excess of rare earths Lu, Eu, Gd, Dy by a factor of 300 to 1000; (d) an increased abundance of Li for some stars.

The regions of anomalous chemical composition seem to concentrate in separate spots and are closely associated with the magnetic field. An analysis of the variations in field strength and in element

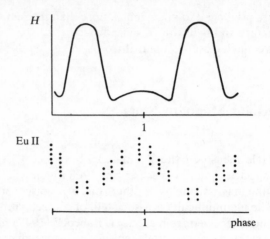

Figure 11.15 Time (phase) dependencies of the effective field on the surface of the α^2 Canes Venatici and the abundance of Eu 11.

abundances testify to this (Figure 11.15). Typical periods of these variations range from 0.5 to 20 days, though there is a group of stars (about 10) with variations of longer period (from 1 to 70 years).

In stellar astrophysics the mean statistical dependence of stellar rotational velocity on spectral type is well known: stars of earlier types rotate faster. The Ap stars do not conform. They rotate more slowly than normal stars of this type. The mean value of the projection of the rotational velocity at the equator onto the line of sight is about 50 km/s for the Ap stars and about 180 km/s for other stars of this spectral type. It is also interesting that Ap stars are more rarely components of binaries than are normal A stars (less than 20% as compared with about 40%).

The main problem in the theory of Ap stars lies in explaining the origin of two, probably closely related, phenomena—a magnetic field and an anomalous chemical composition. Observations show that variations in the magnetic field and the chemical composition have the same period as the stellar rotation. A strong constant component of the field is seen, indicating a predominance of the first (dipole, or dipole and quadrupole) harmonics. In contrast to the Sun and stars of later spectral types that have thick convection zones, A-stars have very thin convective envelopes (of about $10^{-3}R$). The models of magnetic field generation discussed in Section III can therefore not be used for A-stars. The Ap-stars, whose masses vary in the range 1.5 to $6M_\odot$ have convective cores containing, respectively, from 6 to 23% of their masses. Field generation is possible in a convective core (see later). The core is, however, enveloped by a highly conducting blanket, and one must ask, "How can the magnetic field penetrate to the surface?" It is in any case clear that the generation mechanisms and the intrinsic variations of their magnetic fields (i.e. those unconnected with their rotation) operate in Ap stars over longer characteristic times, as compared with stars of solar type.

A naïve extrapolation of the field $2 \cdot 10^{-6}$ Gauss of the interstellar medium with a density $\rho \sim 10^{-24}$ g/cm^3 to mean stellar densities of about 1 g/cm^3, using the law $H \sim \rho^{2/3}$ appropriate to spherical adiabatic compression, yields very strong (in fact, too strong) fields $H \sim 10^{10}$ Gauss. A considerable loss in magnetic flux must have occurred while the protostellar cloud was compressed to form a cold ($T \sim 10$ K) nontransparent core (Chapter 14). This was accomplished simply through the concomitant decrease in conductivity of the protostellar matter. Even more destructive of the magnetic field were the evolutionary stages at which the stellar material became turbulent. This might occur during the protostellar stage (Chapter. 14), or during the so-called "Hayashi phase".

By the Hayashi phase, one means the supposed period (of approximately 10^6 years) in the evolution of a star prior to its transition to the main sequence of the Hertzsprung–Russell diagram. In this period the star is convectively unstable, its contents are thoroughly mixed, and a large-scale magnetic field could be quickly destroyed through turbulent diffusion. On the other hand, at this time conditions are also favorable for a hydromagnetic dynamo to operate in the star, since the turbulence is helical and the rotation of the star is likely to be non-uniform. Schüssler (1975) constructed numerical models for field generation in a star during its Hayashi phase, using the α^2-type dynamo mechanism (without differential rotation). In such a dynamo axisymmetric modes are not necessarily preferred; under certain conditions asymmetric modes are more easily generated. In this way, he could obtain the so-called "oblique rotator"—a dipole field with the magnetic axis inclined to the rotation axis. It is interesting that long period variations of the magnetic fields of Ap-stars with periods exceeding 100 days can be explained by a dynamo process that, like the one proposed for the solar cycle, generates a variable field. It is important to establish from observations whether these stars are in the Hayashi phase at this time i.e. whether they lie above the main sequence.

The oblique rotator model is in fact very crude. It does not explain the difference observed between two successive maxima in the field variation in many Ap stars (see Figure 11.15). Krause (1971) argues that the most realistic model of the magnetic field of an Ap star is symmetric with respect to the equatorial plane. The first two harmonics of the field generated in the dynamo process (an equatorial dipole and a quadrupole) are enough to explain the effective field observed in all well studied Ap stars (Oetken, 1977). Krause does not link the field generation with the Hayashi phase; he assumes that it is taking place in the contemporary stellar core. The time required for this field to spread to the surface of the star by ohmic diffusion is very large (it exceeds the lifetime of the star). The field can, however, rise through nonlinear effects, for example magnetic buoyancy which can operate considerably more quickly.

It is important to explain why some Ap-type stars have a strong magnetic field while others do not. Schüssler (1975) assumes that not all Ap stars have passed through the Hayashi stage, but only those whose masses are less than a certain critical value, $M_{cr}(\approx 2M_{\odot})$. It was noted at the top of p. 233, however, that the masses of the observed Ap stars vary across a much wider range; besides, nonmagnetic A stars with $M < M_{cr}$ are also known to exist. In Krause's opinion (private communication), the Ap stars differ from other stars of A type by having an almost solid body rotation. The differential rotation in the convective core of other stars creates conditions under which the generation of axisymmetric field predominates, and this cannot rise to the surface. There remains the question of why the rotation is solid body (or close to it) in one case, and differential in another. We recall that the Ap stars are slow rotators.

In this discussion, the principal rôle has been assigned to the magnetic field. The anomaly in the chemical compositions was regarded as only a consequence. It has been suggested that the anomaly can be explained by the diffusive separation of elements in a star having a normal composition under the influence of radiation pressure and gravitation. The deficit of helium and other light elements compared with hydrogen is then explained by their slow submergence (over periods of hundreds of years) under the action of gravity. In contrast, the radiation pressure acts more strongly on heavy atoms which have many lines of absorption in the visible and UV spectrum. It should be noted, however, that this mechanism has some defects (for example, it does not explain the excess of He^3, Pm, Li, or the heavy isotopes of mercury). Besides, Ap stars possess rather thin envelopes in strong convection that would quickly mix matter, thus depriving the diffusion mechanism of its charm.

Perhaps it is more attractive, as V. Chechetkin has privately suggested to us, to explain the origin of an Ap star by supposing it was once a member of a binary stellar system whose companion, after having ejected part of its matter to the A star, exploded as a supernova. Heavy elements are formed in the nuclear reactions of the exploding star. The problem is to construct a mechanism for the generation of magnetic field in this rapidly evolving system, and a model that explains the concentration of elements at the magnetic poles.

CHAPTER 12

The Galaxy and Its Magnetic Field: Observations

I The Galaxy

The term "galaxy" has been borrowed from the Greek and stems from "galaktikos" meaning milky. And the Galaxy can really be seen in the night sky as a milky band (the Milky Way), stretching across the heavens almost along a great circle. It is clear from this that Galactic matter is concentrated about a plane. In the first approximation the Galaxy resembles an enormous rotating lens (Figure 12.1) whose diameter is about 30 kpc, with an average thickness of about 2 kpc. As "seen from aloft", the Galaxy would have a more complex form (Figure 12.2). Two tightly twisted spiral arms (about 0.4 kpc wide) would be distinguished, on one of which our Sun is located.

Stars make the major contribution to the mass of the Galaxy. They form its skeleton. The gas contributes only a few percent of the mass,

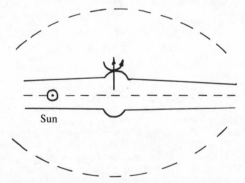

Figure 12.1   The Galaxy as viewed from the edge.

236

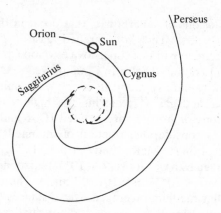

Figure 12.2 Schematic representation of the Galaxy from above.

but it is responsible for the Galactic magnetic field. The gas is also the material from which new stars are continuously generated and, moreover, it is responsible for some active processes, e.g. shock waves. Simultaneously, by evolving and exploding, stars return gas to interstellar space.

The gas is concentrated in the Galactic plane and in the spiral arms but its distribution is not homogeneous.† Large dense gas (or dust-gas) complexes can be distinguished. There are, for example, light nebulae (including the Orion Nebula, discovered by Huygens in 1656) and dark nebulae (including the Coal Sack and Horsehead Nebulae that can be seen with the naked eye, which were found by W. Struve in 1847).

The discovery of the quintruply ionized oxygen lines O(VI) (Jenkins and Meloy, 1974) as well as the observation of soft X-ray radiation from the interstellar medium (Burstein *et al.*, 1977) suggest the existence of vast hot regions in the Galaxy. Such regions arise naturally

---

† The gas distribution can to a first approximation be represented by a thin disk whose semi-thickness $h \sim 10^{-2}R$, where R is its radius. Actually, the gas disk thickens at its edges, and the density isolines have a form close to that of a hyperboloid of evolution, the disk boundary being determined by $z^2/(0.2 \text{ kpc})^2 - r^2/(4 \text{ kpc})^2 = 1$ (Kaplan and Pikelner, 1979). This is due to the decrease in the gravitational field, g_r, perpendicular to the Galactic plane with increasing distance, r, from the Galactic center. In addition, the gas disk is noticeably bent at its periphery: towards the South in the direction of the center and towards the North in the opposite direction (Lozinkaya and Kardashev, 1963).

due to the explosions of supernovae, that occur in the Galaxy once every 20–100 years. The hot rarefied gases ($T \simeq 10^6$ K, $n_e \simeq 3 \cdot 10^{-3}$ cm^{-3}) remain within a supernova envelope as it expands, and their subsequent cooling takes a few million years. During that time another supernova flares up, and its shock wave quickly propagates through the rarefied gas so restoring the high temperature. Vast connected regions of hot gas are formed (Cox and Smith, 1974), which may fill a considerable fraction of interstellar space [of the order of 0.7, according to McKee and Ostriker (1977)]. The remaining volume is occupied by a warm gas ($T \sim 10^4$ K) surrounding cool clouds (of size ~10 pc, $T \sim 10^2$ K, $n = 10$–100 cm^{-3}). The electron density of the interstellar medium, averaged over distances (≥ 100 pc) large compared with the size of the clouds and hot cavities, is

$$\langle n_e \rangle \simeq 3.10^{-2} \text{ cm}^{-3}.$$

This value agrees well with the results of pulsar dispersion measure studies (Terzian and Davidson, 1976). The semi-thickness of the ionized layer of matter, defined as $\int n_e \, dz / \langle n_e \rangle$ where z is the coordinate perpendicular to the plane of the disk, is

$$h \simeq 400 \text{ pc},$$

(Ruzmaikin and Sokoloff, 1977a).

The interstellar gas is in chaotic motions. As was first demonstrated by Kaplan and Pikelner (1970) from a study of the line-of-sight velocities of the interstellar clouds, these motions are turbulent and in the first approximation the velocity distribution on the scale λ is closely consistent with the Kolmogorov spectrum, i.e.

$$v_\lambda \approx (\varepsilon \lambda)^{1/3}, \tag{1}$$

where ε is the turbulent energy dissipating in unit mass per unit time (erg/g sec). Supernova explosions seem to be major contributors to the Galactic turbulent energy budget. We now make a crude estimate of ε. The energy associated with a supernova remnant is about $3 \cdot 10^{50}$ erg (Lozinskaya, 1980). Suppose that a third of this energy is transformed into kinetic energy of the ambient gas. Assuming that the explosions occur every 50 years and that the mass of gas involved is about $10^{10} M_\odot$ (5% of the Galactic mass), we find that $\varepsilon \simeq 3 \cdot 10^{-2}$ erg/g · sec. We can now estimate the characteristic velocity of the random motions on the scale $l \sim 100$ pc, above which the interstellar medium can be considered to be homogeneous. We find [see Eq. (1)]

$$v_l \simeq 10 \text{ km/sec.}$$

In the evolution of large-scale mean fields such as the average temperature or the average magnetic field, the turbulence on scales, λ, small compared with l acts as an effective diffusivity, like a turbulent thermal conductivity or a turbulent viscosity, and

$$\nu_T \approx \tfrac{1}{3}lv \sim 10^{26} \text{ cm}^2/\text{s.} \qquad (2)$$

Moreover the gas participates in the general rotation of the Galaxy. This was reliably established from studies of the Doppler shifts of the 21-cm line (Rougoor and Oort, 1960). The rotation of the Galactic gas disk is inhomogeneous. The angular velocity $\omega(r)$ is large in central regions and decreases outwards (Figure 12.3). Note that regions of solid body rotation are absent.

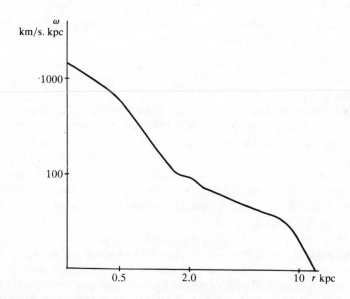

Figure 12.3   Angular velocity of the Galactic gaseous disk as a function of distance, r, from the Galactic center.

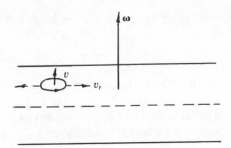

Figure 12.4   Helicity arising in a rotating disk of gas stratified in density. A "rising" vortex expands (i.e. v_r becomes non-zero); Coriolis forces, $2\omega v_r$, cause an additional twisting of the vortex. The direction of twisting is different for rising and falling vortices, but the sign of the quantity $\mathbf{v} \cdot \boldsymbol{\nabla} \times \mathbf{v}$ is the same.

The rotation of the gas and its density distribution (its concentration towards the Galactic plane) imply an absence of reflection symmetry in its random motion (Figure 12.4), and the extent of its absence is measured by the helicity density of the Galactic turbulence. This may be assessed by symmetry and similarity arguments (recall that α is a pseudoscalar that owes its existence to the density gradient vector and the angular velocity pseudovector),

$$\alpha = \tfrac{1}{3}\tau\langle \mathbf{v} \cdot \boldsymbol{\nabla} \times \mathbf{v}\rangle \simeq l^2 \rho^{-1}(\omega \, \mathrm{d}\rho/\mathrm{d}z). \tag{3}$$

Evidently, it is asymmetric $\alpha(z) = -\alpha(-z)$.

In addition to these motions in the Galactic disk, large-scale meridional (quadrupole-type) flows seem to be possible. From the hydrodynamical point of view, such flows may arise just as they do near a rotating disk immersed in a fluid (von Karman, 1921). Observations of the motion of clouds at high latitudes also favor their existence. Estimates by Waxman (1979) show, however, that the meridional velocity is small: $v_M \approx 20$ km/sec. The period of meridional overturning is thus about 10^9 years.

In investigating the Galaxy, special attention should be paid to its central regions ($r < 1$ kpc). This repeats the Galaxy's structure in miniature (Oort, 1977). It too possesses a rotating gas disk ($\omega \approx 10^{-14} \, \mathrm{s}^{-1}$), whose thickness of about 80 pc at the center increases to 250 pc at its periphery. The characteristic velocity of random motions in the disk may reach 20 km/s.

Note that, from an astronomer's viewpoint, the Galaxy is one of a vast class of spiral galaxies. Other examples of such galaxies that resemble our own are M31 and M81 in Messier's catalogue.

II The Magnetic Field of the Galaxy

Stars have magnetic fields. The mean distance between stars is a few parsecs (it is less at the center of the Galaxy and greater at its periphery). Therefore stellar fields are extremely small between stars (in a vacuum they decrease as r^{-3} from their sources) and they cannot explain a large-scale field structure. The observed general Galactic field is certainly associated with the interstellar gas.

Compared with the magnetic fields encountered in laboratory conditions, the Galactic field is tiny in strength (only a few μGauss) but enormous in scale—it extends over lengths of many kiloparsecs. (Recall that $1\,\mathrm{pc} = 3 \cdot 10^{18}$ cm.) It is this large scale that governs the behavior of the field: ohmic attenuation is practically non-existent but the rôle of the motions grows correspondingly in importance. A large-scale field changes insignificantly due to the ohmic dissipation. For example, for a field of scale $l = 100$ pc, the typical time, τ_d, of ohmic dissipation in a plasma at $T = 10^4$ K is very large,

$$\tau_d \approx l^2/\nu_m \sim 10^{26}\,\text{years}, \tag{4}$$

taking $\nu_m = 10^{13} T^{-3/2}\,\mathrm{cm^2/s}$ as the coefficient of ohmic diffusion. It is convenient to characterize this situation by using the dimensionless magnetic Reynolds number

$$R_m = lv/\nu_m,$$

which for $l = 100$ pc, $v = 10$ km/s and $\nu_m = 10^7\,\mathrm{cm^2/s}$ is $3 \cdot 10^{19}$! In fact, diffusion and dissipation in the large-scale magnetic field is determined by the turbulence [see Eq. (2)]. The corresponding effective magnetic Reynolds number,

$$R_m^{\mathit{eff}} = lv/\nu_T,$$

is of order unity.

The magnetic field does not play any significant part in the equilibrium and dynamics of the Galaxy. The publications in which the equilibrium of spiral arms is explained by the magnetic field are now only of historical interest. The magnetic field does, however, play a significant rôle in the propagation of cosmic rays, in gasdynamical processes (shock waves, star formation), and in the mechanism by which cosmic dust is orientated (see below).

A magnetic field of $2 \cdot 10^{-6}$ Gauss allows the maintenance in the Galactic disk of charged particles with energies of up to 10^{18} eV. The Larmor radius of such particles, $r_L \doteq \mathscr{E}/300H$ cm, is less than the semi-thickness of the Galactic disk. The small scale random component of the field affects the diffusion coefficient for cosmic rays (Ginzburg and Syrovatsky, 1963).

It seems that star formation is not possible without a magnetic field. Its rôle consists of transporting angular momentum outwards so that the collapse of the proto-stellar cloud can continue (see Chapter 14).

In addition to the gas, there is much dust in the Galaxy. On being scattered by the dust, star light becomes linearly polarized, through the orientation of the dust particles. The orientation is performed by the magnetic field. It is usually assumed that the rotating dust particles are paramagnetic. On remagnetization in the Galactic magnetic field they line up with their smallest axis orientated along the field (Davis and Greenstein, 1951). The characteristic time of this process of orientation appears to be very great. The mechanism of orientation of dust particles in anisotropic fluxes of gas and radiation is discussed by Dolginov et al. (1979). The magnetic field serves as an orientation axis around which the angular momentum of the dust particles precesses.

The background radiation of the Galaxy is of synchrotron type, i.e. is emitted by relativistic electrons ($n_{er} \sim 5 \cdot 10^{-13}$ cm^{-3}) accelerated by the Galactic magnetic field (Ginzburg, 1953; Shklovsky, 1953). The discovery of the polarization of this radiation was an event of great importance (Razin, 1958; Westerhout et al., 1962).

The magnetic field is an essential ingredient for the interpretation of the observed X-ray radiation of the Galaxy. For example, regions with intensified radiation can be distinguished in the distribution of the soft diffuse X-ray radiation over the celestial sphere. These regions are located within the so-called "spurs" of the magnetic field, where he polarization of the synchrotron radiation is highest (Shklovsky nd Sheffer, 1971; Spoelstra, 1972).

II Local Observations of the Galactic Field

All methods of determining the Galactic magnetic field rely on etailed knowledge of the properties of the observed radiation. The aost direct approach consists in measuring the Zeeman splitting of he hyperfine transition in the 21-cm line of neutral hydrogen. The ifficulty here is that the value of the Zeeman splitting $\Delta\nu = H/2\pi mc \simeq 3 \cdot 10^6 H \sim 10 \sec^{-1}$ is small compared with the Doppler ridth $\Delta\nu = \nu v/c \sim 10^4 \sec^{-1}$ of the lines. Positive results have there- ore been obtained only for a few dense clouds possessing more ntense fields than the Galactic field. The strengths of the field com- onent parallel to the line of sight are given in Table 12.1 for several aterstellar clouds (Verschuur, 1970). From these data an estimate of he strength of the Galactic field can in principle be made by assum- ag that, during cloud formation, the field is intensified adiabatically $H \sim n^{2/3}$). Assuming a Galactic density of $n = 0.7\,\mathrm{cm}^{-3}$, we obtain a aean Galactic field strength of between 1 and 3 μGauss. It is difficult, owever, to obtain any information about the structure of the general ialactic field from these clouds, which are both moving and rotating. Jevertheless, a certain correlation is observed between the orientation f the (non-spherical) dust particles and gas clouds (Shajn, 1955;

able 12.1   Magnetic Fields in Several Interstellar Clouds (Verschuur, 1970).

irection	l^{II}	b^{II}	Velocity km/sec	Field estimate μG	Density cm^{-3}
au A	185	−6	10	−3.5 ± 0.7	14
as A	112	−2	−38	18.0 ± 1.9	193
yg A	76	6	−84	4.0 ± 2.2	2.5
17	15	−1	14	25.0 ± 10	60–100
rion A	209	−19	7	−50 ± 15	680

Verschuur, 1970), and the direction of the magnetic field, which in the first case is determined by the optical polarization of starlight and in the second by the polarization of the continuous radio radiation.

To determine the direction of the magnetic field one may use the data from observations of the polarization of starlight scattered by dust. At the present time such data has been obtained for about 7000 stars, most of which are less than 500 pc distant. The way these data are interpreted depends to a large extent on the orientation of the dust particles relative to the field (Dolginov et al., 1979). If the dust particle (considered as an elongated body) is orientated perpendicularly to the field, as in the mechanism of Davis and Greenstein (1951), then the polarization vector is parallel to the field. On the other hand, the direction of the magnetic field can be determined by the polarization of the background synchrotron radio radiation (originating somewhere not closer than 500 pc) which is determined by the field component perpendicular to the line of sight. A comparison of the optical and radio polarization angles for the radiation coming from the region $120° < l^{II} < 150°$, $-30° < b^{II} < +30°$ (Spoelstra, 1977) strongly suggests that the dust particles are orientated perpendicularly to the magnetic field (the radio and optical angles are perpendicular to each other).

This approach has provided some information about the Galactic magnetic field in the vicinity of the Sun ($\lesssim 500$ pc $- 1$ kpc). Other local methods (an analysis of the synchrotron radiation intensity; the orientation of remnants of old supernovae; the investigation of anisotropy in, and propagation of, cosmic rays, etc.) have been discussed by Verschuur (1970), Heiles (1976) and Spoelstra (1977).

Information about the large-scale magnetic field of the Galaxy can be gained by analyzing the Faraday rotation of discrete sources of polarized radiation.

IV The Large Scale Galactic Field

The Faraday rotation measure of a radio source, usually designated by the two letters RM, determines the position angle of the

olarization plane of the radiation and is directly connected with the
magnetic field (see Chapter 2). We have

$$\phi = RM\lambda^2 + \phi_0,$$

$$RM = \frac{|e|^3}{\pi m_e^2 c^3} \int n_e \, \mathbf{B} \cdot d\mathbf{r} \tag{5}$$

$$= 8.1 \cdot 10^5 (\text{rad/m}^2) \int n_e(\text{cm}^{-3}) \mathbf{B}(\text{G}) \cdot d\mathbf{r}(\text{pc}),$$

here λ is the wavelength of the radiation, ϕ_0 is the initial angle of
inclination of the polarization plane (the intrinsic position angle), and
$_e$ is the electron density of the medium through which the radiation
propagates. Position angles are measured from some standard direc-
on. It should be noted that they are arbitrary to an additive multiple
f π.

By measuring the position angles at several wavelengths (at least
wo), the rotation measure can be found and then the magnetic field
etermined from the electron density and the known distance through
ae medium that the radiation has traversed. This method has the
dvantage that it probes the Galactic magnetic field in the vast spaces
arough which the radio radiation from a source must pass in order
o reach the Earth. Two difficulties are evident here: first, formula
5) contains only the field component parallel to the line of sight;
econd, the field may change repeatedly and greatly over large dist-
nces. Moreover, a contribution to the rotation measure may be made
y the plasma of the source itself, and, in the case of an extragalactic
ource, by the field in the region between the radio source and the
ralaxy. These difficulties can be overcome by using a great number
f radio sources whose radiation permeates the Galaxy from many
ifferent directions, i.e. by exploiting statistical methods.

An analysis of the rotation measures of the Galactic and extragalac-
c radio sources (pulsars, quasars and radio galaxies) established a
ery important and far from obvious fact—the existence of a large-
cale component in the Galactic magnetic field. The first work in this
irection was done by the Australian radio astronomers Gardner and
Whiteoak (1963), Morris and Berge (1964), and Manchester (1972).
eliable and final confirmation was provided by Vallée amd Kronberg
975), Ruzmaikin and Sokoloff (1977b), and Ruzmaikin et al. (1978).

The observed rotation measure of a specific radio source represen
the sum of contributions from the regular Galactic field, from th
fluctuating part of that field, from the source itself, and from th
medium lying between the source and the Galaxy. Each of thes
contributions has its own dependence upon the coordinates. F
example, the rotation measure of the Galaxy must depend strong
on the Galactic angular coordinates (l^{II}, b^{II}), while the contributio
of the extragalactic medium will vary with the distance to the sour
(defined by its redshift). The rotation measures of such extragalact
sources must not show any systematic dependence on the Galact
coordinates (due to the isotropy of the Universe).

Figure 12.5 shows the distribution of the observed rotatio
measures in Galactic coordinates. Data on the coordinates and th
rotation measures of the extragalactic radio sources have been take
from the United Catalogue of Radio Sources by Eichendorf an
Reinhardt (1980) which is based on material drawn from all previo
observations including, for example, those of the well know
catalogues by Mitton (1972), Morris and Tabara (1973), Vallée an
Kronberg (1975), and Haves (1975).

It can clearly be seen that negative rotation measures predomina
in the third quadrant and positive ones in the fourth quadrant. Th
indicates that a large-scale magnetic field exists in the souther

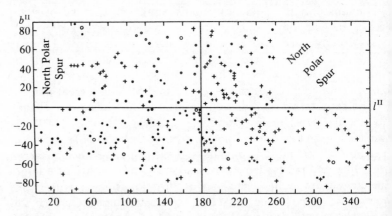

Figure 12.5 Map of the sign distribution of the rotation measures of extragalac
sources, taken from the United Catalogue by Eichendorf and Reinhardt. The posit
RM are marked by crosses and the negative RM by circles.

Galactic hemisphere ($b^{II} < 0$). It is difficult to judge by eye the magnetic state of the northern hemisphere. One can see only that the fluctuations in the rotation measure are rather large (especially in the northern hemisphere). Analyses show (Kuznetsova, 1976; Ruzmaikin et al., 1978) that the irregular part of the rotation measure depends primarily on the Galactic latitude. Hence, it is fluctuations of Galactic origin that play the principal part. The contribution of the Earth's magnetosphere is small and easily extracted from the observations (Verschuur, 1970). Different radio sources may have large and very different intrinsic rotation measures, but the average rotation measure of all sources is zero, and their contribution to the dispersion is imperceptible when seen against that of the Galactic fluctuations.

Faced with these facts, it is natural, in the first approximation, to treat the observed rotation measure as the sum of a contribution from the regular Galactic magnetic field and of a chaotic part arising both from the field fluctuations and those of the Galactic electron density. Thus

$$RM \overset{s}{=} RM + RM_f. \tag{6}$$

The index "s" above the equality sign indicates "statistical" equality. This relation is untrue of course if we consider sources with a very large rotation measure (such sources actually exist as the observations show). Strictly speaking, the equality (6) is a hypothesis to be confirmed below.

Suppose the large-scale magnetic field is homogeneous and directed along (l_0^{II}, b_0^{II}). Since an observer (in the solar system) is situated near the Galactic plane and the rotation measure (5) is proportional to the cosine of angle between the field and the direction to the source, it can easily be shown that (Figure 12.6)

$$RM = K[\cos b_0^{II} \cos b^{II} \cos (l^{II} - l_0^{II}) + \sin b_0^{II} \sin b^{II}]f(l^{II}, b^{II}). \tag{7}$$

Here l^{II}, b^{II} are the known angular coordinates of the source, K is (like l_0^{II}, b_0^{II}) an unknown amplitude which can be expressed in terms of the average electron density (cm^{-3}), the large-scale field $\langle H \rangle \equiv B$ (μGauss) and the semi-thickness of the Galactic gas disk h(pc) $[\equiv \int n_e \, dz / \langle n_e \rangle]$ as

$$K = 0.81 \langle n_e \rangle Bh \quad \text{rad/m}^2. \tag{8}$$

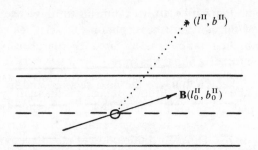

Figure 12.6 Positions of the source (l^{II}, b^{II}) and the large-scale magnetic field (l_0^{II}, b_0^{II}) with respect to an observer situated in the central plane of the Galaxy.

The function $f(l^{II}, b^{II})$ specifies the model. In the simplest case it is assumed that the Galaxy is an infinite plane layer, and then

$$f(l^{II}, b^{II}) = |\sin b^{II}|^{-1}. \qquad (9)$$

In the case of an arm having an elliptical cross-section with semi-axes ratio $\varepsilon\,(\approx 0.5)$ we have, in place of (9),

$$f(l^{II}, b^{II}) = (\varepsilon^2 \cos^2 b^{II} \cos^2 l^{II} + \sin^2 b^{II})^{-1/2}. \qquad (10)$$

Note that for $|b^{II}| \gtrsim 30°$ the formulae (9) and (10) are in approximate agreement.

The fluctuating part of RM is a random function, but we are interested in its dispersion

$$RM_f^2 = 0.81 \int_0^d ds_1 \int_0^d ds_2\, C(s_1, s_2).$$

Here $C(s_1, s_2)$ is the two-point correlation function of $n_e(\mathbf{B} \cdot \mathbf{r}/r)$, the integrals being taken from the point of observation to the boundary of the Galaxy. Under the simplest assumption, that the Galactic turbulence is homogeneous l being then the only relevant correlation scale, $C(s_1, s_2)$ depends only on $s = |s_1 - s_2|$, and we may take (Jokipii and Lerche, 1969; Ruzmaikin and Sokoloff, 1977a)

$$C(s) = C_1 \exp(-s/l).$$

Then we have

$$RM_f = [(d/l) - 1 + \exp(-d/l)]^{1/2} \xi, \qquad (11)$$

where ξ is a random variable with vanishing mean whose dispersion, independent of the coordinates, is given by

$$D_\xi = 2(0.81)^2 l^2 D(n_e B). \tag{12}$$

By recognizing that $d/l \ll 1$ (see the next section) and by expressing l in terms of the disk thickness, we obtain the approximate formula

$$RM_f \simeq [(d/l)f(l^{II}, b^{II})]^{1/2}\xi,$$

which is convenient for practical applications. Note that the contribution from the sources does not depend on the angles l^{II} and b^{II} so it is not included.

We shall suppose that the random variables ξ are independent for a large set of sources. It is interesting to note however that this assumption must become invalid as the number of sources increases since the radiation from neighboring sources will then pass through the same turbulent cells. The angular dimension of the cells is about $0°$ so that, if the sources are homogeneously distributed across the sky, statistical independence of ξ will be realized if there are no more than about 400 sources, and this is approximately the number of observed sources having polarized radio radiation. It is useless to increase this number in order to determine its mean magnetic field, B. To obtain information about the field in a turbulent cell and that of the sources it is necessary to increase further the number of sources of known Faraday rotation.

The mean values of K, l_0^{II} and b_0^{II} are obtained (Ruzmaikin and Sokoloff, 1977b) by regression analysis, which is a generalization of the method of least squares. The essence of this method consists of minimizing the sum,

$$S(K, l_0^{II}, b_0^{II}) = \sum_n \xi^2 \equiv \sum_n \frac{[RM^{(n)} - RM(K, l_n^{II}, b_n^{II})]^2}{(h/l)f(l_n^{II}, b_n^{II})},$$

for the true magnitudes of the mean values sought. Here the index "n" refers to an individual source with coordinates (l_n^{II}, b_n^{II}) and observed rotation measure $RM^{(n)}$; RM is defined by Eq. (7). The extremum condition (the vanishing first derivatives of S) yields the mean values of K, l_0^{II} and b_0^{II}; their confidence limits are obtained from the second derivative of S and the Fisher criterion. The value S_{min} provides an estimate of the strength of the random fields (Section 5). Without going into the details of the calculations we give the

results obtained using the observed rotation measures from the United
Catalogue by Eichendorf and Reinhardt (1980). As the field structure and strength are different for the northern and southern hemispheres, we consider these separately.

(a) *The Southern Hemisphere* ($b^{II} < 0°$; *Figure* 12.5).

Using the ellipsoidal arm model (10), it was found that

$$K = -20 \pm 3 \text{ rad/m}^2, \qquad l_0^{II} = 106° \pm 8°, \qquad b_0^{II} = 0° \pm 8°. \quad (13$$

Parameter values close to those above had already been obtained by
Vallée and Kronberg (1975) who also referred to previous results.
The values in (13) prove to be quite stable under a change in the
number of sources or after other modifications to the model. In
particular, a separate treatment of the high-latitude sources ($b^{II} <$
$-30°$) and the low-latitude sources ($b^{II} > -30°$) had practically no
effect on the values of K and the dispersion D. This should be
contrasted with the results $K = -10 \text{ rad/m}^2$ for $|b| < 30°$ and $K =$
-22 rad/m^2 for $|b| > 30°$ obtained using the plane layer model (9).
The discrepancy here arises from the fact that, in the plane layer
model, the theoretical value of RM is set too high at low latitudes

The result (13) corresponds [see Eq. (8)] to a large-scale magnetic
field of strength

$$B \simeq 2 \cdot 10^{-6} \text{ Gauss} \qquad (14$$

for $\langle n_e \rangle = 0.03 \text{ cm}^{-3}$ and $h = 400$ pc (Ruzmaikin and Sokoloff, 1977a)
We shall now estimate the scale, L, of this field. Although we assumed
that the field was homogeneous, its scale is actually defined by the
maximum distance covered by the radiation of the sources at the
lowest latitudes in the Galactic disk. In the catalogue under consider
ation $b_{min}^{II} \simeq 10°$, for which the number of sources is quite sufficient
and gives

$$L \simeq 2h/\tan b_{min}^{II} \simeq 10 \, h \sim 4\text{kpc}. \qquad (15$$

This scale agrees well with the lengths of the arm on either side of
the Sun. In the transverse direction the field scale seems to be of the
same order of magnitude as the cross-sectional dimensions of the
spiral arm. Note that, allowing for uncertainties, the field directio
is close to that of the local (Orion) arm $l^{II} \simeq 90°$, $b^{II} \approx 0°$ (Figure 12.2

(b) *The Northern Hemisphere* ($b^{II} > 0°$; *Figure* 12.5).

The large-scale component of the magnetic field in the northern hemisphere has the same direction as in the southern hemisphere but proves to be three times weaker. Using the same model (10) of an ellipsoidal arm, it is found that

$$K = -7 \pm 3 \, \text{rad/m}^2, \qquad l_0^{II} = 68° \pm 43°, \qquad b_0^{II} = 49° \pm 23°,$$
$$\tag{16}$$

whence $B \simeq 0.7 \pm 0.3 \, \mu\text{Gauss}$. This interesting fact will be discussed in Section VI.

In general, the observations show that the large-scale magnetic field lies in the plane of the Galactic disk and is directed approximately along the spiral arm. The large helicity of the spiral arm indicates that we are dealing with an azimuthal field. It is important that the field does not change its sign on crossing the Galactic plane. The transverse (poloidal) field components (B_r, B_z) prove to be appreciably weaker. For example, estimates of $0.2 \pm 0.3 \, \mu\text{Gauss}$ and $1.8 \pm 1.5 \, \mu\text{Gauss}$ have been given for the z-component in the Southern and Northern hemispheres, respectively. The sign of this field is, however, the same; this is important and testifies that the poloidal component of the large-scale field is quadrupolar.

In conclusion we note that, using a rather small number of galactic radio sources (pulsars), one obtains parameters for the large-scale

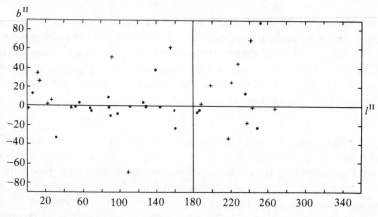

Figure 12.7 Map of the sign distributions of the rotation measures of Galactic pulsars. Spurs are excluded.

field that agree with the statistical results from hundreds of extragalactic sources, namely

$$B = -2.1 \cdot 10^{-6} \, \text{Gauss}, \qquad l = 99°, \qquad b \simeq 0°, \qquad (17)$$

(Manchester, 1974; Ruzmaikin and Sokoloff, 1977a). But it is even more interesting that, by using pulsars, one can estimate the characteristics of the random component of the Galactic magnetic field, due to the fact that a pulsar has a small rotation measure (Manchester, 1972).

V The Random Component of the Galactic Field

Let us examine Figure 12.5 and the analogous figure for pulsars Figure 12.7. We see that the random part of the rotation measure is large. This is confirmed by other methods too, e.g. by the analysis of the polarization of the continuous Galactic radiation which gave $1 < \delta H/H < 3$ (Spoelstra, 1977), with a correlation length in the range 30–100 pc. An independent estimate of the correlation length,

$$l \simeq 100 \, \text{pc}, \qquad (18)$$

was obtained by Ruzmaikin and Sokoloff (1977a) by a statistical study of the pulsar rotation measures. The analysis of the rotation measures of extragalactic radio sources, described in the previous section, with known l and $S_{min} = D\xi$, can be used to obtain the mean square field strength on this scale. Specifically, for the southern hemisphere

$$\overline{(\delta H^2)}^{1/2}/B \simeq 1.7. \qquad (19)$$

Strictly speaking, the contribution of the sources must be separated from the random component of the rotation measure. This can be done by comparing the upper estimate for the RM_f values along the perpendicular to the Galactic plane obtained for the pulsars and the extragalactic sources. It turns out that Galactic fluctuations make the main contribution to RM_f, though residual scatter in the rotation measures of the extragalactic sources is still rather large ($\sim 60 \, \text{rad}/\text{m}^2$).

This is due to the low accuracy with which RM_f can be determined and to the small number of pulsars included.

In addition to estimating the correlation length and dispersion of the random component, it is certainly interesting to obtain data about its spectral distribution. Chibisov and Ptuskin (1982) suggest that the angular variations in the nonthermal radio emission from the Galaxy should be observed in order to obtain information about the random magnetic field. Theoretical estimates for the spectrum of the Galactic magnetic field have been provided by Ruzmaikin and Shukurov (1982); see also Section 8.III.

VI   Asymmetry of the Galactic Field

It can be seen from Figure 12.5 that the distribution of the Faraday rotation measures is much less regular in the northern hemisphere of the Galaxy ($b^{II} > 0$) than in the southern hemisphere ($b^{II} < 0$). The statistical results also confirm this fact (see Section IV). One cause of this asymmetry may be found in the prevalence of the so-called spurs (specific features of radio radiation resembling the spurs of a cavalry-man) in the northern hemisphere. Some spurs were studied by Spoelstra (1972). The greatest of these is the North Polar Spur. The radio radiation and polarization of spurs are abnormally strong. They are believed to be the remnants of supernovae. Their linear polariz-ation is associated with magnetic field loops. The strength of the field is, however, close to that of the mean Galactic field (Spoelstra, 1972, 1977), i.e. the field lines are distorted rather than crushed together.

The asymmetry cannot, however, be fully explained by the spurs. Even when the strongest (North Polar) spur is excluded from the analysis, a notable asymmetry remains in the distribution of RM. The estimated strength of the mean magnetic field is considerably smaller in the northern hemisphere than in the southern hemisphere, and the latitude dependence of the fluctuating component in the northern hemisphere cannot be well described by the simple model (11). A strong large-scale distortion of the mean field seems to have taken place. This can be likened to a local loop in the magnetic field, as

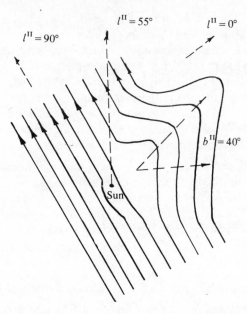

Figure 12.8 The loop of magnetic field in the solar neighborhood and the configuration of the large-scale Galactic field.

suggested by Gardner *et al.* (1969) and by Vallée and Kronberg (1973). An illustrative configuration of the resultant field is shown in Figure 12.8.

Note that the occurrence of asymmetry in the magnetic field distribution for the last one million years or less seems to be a temporary phenomenon associated with supernova explosions. During an interval of time exceeding the period of the Galaxy rotation (250 million years), the magnetic dynamo (see below) would in effect regenerate a mean magnetic field symmetrically about the Galactic plane.

CHAPTER 13

The Galactic Dynamo

We now discuss the subject of the origin and maintenance of the Galactic magnetic field. This is undoubtedly generated within the Galaxy itself. In the sixties the hypothesis was advanced of a pregalactic, relict origin for the field, it did not however receive any support. Numerous investigations of the Faraday rotation measures of extragalactic radio sources did not indicate that intergalactic space possessed an appreciable field (exceeding 10^{-9} Gauss, see Chapter 15). From the cosmological point of view, one can hardly expect that such a field exists. But even if one assumed that, at the time of its formation, the Galaxy possessed a large-scale field, that field would be tangled rather quickly (in fact, in a few Galactic revolutions) through the action of turbulence.

The local sources of field are the explosions of supernovae, stellar winds, and possibly the nucleus of the Galaxy (Chapter 12). A global large-scale magnetic field is formed as a result of the differential rotation and the turbulent spiral motion of the conducting gas. Before we construct the generation model, however, let us estimate the efficiency of these sources.

Sources of Galactic Field

The main factor determining the field dissipation is the turbulent diffusion, which has the characteristic timescale $\tau_0 = l^2/\nu_T$, where ν_T is given by Eq. (12.2). Therefore to maintain (and amplify) a large-scale field with a strength of 2–3 μG in the Galactic disk, power input of the order of

$$\tau_0^{-1} B^2/8\pi \simeq 10^{-28} \text{ erg/cm}^3 \text{ s}$$

255

is required. More power still is needed to maintain the small-scale fluctuating fields, whose energy density in order of magnitude is approximately as large as $B^2/8\pi$. The energy density of the differential rotation is very great (about $6 \cdot 10^{-10}$ erg/cm^3). This energy is, however, not transformed directly into magnetic energy. In the mechanism of magnetic field generation the differential rotation is only a transmission belt. It is important that the differential rotation be maintained at its own level ($\omega \simeq 10^{-15}$ s^{-1} in the vicinity of the Sun; $\omega \simeq 10^{-14}$ s^{-1} near the Galactic center, $\Delta\omega \sim \omega$). As discussed above (see Chapter 12), the energy sources for turbulent motion in the gas disk are the supernova explosions whose power input in the form of kinetic energy is about 10^{-26} erg/cm^3 s (Salpeter, 1976). The expanding H II regions surrounding young hot O- and B-type stars also make an essential contribution (Kaplan and Pikelner, 1970). The power of these sources suffices to produce and amplify the magnetic fields. Since the supernova explosions and stellar winds eject magnetic fields into interstellar space, it is natural to ask, "Can these local sources create by themselves the full Galactic field?" Moreover, in agreement with the original ideas of V. A. Ambartsumian, the nucleus of the Galaxy is, as numerous observations testify (Oort, 1977), in an active state that could provide an additional, and very strong source of field.

Let us evaluate the contribution made by the supernovae (Syrovatsky, 1970). The volume of the Galactic gas disk is

$$V = 2h\pi R^2 = 5.7 \cdot 10^{11} \text{ pc}^3,$$

($h = 400$ pc, $R = 15$ kpc). In accordance with contemporary ideas (Shklovsky, 1976), we shall suppose that one supernova explodes in the Galaxy on average every 30 years. Then the total number of supernovae to have exploded during the Galaxy's existence is about $3 \cdot 10^8$. Hence, to each supernova corresponds a volume of gas of order $2 \cdot 10^3$ pc^3. A reliable estimate of $3 \cdot 10^{-4}$ G has been made for the magnetic field of the Crab Nebula, a young supernova remnant with a radius of about 1 pc. Assuming, for simplicity, that the expansion to $2 \cdot 10^3$ pc^3 follows the adiabatic law ($\sim V^{-2/3}$) we find that the field so created in the interstellar medium is of strength

$$H = 3 \cdot 10^{-4}(1 \text{ pc}^3/2 \cdot 10^3 \text{ pc}^3)^{2/3} \simeq 2 \text{ }\mu\text{G}.$$

The basic scale l of this field will be close to the maximum dimension of the remnants. The radius of a remnant which has existed for several millions years may be as large as 50 pc. Therefore we may assume that $l \simeq 100$ pc. Actually the magnetic field lines will be stretched by the differential rotation, so that the scale will be larger in the azimuthal direction.

A somewhat smaller, but still essential, contribution to the Galactic magnetic field is made by the O- and B-type (and WR) stars of early spectral type (Bisnovaty–Kogan et al., 1973) which have strong stellar winds.

Hoyle (1969) has made the fantastic suggestion that the Galactic magnetic field was generated in a very compact central nucleus. Its size was thought comparable with its gravitational radius, its rotational energy being about $0.1 M_{nucleus} c^2$, and the field on its surface being about 10^9 G. The adiabatic expansion of this field over the entire volume of the Galactic disk would produce a field strength of several μGauss. The existence of such a nucleus with such a strong magnetic field is, for the present epoch, contradicted by observations. In the central region, whose radius is less than 500 pc, the field does not appear to exceed $2 \cdot 10^{-5}$ G and, if the nuclear outflow were an evolutionary event, then the field would have had time to disappear through turbulent diffusion. One may consider Hoyle's idea in a more realistic light by supposing that activity in the Galactic nucleus might have enriched the magnetic field in the central regions of the Galaxy.

Thus local sources are capable of generating a magnetic field of the required magnitude in the Galactic disk. But the loops of field flowing out from stars are orientated at random. Stretching the loops by differential rotation only aligns them parallel to the Galactic plane. In this situation the large-scale field is expected to be small and to decrease as $N^{1/2}$ as the number, N, of supernovae grows; in contrast, the fluctuating field will be practically independent of N.

Hence these sources cannot explain the observed large-scale field. A purely cellular field structure does not agree with the observations which establish the existence of a large-scale average field (Chapter 12). Nevertheless the local sources do play an important part in determining the small-scale structure of the field. It is clear that the supernova explosions, which are the main source of Galactic turbulence, also produce magnetic fields on the small scale. The basic scale of the Galactic turbulence ($l = 100$ pc) agrees well with the correlation scale of the field determined from observations (see Chapter 12).

The explanation of the large-scale field can be found in the theory of dynamos driven by differential rotation and helical turbulence. The dependence of the angular velocity on the distance from the axis of the Galactic gas disk was sketched in Figure 12.3. It can be seen that the rotation is nowhere of solid-body type. In the vicinity of the Sun $r = 10$ kpc, $\omega \simeq 10^{-15}$ s^{-1}; in the central region the angular velocity is a maximum ($r = 1$ kpc, $\omega \simeq 10^{-14}$ s^{-1}) and its gradient is larger.

Figure 13.1 shows the observational curve for the dependence of the differential rotation on distance, r, from the Galactic center (Ruzmaikin and Shukurov, 1981). An important feature here is the deep minimum in the logarithmic derivative of ω in the vicinity of $r = 2$ kpc, that separates the central regions from the outer edge of the disk. It is interesting that, in contrast, the gas density is maximal in the near region 4–6 kpc. We also see again that in the entire Galactic disk the rotation is differential (not a solid body one).

We have already mentioned in Chapter 12 that, under the action of Coriolis forces and with the density decreasing from the central plane to the faces of the disk, helicity develops, i.e. a preferred

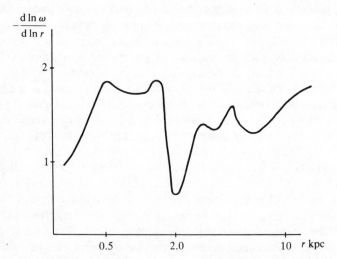

Figure 13.1 Differential rotation of the Galactic gas disk calculated from the observed 21-cm line rotation curve.

direction of screwlike twisting motion takes place (see Figure 12.4) as measured by the pseudo-scalar $\langle \mathbf{v} \cdot \nabla \times \mathbf{v} \rangle$. It is clear that in this case the only available pseudo-scalar is proportional to $\omega \nabla_z \langle \rho v \rangle$, where $\langle \rho v \rangle$ is the mean momentum of the turbulent fluctuations. Dynamo theory requires the quantity $\alpha = -\frac{1}{3}\tau \langle \mathbf{v} \cdot \nabla \times \mathbf{v} \rangle$, which has the dimensions of velocity (τ is the correlation time of turbulent elements). By reasoning based on similarity and dimensionality, we obtain expression (12.3). Of course, to establish the exact form of the function α, a complex hydrodynamic problem has to be solved but, at this stage in the development of Galactic dynamo theory, only qualitative attributes of the mean helicity are important. In particular, the antisymmetrty of this function under reflection ($z \rightarrow -z$) in the Galactic plane, plays an essential rôle. This seems to be a simple consequence of the change in sign experienced by the density gradient on passing through the Galactic plane.

II   A Model of the Galactic Dynamo

The thickness and shape of the Galactic gas disk change with distance, r, from the axis of rotation. In particular, as has already been noted in Section 12.I, the disk expands vigorously and bends towards its rim. But it remains thin practically everywhere, except in a small region near the Galactic center. Assuming its radius to be 15 kpc and the semi-thickness of the ionized layer to be 400 kpc (in agreement with the observed pulsar dispersion measures; see Chapter 12) we obtain

$$h/R \simeq 0.03.$$

The contribution of the differential rotations and helicity to the generation of the large-scale magnetic field is determined by the dimensionless numbers

$$R_\omega = Gh^2/\beta, \qquad R_\alpha = \alpha_0 h/\beta.$$

Here G ($= r\, d\omega/dr$) measures the angular velocity gradient in the interval 3 kpc $\lesssim r \lesssim$ 12 kpc (it is assumed to be $G \simeq -\omega = -10^{-15}\, \text{s}^{-1}$), $\beta \simeq 10^{26}\, \text{cm}^2/\text{s}$ is the coefficient of diffusion of the field (see

Section 12.3) and $\alpha_0 = l^2 \omega / h$ is the amplitude of the function defining the helicity in this interval ($\alpha_0 \simeq 0.8 \cdot 10^5$ cm/s). Hence,

$$R_\omega \simeq -10, \qquad R_\alpha \simeq 1.$$

We may therefore consider the Galactic dynamo to be of the $\alpha\omega$-type with $D \simeq -10$ (Parker, 1971; Vainshtein and Ruzmaikin, 1971). Ruzmaikin, Sokoloff and Turchaninoff (1980) numerically integrated the equations for an $\alpha\omega$-dynamo in a disk [see Eqs. (15)–(16) of Chapter 9], using for $\alpha(z)$ different smooth antisymmetric functions. The results depend weakly on the form of this function. An even azimuthal harmonic of the field is excited with a corresponding poloidal field of quadrupolar type (see Figure 9.1). The critical dynamo-number for $\alpha = \sin \pi z$ is $D_0 = -8$. We estimate the characteristic time of the field growth for $D = -10$ to be

$$\tau = \gamma^{-1} \simeq 5 \cdot 10^8 \text{ years} \sim (1/20) \times \text{the age of the Galaxy.} \tag{1}$$

Thus the Galactic dynamo acts near the excitation threshold, and so we should expect that only the lowest (even) harmonic to be regenerative. The characteristic growth time is rather large, and this raises a question about seed fields (see Section III).

Of all the components of the field produced, the azimuthal one is the largest

$$B_r / B_\phi \simeq (R_\alpha / R_\omega)^{1/2} \simeq 0.3, \qquad B_z / B_r \simeq O(h/r). \tag{2}$$

This fact agrees well with observations (see Chapter 12) according to which the field is nearly parallel to the Galactic plane and orientated approximately along the tightly twisted spiral arms.

Ruzmaikin and Shukurov (1981) used the actual angular velocity, $\omega(r)$, deduced from observation, and also allowed for the decrease in the thickness of the gas disk and the increase in the characteristic turbulent velocity towards the center of the Galaxy. Their calculations confirmed that in most of the disk azimuthal and corresponding quadrupolar poloidal fields are excited, which are even in z, i.e. symmetric with respect to the central plane of the disk and do not oscillate with the time. In the central part of the Galaxy $r < 1$ kpc, however, the dynamo number is so large that the excitation of odd (dipolar) and oscillating modes is also possible. An unexpected result was the impossibility of field generation in the region $r \sim 4$–6 kpc where the differential rotation is very weak (see Figure 13.1). Thus

two isolated generating regions—central and exterior—seem to exist
in the Galaxy. On approximating $\alpha(z)$ by concentrated (δ-type) antisymmetric functions, estimates were obtained for the boundaries of
these regions and the corresponding growth times of field using linear
(kinematic) dynamo theory. These were respectively $r < 1$ kpc, $5 \cdot 10^7$
years and $9 < r < 17$ kpc, $5 \cdot 10^8$ years. The qualitative pattern of the
fields is given in Figure 13.2.

The theory considered is linear, the absolute amplitude of the field
components being undetermined. By taking into account the back-
reaction of the field on the helicity, it is easy to construct a model where
field growth is halted (Vainshtein and Ruzmaikin, 1972). One might
expect the pressure gradient of the steady average field to balance the
Coriolis force, and then

$$B^2/8\pi \sim \rho v \omega l \sim 10^{-13} \text{ erg/cm}^3.$$

It is the Coriolis force that generates the helicity which is so
important to the field generation mechanism. The influence of the

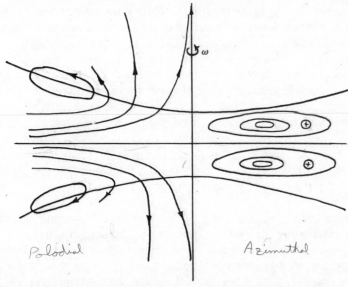

Figure 13.2 Qualitative pattern of the large-scale magnetic field configuration in the
Galactic gas disk. On the left are the lines of force of the poloidal field, on the right
the level lines of the azimuthal field. The directions of the azimuthal field in the central
and outer parts of the disk may be different.

field on the strong differential rotation may be ignored, but the field has a significant effect on the weak chain in the generation mechanism—the helicity. The chaotic component of the field has a somewhat larger energy density, close to the kinetic energy density.

III Seed Fields

The induction equation always has a zero solution. A deeper property is that, like the full equations of magnetohydrodynamics, it is invariant to the interchange $\mathbf{H} \to -\mathbf{H}$, and in particular to $\mathbf{0} \to \mathbf{0}$. For the dynamo to act, one needs an initial seed field to be amplified and changed. Questions about the origin of such a field are not usually asked in applications to planets and stars, since it is assumed that weak fields always exist initially in the medium in which those objects are formed. But the question becomes more relevant for the Galactic dynamo. As has already been mentioned at the beginning of this chapter, no evidence exists of appreciable intergalactic magnetic fields. Let us estimate the strength of the seed field required. We shall assume that the characteristic growth time of the Galactic dynamo is given by Eq. (1). Then, during the period of the Galaxy's existence ($\sim 10^{10}$ years), the observed field ($2\ \mu G$) could be produced by an initial field of $3 \cdot 10^{-13}\ \mu G$. This field is very small but to be relevant it must vary only on a very large scale (a few kiloparsecs). One has to take into account the geometry of the initial field in this way because only the even azimuthal and quadrupolar poloidal components can be amplified.

In considering the origin of seed fields one must take into account the interaction of charged particles. In the cosmological past, when the galaxies were forming and evolving, and the density of radiation was high, weak magnetic fields could develop due to the difference in the Compton interactions of electrons and protons (Harrison, 1970; Mishustin and Ruzmaikin, 1971, see also Chapter 15 below). But the exploding of supernovae and other stars are a much more effective source of seed field. As has been noted above, these explosions mainly produce chaotic fields. Stretching of lines of force by differential rotation orientates them in the plane of the Galactic disk, and as we have seen N randomly distributed loops of magnetic field can, in this way, produce a large-scale field of non-zero strength, probably proportional to $N^{-1/2}$. It is, however, easy to see that this is not the way to produce a field configuration in which B_ϕ has the same sign for all r and z. The observed field does not change sign with z. It is possible, however, that the field direction does change from arm to arm (i.e. with r).

It is interesting that an increase in the number of exploding stars increases the predominance of the random component over the smooth component of field. This eliminates, in fact, the possibility of explaining the entire Galactic field by the supernova explosions without dynamo action. Indeed, if the average field is artificially enhanced to the requisite strength and scale, then too large a fluctuating field is simultaneously produced. According to observations $\delta B/B \leq 3$ (see Chapter 12).

If we accept that stars are the source of the seed field then the question arises, "Can stars be formed in a nonmagnetized medium?" (Hoyle, 1960; Mestel, 1977). In other words, what was the mechanism for expelling the angular momentum of protostars forming in the rotating Galaxy? (See Chapter 14.)

IV   The Rôle of the Spiral Arms

When considering the axisymmetric problem we overlooked a most significant morphological property of the Galaxy—the spiral arms. From the pioneering work by B. Lindblad and C. C. Lin and their colleagues, we know that the spiral arms mark the compressions of density waves propagating around the gas-star disk. [Extensive bibliographies on this subject can be found, for example, in Lin (1967), Kaplan and Pikelner (1974), Friedmann and Polyachenko (1976).] The density wave is associated essentially with the perturbation of the Galactic gravitational field and is, in this sense, different from the usual sound waves where it is associated only with pressure changes. The most impressive feature is the spatial extent of the regular spiral pattern. This indicates unequivocally that spiral density waves are a global property of the Galactic disk (Wielen, 1975) and not merely a pot-pourri of local features.

A large-scale magnetic field of a few microgauss is unlikely to maintain a spiral arm in equilibrium or to influence its global velocity. The magnetic energy density is, however, comparable with that of the kinetic energy in the local perturbations of the wave, and one should therefore take the field into account when considering the gas dynamics. The main peculiarity of these dynamics is the occurrence of a large-scale spiral shock wave (Roberts, 1969). The shock wave arises because the rotational velocity of the spiral pattern is less than that of the disk, though the difference may be small (Suchkov, 1978). Therefore the gas catches up with the shock wave and enters it from the inside. Perturbation of the gravitational potential results in an increase and deflection of the gas velocity from the initial state. A shock wave forms when the perturbation of the gravitational potential and the difference between the angular velocities of the disk and spiral pattern are sufficiently large. From the theoretical point of view the necessity of a spiral shock wave has not yet been definitely established. Pictures of spiral galaxies (similar to our Galaxy) distinctly show dark dust bands at the edges of arms (Linds, 1970), which are interpreted as narrow layers of condensed gas behind shock fronts (Kaplan and Pikelner, 1974). Moreover, the synchrotronic radiation from galactic arms is highly concentrated in narrow bands on their inner sides (Mathewson, et al., 1972). This supports the existence of a spiral shock wave in our Galaxy too.

The tangential component of the magnetic field must (under frozen-in conditions) be enhanced by the shock wave in direct proportion to the density. It would be four times as large in an adiabatic shock wave. In the Galactic shock wave, however, the gas loses considerable energy by radiation, and a ten-fold increase is therefore possible (Roberts, 1969; Kaplan and Pikelner, 1974). It is interesting that, in the presence of a field component normal to the front, amplification remains even immediately after the passage of the shock wave because of the tensions B_n and B_τ (Polovin, 1960; Mishurov et al., 1979). In the first approximation, however, the orbits of the elements of gas in which the field is embedded, are closed. Therefore their repeated passage through the arms does not result in a multiple amplification of the field. On approaching the next arm the previously amplified tangential component returns to its original value, but a refraction of the lines of force takes place (Figure 13.3). It is important that a strong amplification in the tangential component of field occurs in the shock wave (the normal component does not change), i.e. the

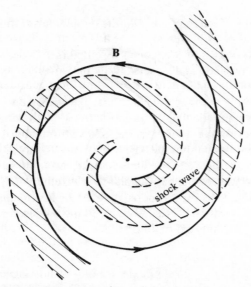

Figure 13.3 Deformation of the azimuthal magnetic field lines by spiral shock waves. On entering the arm, the tangential component is amplified, i.e. the field becomes orientated along the arm.

field becomes orientated along the arm. The arms in the Galaxy are tightly wound, their deflection from the azimuthal direction being only $i \simeq 7°$. Hence over large distances the magnetic field must be located in the arm, a result which is in good agreement with the observed picture (see Chapter 12).

In actual fact, while passing through the shock wave, an element of gas loses, due to radiation, both energy and angular momentum and moves into an orbit that passes closer to the center. According to a numerical calculation by Roberts and Shu (1972) the radial displacement of the element is approximately

$$-\Delta r = ar^3 \sin^2 i, \qquad a \approx 10^{-2} \, \mathrm{kpc}^{-2}.$$

The dependence of Δr on the radius means that gas elements approach each other while moving towards the center in spirals. This should result in an amplification of the azimuthal magnetic field (Ruzmaikin, 1976). The characteristic time of this amplification may be estimated as

$$\tau \simeq \left(\frac{3\Delta r}{r}\right)^{-1} \frac{2\pi}{(\omega - \omega_s)} \simeq \frac{1}{3 \sin^2 i} \left(\frac{10 \, \mathrm{kpc}}{r}\right)^2 \frac{2\pi}{(\omega - \omega_s)}.$$

Due to the small value of $\sin^2 i$ in the denominator, this time proves to be large (about 10^{10} years), so the above effect plays an insignificant rôle in the Galaxy, but it becomes essential in galaxies whose spirals are weakly twisted, and especially in the so-called barred spirals, $i = \frac{1}{2}\pi$.

On the whole we may conclude that the main effect of the spiral arms on the magnetic field is to orientate the field along the arms by the spiral shock wave. Evidently the field, in turn, influences the shock by weakening its amplitude (Roberts and Yuan, 1970). Periodic disturbances of the velocity in the density wave do not exceed the random velocities of the gas (Kaplan and Pikelner, 1974), and in the first approximation they can be ignored in the magnetic dynamo theory. So the axisymmetric $\alpha\omega$-dynamo model described above is a sufficiently good approximation.

V Other Spiral Galaxies

The dynamo theory of magnetic field generation may also be applied to other galaxies similar to our own. Unfortunately data on the thickness of the disks of ionized gas, the dispersion of turbulent velocities, and the characteristic scales of the turbulence are sparse or non-existent. The rotation curves for spiral galaxies, e.g. M31 and M81, are similar to those of our Galaxy. Some spiral galaxies, e.g. the Sombrero hat (NGC 4594), seem, however, to be nearly in solid body rotation in their central regions. Here the α^2-dynamo may prove to be more effective.

Extensive observations made by the Effelsberg and Westerbork radio telescopes provide evidence of magnetic fields in a number of spiral galaxies (M31, M33, M51, M81, NGC253, NGC6946). The structure and strength of the magnetic fields are determined by measuring the linearly polarized radio continuum emission at several wavelengths; see, for example, Beck (1982). An important result is the discovery of a "ring" structure in the field of M31 for $7 \text{ kpc} \leq r \leq 14 \text{ kpc}$ (Beck, 1982) as had been earlier predicted theoretically for both the Galaxy and M31 (Ruzmaikin and Shukurov, 1981).

CHAPTER 14

The Rôle of Magnetic Fields in Star Formation†

I The Modern View of Star Formation

What reason is there to believe that stars are forming at the present time, and that star formation took place in the past? Astrophysics provides the following answer. The Galaxy and a number of other galaxies contain bright massive O-stars and compact gaseous regions composed primarily of ionized hydrogen. Theoretical calculations of stellar evolution show that the characteristic lifetime of the most massive ($M \simeq 50 M_\odot$) O5 stars is of the order of 10^6 years, i.e. very small compared with the ages of galaxies. The brightest compact H II regions are also believed to be less than 10^6 years old. These stars and H II regions are usually associated with known condensations of the interstellar medium (molecular clouds, large gas-dust complexes, e.g. the Orion Nebulae). Taken with some conclusions of gravitational instability theory, this has led to the following concept, now accepted by nearly all investigators: star formation is the result of the contraction (mainly through their own self-gravitation) of inhomogeneities in the interstellar gas in the regions of star formation.

The instability starts to develop when the self-gravitational force becomes greater than the sum of forces opposing the collapse. Self-gravity exceeds thermal pressure gradients in inhomogeneities of scales

$$\lambda > \lambda_J = c_s (G\rho/\pi)^{-1/2} \simeq 3 \cdot 10^{19} (T/n)^{1/2} \text{ cm,}$$

---

† This chapter was written by T. V. Ruzmaikina.

where c_s is the velocity of sound in the medium, ρ the density, n the number density, and T the temperature; λ_J is the well-known Jeans length. The mass contained in a spherical volume of radius $\lambda_J/2$ (the Jeans mass) is

$$M_J = \tfrac{1}{2}\lambda_J^3\rho \sim 20(T^3/n)^{1/2}M_\odot.$$

Typical states for the central parts of dense molecular clouds are $T \simeq 10$ K and $n = 10^4 - 10^6$ cm^{-3}; the Jeans mass is then $M_J \sim 1 - 10 M_\odot$.

To form a star the initial inhomogeneity has to contract drastically. In the case $M = M_\odot$, the average density increases by a factor of about 10^{18} and the temperature by about 10^5 over their respective initial values. So the hydrogen passes from a molecular state to an ionized one. In this process the entropy of the cloud decreases by an order of magnitude due to the radiative cooling of the collapsing protostar. Other obstacles to star formation are the rotation of the molecular clouds and also, in some respects, the magnetic field. These factors are analyzed in some detail in this chapter.

The spatial distribution of massive stars in the Galaxy suggests that their formation starts by the contraction of matter in the shock waves associated with the spiral arms, in supernova shells, and in the collisions of clouds. The formation of (solar-type) stars of small mass apparently occurs in the direct collapse of gravitationally bound fragments (protostars) and does not require any triggering by, for example, shock waves. This view is reinforced by, for instance, the observational fact that T Tauri stars with masses of $0.5-3M_\odot$, which resemble the young Sun and the solar nebula and are often referred to as "pre-main-sequence objects" (with ages of 10^5-10^6 years), are distributed throughout the molecular cloud.

The molecular clouds contain about 40% of the diffuse matter of the Galaxy. Apparently, the clouds are formed through a thermal instability which develops behind the shock fronts that propagate across the interstellar medium. To explain the formation of massive molecular clouds ($\sim 10^3 M_\odot$) one has to consider, in addition, the process by which small clouds adhere together, and also some magnetohydrodynamical instabilities of the large-scale Galactic magnetic field.

The time scale of collapse of a cloud into a star is of the same order of magnitude as the initial free fall time, i.e.

$$t_{ff} = (2\pi G\rho)^{-1/2} \sim 10^{15} n^{-1/2} \quad \text{sec.}$$

In molecular clouds, $n \gtrsim 10^3 \, \text{cm}^{-3}$, so that $t_{ff} \lesssim 10^6$ years. At least $10^9 M_\odot$ of Galactic gas lies within molecular clouds. If all such clouds were in the process of collapsing, one would anticipate a rate of star formation exceeding $10^3 M_\odot \, \text{yr}^{-1}$. In reality it is probable that only about $M_\odot \, \text{yr}^{-1}$ is actually forming new stars, so there must be factors which inhibit the collapse. These can be attributed to the effects of rotation, turbulence and magnetic field.

The broad, two-humped profiles of the spectral lines formed in some molecular clouds provide evidence about their rotation. A wide spectrum of rotational velocities is observed. The directions of the rotation axes of the clouds, as determined by the orientation of their minor axes, are correlated with the Galactic angular velocity. It is evident that the rotation of a cloud will postpone its collapse. The large-scale Galactic magnetic field connecting the cloud to the surrounding interstellar medium will, however, brake the rotation, a process which depends on the ionization of the cloud (Section IV). On the other hand, the magnetic field influences the collapse by itself. A study of the equilibrium configurations of a cloud with a strong frozen-in magnetic field shows that, instead of the $H \propto \rho^{2/3}$ appropriate to spherical collapse, the field, H, depends on the density ρ as $H \propto \rho^{1/2-1/3}$. This may be interpreted by noting that contraction in the central parts of clouds takes place mainly along the field lines (Section IV). This opens the possibility of a collapse in the core of the cloud, followed by its fragmentation into stars. So far the rôle of Galactic turbulence in postponing the collapse of a cloud has been poorly investigated.

It has been observed that there is an intense outflow of matter from the surfaces of T Tauri stars. The large activity and soft X-ray radiation emitted from these stars suggest the presence of a magnetic field. The braking of the rotations of stars of late spectral type during early stages in their evoluion, and the very marked dependence of their rotational velocities on their masses (Table 14.1) can be explained as the magnetic enhancement of the angular momentum loss in their winds (see Section V).

Stellar statistics indicate that more than half of all stars are members of binary systems, thus emphasizing the importance of rotation in the process of protostellar collapse. The recent observational discovery of a shortage of close pairs ($R < 10R_\odot$) amongst the non-evolved binaries, and the relatively slow rotation of pre-main sequence stars

Table 14.1 Mean Rotational Velocities of Main Sequence Stars (McNally, 1965).

Sp	M $(M_\odot)$	R $(R_\odot)$	v_e $(km \cdot s^{-1})$	Ω $(10^{-5} s^{-1})$	P (days)
O5	39.5	17.2	190	1.5	4.85
B0	17.0	7.6	200	3.8	1.91
B5	7.0	4.0	210	7.6	0.96
A0	3.6	2.6	190	10.0	0.73
A5	2.2	1.7	160	13.0	0.56
F0	1.75	1.3	95	10.0	0.73
F5	1.4	1.2	25	3.0	2.42
G0	1.05	1.04	12	1.6	4.55

with $M < 1.5 M_\odot$, show that angular momentum is efficiently removed from the contracting protostars. The possible agents of this angular momentum redistribution are turbulent friction and magnetic field. Circumstellar (proto-planetary) disk formation seems to be a natural consequence of angular momentum redistribution. The hard component of the remnant magnetization of meteorites is a possible indication of a magnetic field of about 1 Gauss in the circum-solar proto-planetary disk (see Section VI).

It is clear from this discussion that magnetic fields can play an important rôle in the formation of stars and circumstellar proto-planetary disks. Many questions have still to be answered, however. The authors hope that this chapter, about magnetic fields in cosmogony, may help to provide a further insight into this subject. Certainly the reader will understand that it is difficult to avoid some hypotheses and speculations in discussing such cosmogenic matters. We shall try to limit their number by maintaining, as far as possible, a strictly physical approach based on fundamental theoretical and observational facts.

II The Observational Background

Evidence for the existence of magnetic fields in regions of star formation has been obtained mainly by two methods: (1) the measurement

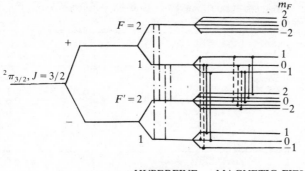

Figure 14.1   Energy-level diagram for the ground-state OH and its expected longitudinal Zeeman pattern of right (dashed)and left (solid) circularly polarized radiation.
The circularly polarized transitions are shown for the $F = 1 \leftrightarrow 1$ (1665 MHz) and
$F = 1 \leftrightarrow 2$ (1612 MHz) radiation. The separations between the two opposite circularly
polarized components of the 1665 doublet is 0.59 km s^{-1}, and of the 1612 components
0.12, 0.37 and $0.61 \text{ km s}^{-1} \text{ mG}^{-1}$ (after Chaisson and Vrba, 1978).

of the Zeeman splitting of the hyperfine lines of the Λ-doublet in the
rotational ground state $(^2\pi_{3/2}, F = \frac{3}{2})$ of OH molecules [in the radio
band at a wave-length of about 18 cm; see Figure 14.1], and (2) the
measurement of either the optical polarization of starlight passing
through the cloud or the infrared polarization of sources within the
cloud. The techniques of magnetic field measurement were discussed
in Chapter 2. Here we present some results that are of interest for
the discussion which follows. A more complete review of magnetic
field detection in clouds is given by Chaisson and Vrba (1978).

Measurements of the energy-level splitting of the hyperfine $^2S_{1/2}$
ground state of the hydrogen atom made possible the detection of
magnetic fields of 3–70 μG in interstellar H I clouds with densities of
10–100 cm^{-3} (Verschuur, 1974). This technique fails, however, in the
case of denser clouds since there the hydrogen is mainly molecular.
By using instead the Zeeman effect for the OH molecules a field of
order 10^{-4}–10^{-3} G can be detected. Definite results are now available
for the strong maser OH radiation associated with regions of active
star formation. These are sources situated within more extended H II
regions. The magnetic field strengths detected are about 10^{-2}–10^{-3} G
(Table 14.2).

Table 14.2   Magnetic Field Strength in Maser Sources (Chaisson and Vrba, 1978; Crutcher *et al.*, 1975; Lo *et al.*, 1975).

Source	ΔV (km · s^{-1})	$\Delta\nu$ (kHz)	H (10^{-3} G)
W 3 (OH)	0.7	4.0	6.0
W 51	0.9	5.2	8.0
NGC 6334 (N)	0.8	4.6	7.0
NGC 7538 (N)	0.4	2.3	3.5
Sgr B2	1.0	5.7	9.0
V 1057 Cyg	0.3	1.7	2.5
Ori A OH			3.5 ± 4.5

The estimated density in the OH masers under pumping conditions is 10^{6}–10^{9} cm^{-3}. One may be tempted to use these values in conjunction with a compression law such as $H \sim n^{2/3}$ to estimate the magnetic field strengths. It is necessary to bear in mind, however, that the radiating maser domains are excited by the passing of shock waves which themselves affect the magnetic field. On the one hand, the compression of matter results in the amplification of the component of field parallel to the shock front while, on the other, the turbulence excited behind the front enhances the decay of magnetic field (see also Section 13.IV).

A number of OH sources do not exhibit any distinct Zeeman splitting but show instead strong circularly-polarized features in their spectra. Theory suggests that such features are characteristic of sources having a milligauss magnetic field and a slight velocity gradient along the line of sight (Varshalovich and Burdjuzha, 1975). For the less dense molecular clouds (10^{3}–10^{5} cm^{-3}) this provides only upper limits for the magnetic field of 0.05–0.4 mG (Chaisson and Vrba, 1978).

The polarization of starlight is known to be due to non-spherical interstellar grains aligned by the ambient magnetic field (Chapter 2). Measurements give information about the structure of the magnetic field but not its strength, because the magnetic properties of the grains are too poorly known. The magnetic field structures in a number of massive dust clouds have now been traced. Measurements were made of the polarization of starlight passing through the peripheral regions of clouds and of the 2.2 μm radiation from the infrared sources inside the clouds (Vrba *et al.*, 1976). These observations show that the magnetic fields of the massive molecular clouds Rho Ophinchi,

Lynds 1630 and NGC 1333 have large-scale structures similar in form to the clouds themselves.

Magnetic fields in the neighborhood of some regions of star formation (the young stellar cluster Per OB3, the Orion molecular cloud complex, and the dark cloud R Coronal Austrina) are specially disturbed. It appears that these objects are suspended in pockets of interstellar magnetic field (Appenzeller, 1974).

The magnetic field in H II regions is detected by Zeeman splitting of radio-recombination lines of hydrogen. Troland and Heiles (1977) determined, by measuring the splitting of the recombination line H90α, that the magnetic field strength in the Orion Nebulae is 0.4 ± 0.1 mG. The corresponding density of the radiating regions is 10^3–10^4 cm^{-3}.

It may be easily verified from the available observational data that the magnetic field is intensified during the formation and contraction of the molecular clouds. The dependence of the strength, H, on the density, ρ, appears however to be different from the $H \sim \rho^{2/3}$ following from isotropic compression of a frozen-in field; it appears that H increases more gradually with increasing ρ. A possible reason for this is discussed in the next section.

III The Evolution of Magnetic Clouds

Consider a homogeneous diffuse cloud in the vicinity of the Sun of the following (typical) character: $T = 50$ K, $n_i = 10$ cm^{-3}, $R_i = 10$ pc, $M = 1500 M_\odot$, $H_i = 10^{-6}$ G, and co-rotating with the Galaxy, (the local angular velocity is $\omega_i = 10^{-15}$ s^{-1}). The gravitational, thermal, magnetic and rotational energies of the cloud are respectively,

$$E_G = 3GM^2/5R_i \simeq 1.1 \cdot 10^{46} \text{ erg,}$$

$$E_T = (kT/\mu m_H)M = 1.3 \cdot 10^{46} \text{ erg,}$$

$$E_H = H_i^2 V_i/8\pi = H_i^2 R_i^3/6 = 0.6 \cdot 10^{46} \text{ erg,}$$

$$E_R = M\omega_i^2 R_i^2/5 = 0.6 \cdot 10^{45} \text{ erg.}$$

The average angular momentum density is $j = 2\omega_i R^2/5 = 4.3 \cdot 10^{23}\,\text{cm}^2\,\text{s}$. One can see that $E_H \sim E_T \sim E_G > E_R$, so that the magnetic field is of importance for the structure and evolution of such interstellar clouds.

The electrical conductivity of the diffuse interstellar medium and even of dense molecular clouds is rather high, so the magnetic field can be considered as frozen-in during the early stages of the collapse (up to densities $\sim 10^4\,\text{cm}^{-3}$). The field is disturbed by motions in the cloud and in turn modifies those motions. In papers dealing with the influence of the magnetic field on the collapse, fragmentation and rotation of the cloud, the initial field is assumed as a rule to be homogeneous. This assumption does not reflect the real situation in the interstellar medium, where the clouds partake in the prevailing interstellar turbulence (Kaplan, 1966; Larson, 1978). This turbulence is apparently magnetohydrodynamical with a rather flat spectrum $B \sim k^{-1/4}$ over wave numbers k on the scale 0.1–100 pc (Ruzmaikin and Shukurov, 1981; see also Chapter 18), which includes the typical sizes of the molecular clouds. The characteristic correlation time depends on the scale. While it is great ($\sim 10^7$ years) for $k_0^{-1} \simeq 100$ pc, it is considerably less than the lifetime of a cloud on its scale. The magnetic fields can affect the pressure and heating of the clouds. Moreover, one should not neglect possible magnetic field transfer and generation in clouds.

Detailed reviews of the theory of magnetic cloud evolution in the interstellar medium are given by Mestel (1977) and Mouschovias (1978). The magnetic field of a cloud is thought to originate from the large-scale Galactic magnetic field, intensified by differential rotation and compression. Ambipolar diffusion is the principal mechanism by which the constraint of frozen-in magnetic fields is broken.

Studies of the influence of a magnetic field on gravitational (Chandrasekhar and Fermi, 1953a) and thermal (Field, 1965) instabilities show that the field suppresses the growth of small-scale perturbations. The critical wavelength of the perturbations that can grow through gravitational forces in a direction perpendicular to the large-scale magnetic field is $(1 + v_A^2/v_S^2)^{1/2}$ times the Jeans wavelength (v_A is the Alfvén velocity, and v_S the velocity of sound). The magnetic energy of the interstellar medium is comparable with its thermal energy $v_A \sim v_S$, so the magnetic field increases the critical mass for interstellar cloud formation to $M_{cr} = M_J(1 + v_A^2/v_S^2)$, where M_J is the Jeans mass.

In the limiting case $v_A \gg v_S$, the critical mass can be expressed through the magnetic flux, Φ, across the equatorial plane of the cloud and the gravitational constant G,

$$M_{cr} = \beta^{1/2} \Phi / G^{1/2}, \tag{1}$$

(Mestel, 1965). The numerical factor β changes from $5/9\pi^2$ for a spherically-symmetric cloud to $1/7\pi^2$ for a cloud that is highly flattened in the direction of the magnetic field (Strittmatter, 1966). When $M < M_{cr}$ the cloud reaches an equilibrium state in which the gravitational force is balanced along the field by a thermal pressure gradient and across the field by the magnetic stresses.

As we noted in Section I, many molecular clouds do not collapse even though the thermal pressure gradient is small compared with the self-gravitational force. (This is supported by the low efficiency of star formation in the Galaxy.) The magnetic field is one of the means by which the gravitational compression of clouds is prevented. Some important features were brought to light by a numerical study of the nonhomologous (i.e. those not belonging to a self-similar sequence of equilibrium configurations) contraction of self-gravitating isothermal clouds, of mass $M < M_{cr}$, embedded in a hot diffuse interstellar medium (Mouschovias, 1978). In general, the clouds became highly centrally condensed and flattened along the magnetic field; Figure 14.2. The calculations predicted that the magnetic field strength, H, and the gas density, n, are related by

$$H \sim n^q,$$

where the exponent q depends on position in the cloud as well as on the relative sizes of the initial magnetic and gas pressures. It was found that at the center of the cloud $\frac{1}{3} \lesssim q \lesssim \frac{1}{2}$. Such a weak dependence of the field strength on n implies that matter contracting at the center of the cloud moves mainly along the magnetic lines of force. As the strength of the external field diminishes, q increases and approaches the isotropic value $\frac{2}{3}$ in the limit $H \to 0$.

The matter flowing along the magnetic lines of force in the central parts of the cloud causes a decrease in the magnetic flux through a cross-section containing the center. The critical mass (1) decreases correspondingly, $M_{cr} \simeq 2 \cdot 10^4 (B_i/n_i^q)^3 n^{-2+3q} M_\odot$. In the numerical models mentioned above, the critical mass was shown to be as small

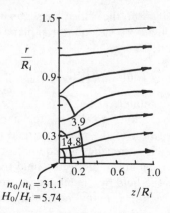

Figure 14.2 Equilibrium state of a magnetic cloud of mass $M = 1554M_\odot$ and temperature $T = 50$ K embedded in a hot and tenuous intercloud medium of pressure $P_{ext} = 3 \cdot 10^{-13}$ erg cm^{-3}. Both axes are labeled in units of the radius of the cloud, R, in its initial uniform, spherical state. The field lines are denoted by thin solid lines bearing arrows. The iso-density contours are the thick solid curves labeled in units of the total uniform density of the initial state, n_i. The initial state is characterized by $n_i = 9.2$ cm^{-3}, $R_i = 10.9$ pc, $H_i = 0.9$ μG. For the state presented in the figure (at the verge of gravitational collapse) the cloud parameters are $n_{center} = 285$ cm^{-3}, $n_{surface} = 36$ cm^{-3}, $R_{equator} = 7.7$ pc, $R_{polar} = 2.8$ pc, $H_{center} = 5.2$ μG (after Mouschovias, 1978).

as 15% of that predicted by the virial theorem under identical conditions. Thus, nonhomologous quasi-equilibrium contraction can result in a free collapse of the central regions, which contain only a small part of the cloud mass. It can be argued (Mouschovias, 1978) that the tension in the field lines can support the extended envelope of the cloud. This is contrary to the notion (valid in the case of a spherically symmetric contraction) that, if the gravitational forces induce a contraction by exceeding the magnetic force initially, they will continue to exceed the magnetic forces at all later times.

These features can explain the observed low efficiency of star formation. Apparently, star formation takes place only in the central collapsing condensation of a cloud. The extended envelope remains diffuse and can easily be destroyed by the strong winds flowing from stars of early types or by supernova explosions.

The change of the critical mass in the central parts of a contracting magnetic cloud promotes its fragmentation. In an isotropic contraction ($q = \frac{2}{3}$) the magnetic energy grows as $n^{4/3}$, i.e. it follows the same law as the gravitational energy, so that the ratio of these energies remains constant. Hence, if the mass of some part of the cloud is initially less than the critical Jeans mass, an instability that results in fragmentation will not develop. A reduction in M_{cr} opens up the possibility of the fragmentation and collapse of clouds of stellar mass (protostars).

So far it has been assumed that the magnetic field in a cloud is frozen to the matter. In reality, as a cloud contracts the degree of ionization falls and the frozen-in condition can be broken through ambipolar diffusion. This diffusion occurs because the magnetic field lines are attached to ions but not to neutral atoms or molecules. Due to collisions, the drift of the magnetized ions through the neutrals results in a reduction of the magnetic energy. The characteristic diffusion time is

$$\tau_d = 8\pi R^2 X n^2 \mu_{in} \langle \sigma v \rangle_{in} H^{-2}, \qquad (2)$$

(Spitzer, 1968), where R is the radius of the cloud, X the degree of ionization, μ_{in} the reduced ion-neutral mass, and the quantities σ and v are, respectively, the collision cross-section between neutrals and ions and the random speed of neutrals relative to ions.

The degree of ionization in molecular clouds of moderate density is determined by a balance between the rate of ionization due to cosmic radiation ($\xi_{CR} \sim 10^{-17}$ s^{-1}; Nakano and Tademaru, 1972) and the rate of recombination on the dust grains (Consolmagno and Jokipii, 1978). Ionization in the very dense parts of clouds and protostars, which have surface densities greater than 10^2 g $\cdot$ cm^{-2} so that they are opaque to cosmic radiation, can be brought about only by their inner radioactive sources. The minimum ionization rate in the solar neighborhood, estimated according to the ^{40}K abundance, is $2 \cdot 10^{-21}$ s^{-1} (Cameron, 1962). Pollution of matter by radioactive isotopes of short half-life originating in supernova explosions can significantly increase the ionization rate in the regions under consideration. When ions recombine on dust grains of typical radius $\sim 10^{-5}$ cm and of number density $n_g \sim 10^{-12} \, n$, the degree of ionization is given by

$$X \sim 10^{17} n^{-1} T^{-1/2} \xi,$$

where T is the temperature in the cloud, and ξ (s^{-1}) is the total ionization rate.

The possibility that the state of frozen-in magnetic flux is unrealized must be faced whenever the diffusion time is less than the time scale of the process under consideration (see Section 3.II). In application to contracting interstellar clouds, magnetic flux decreases because ambipolar diffusion becomes important when τ_d is comparable with the time scale of the contraction. The equality $\tau_d = \tau_c$ defines a characteristic magnetic field strength

$$H_{cd} \sim 10^{-16} X^{1/2} n R \tau_c^{-1/2} \text{(G)}. \qquad (3)$$

If, rather crudely, we take $HR^2 = \text{const.}$, then $H \sim n^{1/2}$. More accurate arguments give

$$H_{ed} \sim 10^{-11} \xi^{2/5} R_0^{2/5} H_0^{1/5} n^{2/5} \text{(G)}, \qquad (3')$$

where H_0 and R_0 are the strength and the length scale of the magnetic field at the time when ambipolar diffusion destroys its frozen-in state. Numerical calculations by Black and Scott (1982) give $H \sim n^{0.44}$ for $X \sim n^{-1}$.

The time scale for the collision of electrons with ions and neutrals decreases with increasing density so that Ohmic dissipation becomes important. Ohmic diffusion exceeds ambipolar diffusion when (Figure 14.3)

$$n > n_{cr} \equiv (\pi c)^{-1} \eta^{-1/2} (m_e m_p)^{-1/4} H \sim 0.5 \cdot 10^{14} H \text{(cm}^{-3}),$$

where η is the polarizability of the H_2 molecule ($= 0.8 \cdot 10^{-24}$ cm^3), and m_e and m_p are the masses of the electron and proton respectively

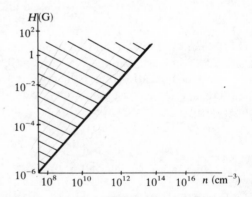

Figure 14.3 Solid line H versus n corresponds to the equality of the Ohmic and ambipolar diffusions. The region in which the magnetic field behavior is determined by ambipolar diffusion is shaded.

For $H = H_{cd}$ determined by (3'), $n_{cr} \sim 10^{26} H_0^{1/3} \xi^{2/3} (\text{cm}^{-3})$, and for $R_0 = 10^{17}$ cm, $H_0 = 10^{-3}$ G and $\xi = 10^{-19}$ s^{-1} this gives $n_{cr} \sim 10^{12}$ cm^{-3} and $H_{cd}(n_{cr}) \sim 10^{-2}$ G. For $n > n_{cr}$, the magnetic field at the center does not grow as the density increases, and $H = H_{ed}(n_{cr})$. Ohmic diffusion is important until the temperature rises to 1600 K (and the density correspondingly to about 10^{-8} g cm^{-3}) at which dust evaporates, molecular hydrogen dissociates, and the alkaline metals thermally ionize; a rapid increase in the degree of ionization follows. The magnetic field can then again be considered to be frozen-in, and will intensify during the dynamical contraction initiated by the dissociation of the H_2 molecules. At this time a starlike core forms which is in hydrostatic equilibrium and has an initial density of 10^{-2} g cm^{-3}. The magnetic field in the core will be of order $10^4 H_{cd}(n_{cr})$, i.e. 10^2G for the numerical values taken (Ruzmaikina, 1981a). This field can redistribute angular momentum in the core. We shall focus attention in Section VI on this process, which is important in cosmogony.

In conclusion let us discuss the magnetohydrodynamical instability of a galactic gas disk having a large-scale magnetic field (Parker, 1966), and which may be responsible for the formation of massive $(M \sim 10^3 M_\odot)$ interstellar clouds.

The interstellar gas, which is supported by the magnetic field, and the thermal and cosmic-ray pressure gradients against the component of the Galactic gravitational field perpendicular to the disk, may be unstable to magnetic field line deformations. Let the field be parallel to the plane of the disk. If a disturbance moves some domain out of the Galactic plane and so bends the magnetic field lines, then the gas

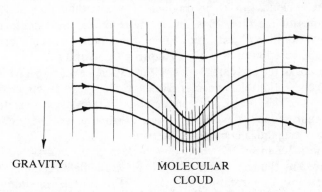

GRAVITY   MOLECULAR
                              CLOUD

Figure 14.4 Configuration of the magnetic field lines resulting from Parker's instability.

tends to accumulate in those "valleys" of field lines under the action of the gravitational field (Figure 14.4). This reduces the gravitational burden on the distorted domain, so permitting the magnetic field (and the cosmic-ray gas) to expand further away from the Galactic plane, thereby exhancing the flow of gas along the magnetic field lines into the valleys. This instability is similar to the Rayleigh–Taylor instability. The accumulation of gas in the magnetic field valleys increases its opacity and cooling rate. In the Galactic disk, however, there are random gas motions additional to the thermal and cosmic-ray pressures. The magnetic Rayleigh–Taylor instability apparently becomes effective as soon as the kinetic energy of these motions is less than the magnetic energy.

IV Angular Momentum Transport

Many molecular clouds show the following features (Hopper and Disney, 1974; Heiles, 1976): (1) stretched forms; (2) a tendency for most stretched clouds to be parallel to the Galactic plane; (3) a variation in the line-of-sight velocities, as determined by the spectral lines, across the cross-section of a cloud. These features all provide evidence that the clouds rotate.

Interstellar turbulence (Cameron, 1962) and Galactic differential rotation (Hopper and Disney, 1974) are considered to be the basic mechanisms in cloud rotation. The angular momentum per unit mass in these clouds (or their fragments) is less than that of spherical volumes of the same mass in the more diffuse interstellar medium but greater than the intrinsic angular momentum of a star. This implies that the rotation of both the collapsing cloud and protostar has been reduced. In the early stages of star formation, i.e. during the formation and contraction of molecular clouds, braking can be effected by the stresses of the frozen-in magnetic field. In the final stages, during its pre-main sequence evolution, a star of small mass is very effectively braked magnetically by its stellar wind.

The mechanism by which the rotation of an interstellar cloud is reduced by the large-scale Galactic magnetic field is simple. A cloud in a steady state, rotating relative to the interstellar medium with an angular velocity Ω, is coupled to it by the frozen-in magnetic field via changes induced in the azimuthal component, H_ϕ, of the field, which obeys

$$\partial H_\phi / \partial t = s(\mathbf{H} \cdot \nabla)\Omega, \qquad (4)$$

where s is distance from the rotation axis. In turn this ϕ-component helps to redistribute the angular momentum, $\mathbf{J}$, of the cloud, according to

$$\rho \, \partial(\Omega s^2)/\partial t = \mathbf{H} \cdot \nabla(s H_\phi / 4\pi). \qquad (5)$$

Combining (4) and (5) we obtain a wave type equation

$$4\pi\rho \, \partial^2 \Omega / \partial t^2 = s^{-2} \mathbf{H} \cdot \nabla(s^2 \mathbf{H} \cdot \nabla\Omega). \qquad (6)$$

In the simplest and most transparent case ($\mathbf{H} \| \mathbf{J}$) of a uniform poloidal field directed along the z-axis, Eq. (6) reduces to

$$\partial^2 \Omega / \partial t^2 = v_A^2 \, \partial^2 \Omega / \partial z^2, \qquad (7)$$

Ebert *et al.* 1960; Mouschovias, 1978), where $v_A = H(4\pi\rho)^{-1/2}$ is the Alfvén speed. Thus, the rotation of the cloud relative to the background creates moving magnetic disturbances resembling Alfvén waves. These carry away the angular momentum of the cloud, thus smoothing the difference between the angular velocity of the cloud and that of its environment. The characteristic time for the loss of angular momentum by a spherical cloud of radius R is

$$\tau_{\mathbf{H}\|\mathbf{J}} = 8(\rho_{cl}/\rho_{ext})^{1/2} R/15 v_{Aext}, \qquad (8)$$

where ρ_{cl} and ρ_{ext} are respectively the densities within and outside the cloud, and $v_{Aext} = H(4\pi\rho_{ext})^{-1/2}$. When the cloud is flattened along the rotation axis (the magnetic field direction) R is the polar radius of the cloud. Formula (8) can be interpreted as follows: $\tau_{\mathbf{H}\|\mathbf{J}}$ is the time taken by an Alfvén wave propagating along the rotation axis to travel the distance h across a disturbed region whose moment of inertia is equal to that of the cloud,

$$h/R \simeq \rho_{cl}/\rho_{ext}. \qquad (9)$$

When the initial magnetic field is perpendicular to the rotation axis the propagating waves entrain matter further and further from the rotation axis, i.e. matter with larger and larger moments of inertia. It is clear that the braking of the rotation is more effective in this case. By assuming that the cloud has a "cylindrical" form it is easy

to estimate the radius, r, of the region whose moment of inertia is
equal to that of the cloud by

$$r/R \simeq [1 + (\rho_{cl}/\rho_{ext})]^{1/4}. \qquad (10$$

It is evident that $r - R \ll h$ [see Eq. (9)]. Detailed calculations (Mouschovias and Paleologou, 1980) confirmed the greater efficacy of
angular momentum loss in the case $\mathbf{H} \perp \mathbf{J}$ (Figure 14.5). The ratio
$\tau_{\mathbf{H}\|\mathbf{J}}/\tau_{\mathbf{H}\perp\mathbf{J}}$ is about 5 for $\rho_{cl} = \rho_{ext}$, and increases with ρ_{cl}/ρ_{ext}. The
characteristic time $\tau_{\mathbf{H}\perp\mathbf{J}}$ diminishes as the cloud contracts. For
instance,

$$\tau_{\mathbf{H}\perp\mathbf{J}}/\tau_0 \lesssim \begin{cases} 0.1 \\ 0.06 \\ 0.02 \end{cases} \text{ for } \rho_{cl}/\rho_{ext} = \begin{cases} 1, \\ 10^2, \\ 10^3, \end{cases}$$

where τ_0 is the characteristic time of the Alfvén wave propagation
across a spherical cloud of the same mass when $\rho_{cl}/\rho_{ext} = 1$.

The real configuration of the magnetic field threading contracting
clouds to the interstellar medium is, of course, more complicated.
The disturbance of the field was taken into consideration by Gillis et
al. (1979) in a simple model of spherically-symmetric homologous
contraction. The rate of angular momentum loss was found to have
a value intermediate to those for $\mathbf{H}\|\mathbf{J}$ and $\mathbf{H} \perp \mathbf{J}$. It is worth mention-
ing an interesting effect. The decrease in angular velocity (for $\rho_{cl} \neq \rho_{ext}$)
proceeds, not monotonically, but through oscillations of diminishing
amplitude. This may explain why some clouds are observed to rotate
in the retrograde sense with respect to the Galactic rotation. The total
time taken for the angular momentum to be transferred out of the
cloud is of order τ_0 in all cases considered. Detailed three-dimensional

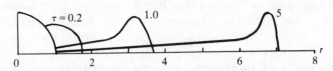

Figure 14.5 Behavior of a field line which initially coincided with the line $\phi = 0$
polar coordinates (r, ϕ) for perpendicular $\mathbf{H}$ and $\mathbf{\Omega}$, and $n_{\text{cloud}}/n_{\text{ext}} = 10$. Radial distance
is normalized to the cloud radius. The time τ is normalized to that taken by an Alfvén
wave to travel across the cloud, using for the Alfvén velocity that value appropriate
to conditions just outside the cloud surface (after Mouschovias and Paleologou, 1980).

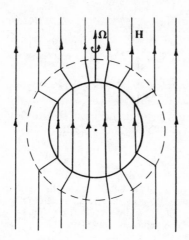

Figure 14.6   The Gillis *et al.* (1979) model of the magnetic field structure of a cloud.

calculations of rotational braking for interstellar clouds in quasi-equilibrium with a frozen-in magnetic field have been performed by Dorfi (1982).

The problem of angular momentum transport from a collapsing cloud of mass greater than M_{cr}, an important aspect of star formation, was studied by Mestel and Paris (1979) using the model of a spherically-symmetric cloud. The magnetic field, both in the cloud and in the remote unperturbed interstellar medium, was supposed uniform and parallel to the rotation axis. In the intermediate region the field was approximated by a radial field $H_r \sim R^{-2}$ (Figure 14.6). The initial angular velocity of the cloud was assumed to be equal to that of the background so that $\partial\Omega/\partial t = 0$, and there was therefore no magnetic braking. The subsequent evolution was determined by the ratio of the centrifugal and gravitational forces. A typical dependence of this ratio on the contraction factor η ($\equiv R/R_0$) is illustrated in Figure 14.7. It is close to being linear until an equilibrium state is set up by the amplification in the magnetic field, H_ϕ, created by the differential rotation which is itself the result of the contraction of the cloud. The characteristic time of contraction is determined subsequently by the efficiency of the magnetic braking.

If the initial state of a cloud is one of equilibrium and the cloud then begins to contract due to a decrease in the magnetic flux that

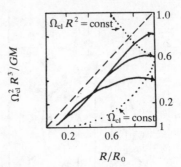

Figure 14.7 Typical curves (solid lines) showing the variation of the centrifugal to gravitational forces ratio $\Omega^2 R^3/GM$ as a function of the degree of cloud contraction R/R_0 for the case $M = M_{cr}(1)$ and $\rho = \rho_0(R_0/r)^3$ in the region $R_{ce} < R < R_0$ where $H \sim R^{-2}$. The dashed line represents magneto-centrifugo-gravitational equilibrium defined by the equation

$$1 - \Omega^2 R^3/GM - (M_c/M)^2(1 - R/R_0) = 0.$$

The dotted curves representing the corotation (Ω = constant) and the angular momentum conservation (ΩR^2 = constant) are shown for the initial ratio $\Omega^2 R^3/GM = 0.6$.

threads it, it will continue to co-rotate with its background until ambipolar diffusion begins to act, after which a rapid loss of its angular momentum commences. Angular momentum transport essentially ceases when the characteristic time (2) of ambipolar diffusion becomes comparable with the time-scale of magnetic braking. For the core of a molecular cloud with $M = 200 M_\odot$ and $T = 50$ K the transport of angular momentum by a frozen-in magnetic field, parallel to the rotation axis and proportional to $n^{1/2}$, ceases when the density is $n_{cr} \simeq 2 \cdot 10^3$ cm^{-3}. This critical density depends on the characteristic time of angular momentum transport as (Mouschovias, 1978),

$$n_{cr} \sim \tau_J^{-3/2}.$$

The larger n_{cr}, the more effective is the angular momentum transport in the case $\mathbf{J} \perp \mathbf{H}$. The angular momentum per unit mass of a cloud of mass M rotating at the Galactic angular velocity $\Omega = 10^{-15}$ s^{-1} is

$$j \sim 10^{37} \Omega n^{-2/3}(M/M_\odot)^{2/3} \text{ cm}^2 \text{ s}^{-1}.$$

The cloud can co-rotate up to $n \sim 10^4$–10^6 cm^{-3}, and the fragments of stellar masses within the cloud then have angular momenta comparable with those of binary stars or with the initial angular momentum of the solar system. This applies only to clouds whose angular momentum and magnetic field enjoy this favorable relative orientation (Mouschovias, 1978). There is little doubt that in such cases magnetic braking can, during the initial stage of contraction, carry the surplus angular momentum away from the cloud.

The magnetic field is not important when ambipolar diffusion and Ohmic decay dominate. As was noted in the previous section, however, the magnetic field in the central regions of protostars becomes frozen-in again once the thermal ionization at $T = 1600$ K has taken place; it intensifies further when the dissociation of molecular hydrogen occurs. The field may then provide an effective means of redistributing angular momentum in the forming stellar-like core, and prevent that core from dividing to form a binary system. It is natural to relate this picture to the observed deficit of close pairs (i.e. those separated by less than $10R_\odot$) among the non-evolving binary stars (Krajcheva et al., 1978).

V Braking of Stellar Rotation

There has been no direct detection of a magnetic field in any pre-main sequence stars. There are indications, however, that magnetic fields play an essential rôle in stellar activity. The HEAO-2 observations of the Orion complex showed that a number of pre-main sequence stars are sources of soft X-rays (Chanon et al., 1979). The X-ray coronal luminosity ranges from 10^{-5} to 10^{-6} of the optical luminosity of these stars. The existence of stellar coronas can be explained by magnetic fields (see Chapter 11). Coronal heating apparently occurs both through the dissipation of magnetohydrodynamical waves and through flare activity associated with the reconnection of magnetic field lines. Moreover, the short-period variations in the brightness of the T Tauri stars, often referred to as pre-main sequence objects, can be interpreted as the result of the periodic coverage of large areas of their surfaces by stellar spots similar to sunspots (Gershberg, 1975).

The existence of magnetic fields in low mass stars was early invoked to explain their slow rotation. It is known there is a marked difference between the angular momenta per unit mass of the early and late type main sequence stars. The rotational velocities for stars of spectral types later than F diminish sharply with decreasing stellar mass, i.e. with the progression to later spectral type (see Table 14.1). It must be emphasized that the mass dependence here is stronger than would be the case had the stars of different masses formed from a medium of uniform angular momentum per unit mass. In that case the mass dependence of the angular momentum would have been $J \sim M^{2/3}$. It follows that a mass-sensitive mechanism for braking stellar rotation acted during the formation and early evolution of stars with $M <$ $1.5M_\odot$.

Kraft (1967) was the first to provide observational evidence for the gradual slowing-down of the rotation of solar type stars. Further observations showed that the mean angular velocity decreased as $t^{-1/2}$ over a time interval of between 10^8 to 10^9 years (Skumanich, 1972). An explanation of this angular deceleration of late type stars was given by Schatzman (1962). His idea relied on the presence of a stellar wind. If the magnetic field is strong enough, the outward flowing ionized gas in a wind moves along the field lines, in co-rotation with the star, until it reaches the distance, R_A, where its radial velocity becomes equal to the Alfvén speed. Subsequently the gas no longer co-rotates with the star but approximately conserves its angular momentum. When the magnetic energy density at the stellar surface is comparable with the thermal energy density, the angular momentum carried away by a unit mass wind proves to be much greater, by a factor of up to $(R_A/R)^{1/2}$, than it had when at the stellar surface. Thus, by the ejection of a small part of its mass, the star can markedly decrease its angular momentum. The efficiency of this mechanism is determined by the intensity of the convection in the upper envelope of the star. It is the convection zone that is responsible for both the generation of the magnetic field and the origin of the stellar wind. The duration of well known Hayashi phase of convective mixing, and the depth of the convection zone of the star when on the main sequence (Iben, 1965) both increase as the mass of the star decreases.

Detailed calculations of rotational braking by a stellar wind (Mestel, 1968) show that the outflow is greatly influenced by a strong magnetic field and by centrifugal forces. The stellar wind flows only within cones of open field lines surrounding the symmetry axis of the poloidal magnetic field in both northern and southern magnetic hemispheres. The boundaries of the cones are determined by equating the energy density of the stellar wind to that of the unperturbed field at the magnetic equator. In fact, fluid elements of smaller energy density cannot leave the stellar magnetosphere. The cones become narrower as the ratio of the magnetic energy density at the base of the corona to the thermal energy density increases. This offsets the concomitant increase of R_A. In the limit of large field strengths, the rate of angular momentum loss becomes independent of the magnetic field. The loss rate for rapidly rotating stars with cool coronas proves to be strongly dependent on their angular velocities. The dynamics of the stellar wind are then strongly influenced by the centrifugal forces. In this case it is the centrifugal energy transported to the coronal gas by the magnetic field stresses, rather than the thermal conductivity, that is the source driving the stellar wind.

The angular momentum transport from stars of late spectral types can be highly efficient during their pre-main sequence evolution, during which they are thought to be T Tauri stars. The intense stellar wind observed from some T Tauri stars and the strength of their

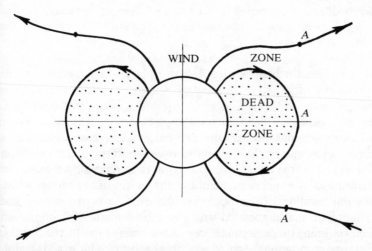

Figure 14.8   Field lines for a rotating magnetic star with stellar wind are shown. The magnetic and rotation axes are parallel. The points A determine the equality of the wind speed and the local Alfvén speed. Within A the field maintains approximate corotation. Beyond the gas effectively moves free of torque. For the limiting field line, which separates the wind zone from the dead zone (dotted region), the point A is at the equator. Inside the dead zone the gas is prevented from expanding by the magnetic field pressure.

magnetic fields (10^3 G, according to the theoretical speculations of Levy and Sonnet, 1978) could lead to a rapid decrease in the angular momenta of these stars at this stage in their evolution.

Careful measurements of the rotational velocities of pre-main sequence stars in the Taurus–Auriga cluster and in NGC 2264 (Vogel and Kuhi, 1981) reveal that stars of mass less than $1.5M_\odot$, which follow a non-convective evolutionary track (i.e. evolve in radiative equilibrium), generally rotate at less than 25 km s^{-1}, while well over half the stars of greater mass have rotational velocities in the range 25–150 km s^{-1}. For most stars on convective tracks $\langle v \sin i \rangle <$ 35 km s^{-1}. This suggests that most of their angular momenta was lost during the early hydrostatic stages of their evolution and before these pre-main sequence stars showed any photospheric activity.

It appears that, in order to quench the rotation of stars that are born fast rotators, strong stellar winds are needed for the first few hundred thousand years after their hydrostatic cores formed. The other possibility, that nascent stars lose angular momentum due to rotationally driven outflow of matter from their equators, is not excluded, and is examined further in the following section.

VI Magnetic Fields and Cosmogony

Magnetic fields are of interest in the cosmogony of the solar system and more generally in the cosmogonies of other planetary systems primarily because of their ability to redistribute angular momentum. A fascinating feature of the solar system is its highly nonuniform distribution of angular momentum per unit mass. That associated with the rotation of the Sun ($j \sim 10^{15}$ cm^2 s^{-1}) is about 10^5 times smaller than that associated with the orbital motion of the planets. Even assuming that the young Sun rotated more rapidly, the disparity still remains as large as two orders of magnitude. This has invited the hypothesis that a proto-planetary cloud was captured by the young Sun e.g. that the solar nebulae acquired its angular momentum through tidal interactions with another protostar. This hypothesis even in its most developed form (Kobrick and Kaula, 1978), cannot be regarded as the only, or even physically the most likely, solution

(Chemical and isotopic abundances give no support to the idea that solar and planetary matter has an independent origin.) A configuration with a mass concentrated in the central body and with an angular momentum shared with a smaller mass in orbital motion— just like the Solar system—is energetically preferable (Lynden-Bell and Pringle, 1974). Therefore, the formation of a star accompanied by a planetary system may be the natural result of the evolution of a rotating protostar, provided a sufficiently efficient transfer of the angular momentum from the central protostar to the periphery of the nebula can occur.

It was Alfvén (1954) who first pointed out that the magnetic field might play a crucial rôle in the evolution of the solar nebula. The first physical model of the simultaneous formation of the Sun and a protoplanetary cloud from a single rotating nebula was constructed by Hoyle (1960) along the lines sketched by Kant and Laplace. Hoyle estimated the angular momentum of the solar nebula to be $3 \cdot 10^{51} \, \text{g cm}^2 \, \text{s}^{-1}$ (the matter in the planets was complemented by volatiles to reconstruct the Solar abundances). This value represents a lower bound on the angular momentum of the Solar Nebula; that of the present solar system is ten times less, viz. $3 \cdot 10^{50} \, \text{g cm}^2 \, \text{s}^{-1}$. Assuming that a solar nebula of this angular momentum condensed uniformly, Hoyle found that its radius could become as small as the present radius of Mercury's orbit ($R \sim 10^{12} \, \text{cm}$). After that the nebula becomes rotationally unstable. The outflowing gas gradually forms a quasi-Keplerian disk, a frozen-in magnetic field connecting the disk to the central body. Magnetic stresses carry angular momentum from the proto-Sun to the disk, and from the inner to the outer part of the disk. This causes the disk to grow and, when later the rotational instability ceases, the inner edge of the disk moves outwards from the solar surface to the accompaniment of a dimunition in the Sun's angular velocity. For these processes to be effective over a period of 10^7 yr, the poloidal field strength and the degree of ionization should be of order 1 G and not less than 10^{-7}, respectively.

The hypothesis proposed by Hoyle aroused widespread interest, although some of its aspects were later criticized (Safronov, 1969). Hoyle's general idea is, however, tenable provided some modifications, inspired by recent progress in understanding protostellar collapse, are incorporated. Simulations of the collapse of a protostar (Larson, 1969, 1978; Tscharnuter, 1978) have shown that the condensation is initially highly inhomogeneous, with the concentration increasing towards the center. The evolution of the central regions depends crucially on the angular momentum per unit mass. If the angular momentum J of the protostar exceeds $10^{53}(M/M_\odot)^{5/3}$ g cm^2 s^{-1}, a disk or a ring structure forms that is very unstable to fragmentation. Perhaps a vigorous turbulence can suppress this instability so that a single star can form at the center of the disk (Tscharnuter, 1978). If J does not exceed $10^{52}(M/M_\odot)^{5/3}$ g cm^2 s^{-1}, a single star-like core is formed whose initial parameters do not differ substantially from those in the core of a non-rotating protostar ($M \sim 10^{-2}M_\odot$, $\rho \sim 10^{-2}$ g cm^{-3}, $T \sim 10^4$ K); see Ruzmaikina (1981b). Such a core is in a state of (approximate) hydrostatic equilibrium, and its subsequent evolution is governed by the accretion of an envelope and a redistribution of the angular momentum within it. The interstellar magnetic field, amplified during the collapse, is capable of redistributing the angular momentum and of setting up rigid body rotation throughout most of its mass (Ruzmaikina 1981a); see also Section III.

In a protostar with $J \sim 10^{52}-10^{53}(M/M_\odot)^{5/3}$ g cm^2 s^{-1}, during the growth of the core the rotational equatorial velocity reaches its threshold value, and the subsequent onset of the outflow of gas sets up a proto-planetary disk. The redistribution of angular momentum within the disk results in an increase of its radius. Presumably, this scenario represents one possible way to form a star from a proto-planetary disk. It is of special interest in the context of the solar system, because the estimated angular momentum of the solar nebula lies in the range of angular momenta in which theory predicts that this mechanism can function (see Safronov and Ruzmaikina, 1978).

The angular momentum of a star with a disk formed in this way should lie somewhere between that of single and binary stars. A collapsing protostar, with an angular momentum too low to fragment during dynamical collapse but too high to form a single star, gets rid of its excess angular momentum through the formation of a disk. Further evolution of the disk results in the formation of planets. The dynamics of accretion of the envelope and the formation of the disk have been investigated by Cassen and Moosman (1981) and Ruzmaikina (1982).

The magnetic field should obviously be important in the early stages of disk formation, particularly in its inner parts where the degree of thermal ionization is high. A dynamically significant magnetic field may, however, permeate even larger volumes of the disk. This can be inferred from the data on meteorite magnetism. The specific magnetization of a meteorite depends upon its genetic type (Guskova and Pochtarev, 1967). It has been found that meteorites possess a stable magnetization, i.e. one that is hard to remove by demagnetization, which is the firstborn component of its total magnetization. Stacey and Lovering (1959) proposed that this component may impart information to us about the origin of the Solar System. For this purpose, the magnetization of carbon chondrites are most valuable. These meteorites include those that have undergone only minor chemical and mineralogical changes from the time of their formation in the Solar System (Wasson, 1974). Refined measurements of some carbon chondrites indicate that their magnetization may have been acquired during cooling in a magnetic field of strength 1–10 G, assuming that the direction of the magnetic field did not change in that time (Levy and Sonett, 1978). The origin of such a field is unclear. It is possible that it was generated in rather massive parent bodies, the fragmentation of which gave rise to meteorites. But the possibility that such a field permeated the proto-planetary cloud in the region of meteorite formation (at a heliocentric distance of 3 a.u.) cannot be ruled out. The generation of a magnetic field in the disk through dynamo action or the dragging out by the solar wind of a solar field, generated during the Hayashi phase, come to mind as possible mechanisms for the origin of this field (Levy and Sonett, 1978).

The rôle of the magnetic field in the evolution of the solar nebula is not yet fully understood. Recent developments in the theory of star formation, cosmogony and the theory of the hydromagnetic dynamo provide, however, a firm foundation for future progress.

CHAPTER 15

Magnetic Fields in Cosmology

I General Considerations

In Book III of his Physics, Aristotle (200b) states that†

"There is that which is in actuality only, and that which is in possibility and in actuality, on the one hand that which is a particular individual, on the other that which is a quantity, a quality, and similarly with the other categories of that which is".

We take this passage to mean that we should not merely investigate things that actually exist but also consider phenomena that we can conceive of in possibility, and of the latter not merely objects that we can quantify but also concepts that have only a qualitative character. The last situation is exactly relevant to the subject of intergalactic magnetism, and we shall now discuss, in the spirit of Aristotle's advice, the nature of this magnetism, and its rôle in cosmological evolution.

At first, the existence of an intergalactic field seemed to be the inevitable consequence of an influential hypothesis by Hoyle (1958), that galactic fields are created by the compression of a pre-existing relict intergalactic field (see Chapter 3). The leitmotif of the

---

† The authors are grateful to W. Charlton of the University of Newcastle upon Tyne for providing this translation of the passage from Aristotle, as given in the text of Ross (1936).

present book is, however, the belief that magnetic fields evolve through dynamo processes from weak seed fields, not of a relict nature. The hypothesis of a relict field must today also be rejected on the basis of observational data and fundamental theoretical principles.

II Observational Limitations on Intergalactic Magnetic Fields

Let us see what limits are placed on the strength of an intergalactic magnetic field by the available observations. First of all, we shall estimate what size of intergalactic field is necessary to produce the observable galactic fields ($B_g \simeq 2 \cdot 10^{-6}$ Gauss, see Chapter 12) during the compression of the intergalactic medium that preceded the birth of the galaxies. By assuming, for simplicity, the compression to be isotropic we obtain

$$B_0 = B_g(\rho/\rho_g)^{2/3} \simeq 10^{-9}\Omega^{2/3} \text{ Gauss,} \qquad (1)$$

where $\Omega = \rho/\rho_{cr}$, $\rho_{cr} \simeq 2 \cdot 10^{-29}$ g/cm^3, and $\rho \simeq 10^{-24}$ g/cm^3 is the mean density of the Galaxy. The energy density of such a field, $B_0^2/8\pi \simeq 10^{-19}\Omega^{4/3}$ erg/cm^3, is much less than that of the relict background radiation measured today, $\varepsilon_\gamma \simeq 4 \cdot 10^{-13}$ erg/cm^3 and that of the matter $\varepsilon \simeq 2 \cdot 10^{-8}$ erg/cm^3 (Zeldovich and Novikov, 1975). Thus, the required field strength is rather small. To make this estimate, we needed to know the density of the intergalactic medium which is an intrinsically difficult problem (Longair and Suryaev, 1971).

We list here the three main limits on the strength of the mean intergalactic field (assumed uniform):

(1) An upper limit on the field strength at the present time can be obtained from measurements of the anisotropy of the relict background radiation. A direct measurement is made of the relative deviation, $\Delta T/T$, of the temperature of this background from the mean (isotropic) one, $T_0 \simeq 2.7$ K. A large-scale anisotropy in the background radiation is a result of the anisotropy of the expansion (Zeldovich and Novikov, 1975),

$$\Delta T/T \simeq (a-b)/a|_{z=z^*},$$

where $a(t)$ and $b(t)$ are the scale factors of the expansion with respect to different axes [the axis for $b(t)$ is directed along the field], and $z^* = 10$ is the red shift corresponding to the time at which the expanding intergalactic gas starts to become transparent to the background radiation. The value of z^* can easily be obtained by equating the characteristic time of the Compton interaction, $(\sigma_T n_e c)^{-1} \simeq 10^{19}(1+z)^{-3}$ sec, with the cosmological expansion time $\tau \simeq 3 \cdot 10^{17}(1+z)^{-3/2}$ sec. Assuming the simplest quasi-Euclidean model, the anisotropy of expansion is due entirely to the magnetic field (Zeldovich, 1965), and it may be shown (Doroshkevich, 1965) that

$$b/a = 1 - 12\varepsilon_B/\varepsilon,$$

where ε_B is the energy density of the magnetic field. Thus,

$$\frac{\Delta T}{T} \simeq 12 \left(\frac{\varepsilon_B}{\varepsilon} \right)_{z=z^*} = 12 \left(\frac{\varepsilon_B}{\varepsilon} \right)_{z=0} (1+z^*).$$

Substituting for ε_B yields

$$B_0^2 \simeq \tfrac{2}{3}\pi\varepsilon_0(1+z^*)^{-1}\Delta T/T.$$

The upper limit for the large-scale anisotropy of the background radiation is no more than $3 \cdot 10^{-3}$ (see Table XVII in Zeldovich and Novikov, 1975). From this we obtain

$$B_0 < 3 \cdot 10^{-6} \text{ Gauss.} \tag{2}$$

(2) Another estimate of B_0 follows from the fact that, during cosmological expansion, the quantity $B^2/8\pi\varepsilon_\gamma$ must be nearly constant, possibly to within a logarithmic factor (Zeldovich and Novikov, 1975). Since the chemical composition of most matter present in the Universe today agrees with that predicted by the homogeneous and isotropic model of Friedmann (about 30% hydrogen and 70% helium) it is natural to suppose that the above quantity was small in the epoch $(1 < t < 100 \text{ sec})$ after the big bang during which nuclear reactions mainly occurred. This immediately gives an upper bound on the field today:

$$B_0 < 3 \cdot 10^{-7} \text{ Gauss.} \tag{3}$$

Unfortunately both the upper limits (2) and (3) prove to be very large compared with (1).

(3) The strictest estimate of B_0 has been obtained through an

analysis of the observed rotation measures (RM) of remote radio
sources (radio galaxies and quasars) by a number of authors
(Kawabata *et al.*, 1969; Reinhardt, 1972; Vallée, 1975; Kolobov *et
al.*, 1976; Kuznetsova, 1976; Ruzmaikin and Sokoloff, 1977b). The
rotation measure obtained by measuring the position angle of the
linearly polarized radio radiation on several wavelengths determines
the product of the electron density and the component of magnetic
field along the line of sight (see Chapter 2).

Intensive radioastronomy observations in recent years have given
the rotation measures of several hundred extragalactic radio sources
(radio galaxies and quasars). It is well-known that a rotation measure
contains immediate information on the magnetic fields in the space
between the source and the observer (see Chapter 2). The main
difficulty of interpretation stems from the fact that an observed RM
is the sum of contributions from the source, from the Galaxy and
possibly from the intergalactic medium. The contribution from the
gas and fields between the galaxies of a cluster must also be taken
into account. Several attempts were made to separate the contributions from the Galaxy, from the intergalactic field and from the radio
sources. Ruzmaikin and Sokoloff (1977b) constructed a model
through which these three main contributions to the observed RM
could be assessed. The possibility of extracting such information is
based on the individual character of each contribution. The rotation
measure of the Galaxy, for instance, depends on the angle between
the direction to the source and the orientation of the Galactic magnetic
field; it does not depend on the redshift z of the source, or on the
mean density of the intergalactic medium (i.e. on $\Omega = \rho/\rho_{cr}$); the RM
of a source depends on z but not on its angular Galactic coordinates;
the contribution from the hypothetical intergalactic magnetic field
must depend on a certain angle (the direction of field) and must
increase with z.

We represent the observed rotation measure of a radio source by
a sum of the individual rotation measures from the Galaxy, the
intergalactic medium and the source:

$$RM = RM_g + RM_{ig} + (1+z)^{-2} RM_s. \tag{4}$$

The contribution from the Galaxy can be approximated, on average,
by three quantities: an amplitude K and the angles l_0^{II} amd b_0^{II}
determining the direction of the Galactic field. This contribution,

together with its attendant fluctuations, has been studied separately in Chapter 12. The factor $(1+z)^{-2}$ in the third term allows for the fact that the angle of rotation of the polarization plane is proportional to the square of the wavelength λ, which suffers a redshift, $\lambda = \lambda_s(1+z)$. Thus, RM_s is the intrinsic rotation measure of the source. Even after extracting this factor, RM_s can still depend on z, as noted by Komberg (private communication), because of evolutionary and selection effects (the larger the z, the brighter and more compact the sources must be to be seen, and therefore the greater their RM). Since the exact dependence of $RM_s(z)$ on z is unknown, the available observational data was divided into discrete zones each spanning Δz, in each of which RM_s is supposed to be constant. When assembled, the values of RM_s obtained in these zones gave a piecewise linear approximation to $RM_s(z)$ that approximately characterized the evolutionary and selection effects.

It was natural to suppose that RM_s is independent of angle, that at every fixed z the mean, $\langle RM_s \rangle$, of the distribution function of the rotation measures of the sources is zero, and that there is a dispersion in RM_s about that mean. The last assumption could be checked immediately after the parameters have been numerically calculated. The dependence of the dispersion of the rotation measures of the sources on their redshift characterizes the evolutionary and selection effects.

An expression for the intergalactic rotation measure is easily obtained by noting that in the Friedmann model the electron density, magnetic field and wavelength obey

$$n_e = n_{e0}(1+z)^3, \qquad B = B_0(1+z)^2, \qquad \lambda = \lambda_0(1+z)^{-1}.$$

The path travelled by the radio radiation is related to the redshift by (Zeldovich and Novikov, 1975)

$$dr = c \ dt = c[\mathcal{H}(1+z)^2(1+\Omega z)^{1/2}]^{-1} \ dz,$$

where c is the velocity of light, and $\mathcal{H}$ the Hubble constant. Substitution of these relations into the integral $\int n_l \mathbf{B} \cdot d\mathbf{r}$ and its subsequent integration yield

$$RM_{ig} = A\Omega^{-2}[(1+\Omega z)^{3/2} - 3(1-\Omega)(1+\Omega z)^{1/2} + 2 - 3\Omega] \cos \theta, \quad (5)$$

where

$$A = 8.1 \cdot 10^5 \frac{2c}{3\mathcal{H}} n_{e0} B_0 \approx 10^{10} \left(\frac{\mathcal{H}_0}{50 \text{ km/s Mpc}}\right) \left(\frac{B_0}{1G}\right) \Omega \text{ rad/m}^2.$$

The angle θ between the direction of the source and the hypothetical intergalactic field may be expressed in terms of the angular Galactic coordinates of the source (l^{II}, b^{II}) and the field (l_1^{II}, b_1^{II}) by

$$\cos\theta = \cos b^{II} \cos b_1^{II} \cos(l^{II} - l_1^{II}) + \sin b^{II} \sin b_1^{II}.$$

Eq. (5) assumes a very simple form in limiting cases

$$RM_{ig} \simeq \begin{cases} \frac{3}{2}Az, & \Omega z \ll 1, \\ A[(1+z)^{3/2} - 1], & \Omega = 1. \end{cases} \tag{6}$$

To obtain the parameters K, l_0^{II}, b_0^{II}, l_1^{II}, b_1^{II}, Ω and A a regression method was developed and numerically implemented (Ruzmaikin and Sokoloff, 1977b; Ruzmaikin et al., 1978). In essence it consists of minimizing the likelihood,

$$S(K, l_0^{II}, \ldots, A) = \sum_n (1 + z^{(n)})^4 [RM^{(n)} - RM_g^{(n)} - RM_{ig}^{(n)}]^2,$$

where $RM^{(n)}$ and $z^{(n)}$ are the rotation measure and the redshift of the nth radio source. The mean parameter values were obtained by minimizing S, and confidence intervals were deduced by estimating the second derivatives of S at this minimum. In our opinion, this method, being in fact a generalization of the method of least squares, provides a simple and adequate technique for this problem. Statistical methods used by other authors (for example, the calculation of various correlation coefficients) do not always lead to reasonable results. The best estimate of the magnetic field strength obtained in this way is (Ruzmaikin and Sokoloff, 1977b)[†]

$$B_0 < 5 \cdot 10^{-10} \Omega \text{ Gauss.} \tag{7}$$

The limit (7) obtained in the previous paragraph applies strictly only when the intergalactic field is uniform: it should be reliable if it varies over lengths exceeding the distances to remote sources ($z \simeq$ 2–3). Observational bounds for fields of smaller scales are less restrictive (Reinhardt, 1972; Kolobov et al., 1976; Kronberg et al., 1977). Fields of galactic scale and, possibly on the scale of galactic clusters may exist.[‡]

[†] According to some estimates of visible matter in the Universe, $\Omega \simeq \frac{1}{40}$ (Longair and Sunyaev, 1971).

[‡] A possible observational test, to reveal magnetic fields on scales larger than the galactic scale, is discussed by Komberg et al. (1979) and by Kronberg and Perry (1982).

III   The Rôle of Magnetic Fields in Galaxy Formation

A most important problem of modern cosmology is to develop a theory of galaxy formation that would explain the observed distributions of their mass, angular momentum, type and structure. Zeldovich and Novikov (1975) amongst others have reviewed the existing theories of galaxy formation. Some new ideas on the detection and study of the cellular structure of the Universe are presented in papers by Zeldovich et al. (1982) and Shandarin et al. (1983).

Let us discuss the rôle of relict magnetic fields in the birth and early evolution of galaxies. The fields, even though very weak today, become stronger if we go back in time, and may have been essential for the physical processes that occurred during the early stages of galactic evolution. In this section we shall assume the existence of initial seed fields. The question of magnetic field generation in an initially unmagnetized plasma is discussed in the next section.

Models of the early Universe are required today to be consistent not only with the high degree of isotropy exhibited by the background radiation, but must also not contradict the observed chemical composition of the matter in the Universe that has not experienced nuclear synthesis in stars. These criteria are well met by the homogeneous isotropic Friedmann model. Galaxies are formed through the growth of small perturbations in the Friedmann solution that are either adiabatic or entropic by nature. An alternative theory, a major portion of which is devoted to vortex perturbations (Ozernoy and Chernin, 1967), has also been developed. In this theory, however, strong deviations from the Friedmann solution inevitably appear at an early stage.

A uniform magnetic field, B, has an effect on the growth of perturbations (Chandrasekhar and Fermi, 1953b). In the Friedmann model at the later stages of expansion, on the one hand, the scales of galaxies and their clusters are so large that the unperturbed solution may be considered to be "dust-like" ($p = 0$); on the other hand, these scales are much less than the event horizon ct, so that Newtonian theory may be applied. A critical wavelength

$$\lambda_B \simeq \tfrac{1}{2}G^{-1/2}B_0[\rho_0(1+z)]^{-1}$$

(8)

appears naturally. In (8), $G = 6.7 \cdot 10^{-8} \, \text{cm}^3/\text{g s}^2$ is the gravitational constant, z is the redshift, B_0 and ρ_0 are the field strength and density in the contemporary epoch.

The component of field perpendicular to the velocity perturbations affects their evolution. When $\lambda > \lambda_B$ the density perturbations grow asymptotically $(t \to \infty)$ as

$$\delta\rho/\rho \sim t^{\alpha}$$

(Ruzmaikina and Ruzmaikin, 1970), where the factor α would be $\frac{2}{3}$ were the magnetic field zero (Zeldovich and Novikov, 1975), but which is smaller when $B_0 \neq 0$ by an amount $(\frac{2}{3})k^2 B_{0\perp}^2/4\pi\rho_0$ (k is the perturbation scale). Thus, as one would expect, the field inhibits the growth of perturbations in directions perpendicular to itself. For $B_0 = 10^{-9}$ Gauss and $\rho_0 = 10^{-29} \, \text{g/cm}^{-3}$, we have $\lambda_B \simeq 10^{23}(1+z)^{-1}$, i.e. the wavelength exceeds the Jeans length for density perturbations in an epoch close to that of recombination. Taking into account the pressure of the matter, one obtains a critical mass

$$M_B \sim \rho\lambda_J\lambda_B^2 \sim 10^6 - 10^7 M_\odot,$$

for which the density perturbations can grow. Here $\lambda_J = 2\pi u_s(4\pi G\rho)^{-1/2}$ is the Jeans wavelength, and the Jeans mass is

$$M_J \sim \rho\lambda_J^3 \sim 10^5 M_\odot.$$

This result should be borne in mind when constructing an entropic theory of galaxy formation.

At the stage where nonlinearities affect the growing perturbations in the expanding dust-like matter, the solution is usually written in Lagrangian form (Zeldovich, 1970)

$$\mathbf{r} = a\mathbf{q} + \mathbf{s}(\mathbf{q})b,$$
$$\rho(\mathbf{r}, t) = \rho_0(\mathbf{q})/(a - b\lambda_1)(a - b\lambda_2)(a - b\lambda_3), \tag{9}$$

where $\mathbf{r}$ and $\mathbf{q}$ are the Eulerian and Lagrangian coordinates of a particle; $\lambda_1, \lambda_2, \lambda_3$ are the eigenvalues of the deformation tensor, $\partial s_i/\partial q_k$; and $a(t)$ and $b(t)$ are known powers of t (b growing faster than a). Under frozen-in conditions, when the field has no effect on the compression dynamics, one can also write the Langrangian solution for $\mathbf{B}$ as

$$B_i(\mathbf{r}, t) = B_{0k}(\mathbf{q})[\rho(\mathbf{r}, t)/\rho_0(\mathbf{q})] \, \partial r_i/\partial q_k,$$

where $\mathbf{B}_0(\mathbf{q})$ is the initial field given by the linear analysis (for example, it might be a relict field). By substituting in (9), one obtains the projections of $\mathbf{B}$ onto the major axes of the deformation tensor

$$B_1 = B_{01}/(a - b\lambda_2)(a - b\lambda_3),$$

$$B_2 = B_{02}/(a - b\lambda_1)(a - b\lambda_3),$$

$$B_3 = B_{03}/(a - b\lambda_1)(a - b\lambda_2).$$

When there is no magnetic field and one of the multipliers in (9) is zero, the density grows indefinitely along the corresponding axis forming a "pancake" configuration. If a field exists and is not directed along the axis of compression, the collapse stops before large densities are reached. This occurs at a characteristic distance of order (8),

$$B/G^{1/2}\rho \simeq 100(1 + z)^{-1} \text{ kpc},$$

for $B_0 \simeq 10^{-9}$ Gauss, i.e. the field, when large enough, is capable of making a sphere-like configuration. This discussion again highlights a need to estimate the strength of the intergalactic magnetic field more precisely (see Section II). The existence of pancake configurations of matter in the contemporary epoch places a severe limitation on the intergalactic field strength.

IV Excitation of Magnetic Fields in a Radiation Plasma

We shall now consider what magnetic fields can actually arise in an expanding Universe in the early stages of its evolution. The discovery of the background radiation with a mean temperature of 2.7 K has indicated that the number of photons per unit volume of space is about 10^8 times the number of baryons.[†] Therefore, an expanding cosmological plasma is radiative. The density of the photon

---

† It seems there are a considerable number of neutrinos in the Universe. Except for the very early stages of the expansion, however, their interaction with the plasma electrons is weak.

component, ρ_γ, is at present much lower than that of the matter, ρ, but this was not always so. The evolution of ρ and ρ_γ is governed by

$$\rho \sim a^{-3}(t), \qquad \rho_\gamma \sim a^{-4}(t), \tag{10}$$

where $a(t)$ is the scale factor of the expansion. Thus, in times past, close to the epoch of recombination of the relict helium-hydrogen plasma, these densities were equal, and prior to that time the plasma was radiation-dominated, i.e. $\rho_\gamma > \rho$.

Harrison (1970) proposed a mechanism for the generation of a magnetic field in the *expanding* radiation-dominated plasma. The physical idea behind this mechanism can be illustrated using the example of a uniformly rotating, expanding vortex of radius $a(t)$, consisting of radiation with density $\rho_\gamma(t)$ and a nonrelativistic plasma (electrons and protons) with density $\rho(t)$.[†]

While the vortex is expanding the "masses" ρa^3 and $\rho_\gamma a^4$ are separately conserved [see Eq. (10)], and when there is no interaction between them the angular momenta of the matter and the radiation, $\rho \omega a^5$ and $\rho_\gamma \omega_\gamma a^5$, are also conserved. Hence,

$$\omega \sim a^{-2}(t), \qquad \omega_\gamma \sim a^{-1}(t), \tag{11}$$

i.e. because of the expansion, the angular velocity of the matter and radiation will evolve in different ways. Now consider the interaction. The Thompson cross-section for the scattering of electrons and γ-quanta is $(m/m_p)^2$ times as large as that for proton scattering with γ. The electrons in this radiation-dominated plasma are therefore swept round with the radiation. The medium thus consists of two fluids: one of positively charged protons with density ρ and the other of photons and negatively charged electrons with density ρ_γ. The vortex expansion produces a difference in the angular velocities of these fluids [see Eq. (11)], and hence a vortex current and a magnetic field are created. The decay of magnetic field caused by the Coulomb interaction between electrons and protons is not potent on galactic (or larger) scales. Protons and electrons are bound only by the induced electric field, which tends to equalize the difference in the angular

---

[†] Electrons become non-relativistic when the temperature in the expanding Universe drops to $mc^2/k = 6 \cdot 10^9$ K. At higher temperatures lepton-nucleon interactions occur and the mechanism considered here ceases to function.

velocities. The magnetic field having been generated, the angular momenta of the matter and the radiation are no longer separately conserved. As in the case of the well-known Einstein–de Haas effect, however, the angular momentum associated with the "total" vorticity $\xi \equiv \omega + (e/m_p c)B$ will be constant in time, i.e.

$$\omega + (e/m_p c)B = (a_1/a)^2 \omega_1.$$

Here ω_1 is the vorticity at the initial time, t_1, when the scale factor (the vortex radius) is a_1. Clearly, $\omega_e \simeq \omega_\gamma = \omega_1 a_1/a$. If the vortex scales are such that the viscosity may be ignored, then on the right-hand side to leading order $a_1 \omega_1/a \simeq \omega$, and hence

$$B = -\frac{m_p c}{e}\left(1 - \frac{a_1}{a}\right)\omega \simeq -10^{-4}\left(1 - \frac{z}{z_1}\right)\omega,$$

where the redshift, $z \equiv a_0/a(t) - 1$, has been introduced, a_0 being the vortex radius at the moment $z = 0$. It may be seen that the magnetic field generated is maximal when $z \ll z_1$, and directly "follows" the angular velocity $B_{max} = -(m_p c/e)\omega$. For scales on which the viscosity is essential, one may assume that asymptotically $\omega \to 0$, and therefore

$$B \simeq \frac{m_p c}{e}\frac{a_1}{a}\omega = \frac{1+z}{1+z_1}\frac{m_p c}{e}\omega.$$

Thus, in this case the field is $(1+z)/(1+z_1)$ times smaller than the maximal field, and is in the opposite direction.

Cosmology deals with not just one vortex but with a large set of interacting vortices, i.e. with a turbulent radiation-dominated plasma (Gamov, 1952; Ozernoy and Chernin, 1967). The generation of a magnetic field in cosmological turbulence has been studied by Ozernoy and Chibisov (1970), Harrison (1973) and Baierlein (1978). This problem may be reduced to an investigation of a total vorticity equation

$$\partial(a^2\xi)/\partial t = a^2(\xi \cdot \nabla)v - (v \cdot \nabla)a^2\xi + \nu_\gamma \nabla^2(a^2 e B/m_p c). \quad (12)$$

By solving this equation for given vorticity ω, one can obtain information about the magnetic field. The quantity ν_γ is the radiative kinematic viscosity determining the characteristic dissipation scale.

This problem is an analog of the magnetic field generation problem in mirror-symmetric turbulence (see Section 8.IV). An important difference is that the coefficients in Eq. (12) depend on time. The magnetic Reynolds number of the cosmological plasma is very large, so both the conditions $R_m \gg R_e$ and $R_m \gg (R_m)_{cr}$ are satisfied. Since the coefficients are nonstationary, the growth of the field will be slower than exponential. Such an effect, when an exponential instability is replaced by an algebraic one, is well-known in the theory of the growth of gravitational perturbations in the expanding Universe (Zeldovich and Novikov, 1975). A crude estimate of the growth law can be made by replacing $\exp(\gamma t)$ with $\exp(\int \gamma \, dt)$. The maximum growth rate is $\gamma_{max} = 4v/5l$ (see Section 8.IV). On the largest scale $v/l \cong t^{-1}$ (Harrison, 1973), and hence we obtain $B \sim t^{4/5}$ i.e. the growth is very slow. But there is some doubt as to the existence of cosmological turbulence. Zeldovich and Novikov (1975) have pointed out the difficulties associated with the vortex theory. Therefore we shall not discuss further the consequences of this magnetic field growth for the pregalactic stage of the Universe's evolution.

According to the modern view, galaxies started to form after the recombination of the relict plasma. At that time the expanding Universe consisted mainly of neutral atoms of hydrogen and helium of density n_H, free electrons and protons of densities n_e and $n_p [(n_e/n_H) = 3 \cdot 10^{-4} – 3 \cdot 10^{-5}$, being the residual degree of ionization in the absence of secondary heating of the matter], as well as relict radiation with density n_γ. In fact, immediately after the recombination $(z > z_r)$, the matter becomes transparent to radiation on scales corresponding to the size of future galaxies. The motion of the gas does not advect the radiation and, the latter may therefore be considered as a homogeneous background.

Without going into specific details of galaxy formation for the time being, let us simply imagine that a protogalaxy, i.e. some gas cloud which is a candidate for conversion into a galaxy, rotates with a given angular velocity ω relative to the radiation background. If the protogalaxy continues to participate in the Hubble expansion, or if the motion is still unsteady, then $\omega = \omega(t)$. The rotation of the protogalaxy may be caused either by the tidal interactions between protogalaxies (Peebles, 1969), or by large-scale shock waves (Doroshkevich, 1973; Jurevich and Chernin, 1978), or it may be a consequence of the initial vortex perturbations.

Because of the presence of the angular velocity, ω, a magnetic field must be excited by the Compton interaction mechanism (Mishustin and Ruzmaikin, 1971). But, in contrast to the Harrison mechanism, there are not now two fluids (the protogalaxy is "transparent" to radiation!) and expansion does not play an important part. The process can be described as follows.

The collisions of the protons with the much heavier neutral atoms make the former rotate with the same angular velocity ω. Protons do not significantly interact with the radiation, unlike electrons which do so much more efficiently than with neutral atoms or protons. The radiation brakes the rotation of the electrons, so producing a difference in the electron and proton angular velocities, i.e. a current and, hence, a magnetic field. The electrons and protons find themselves bound by the induced electric field that balances the radiation friction. Their radial velocities become equal through the action of this potential field.

The mechanism of field excitation just described can easily be confirmed by approximate calculations. The equation of motion for the electrons is

$$0 = -(e/mc^2)(c\mathbf{E} + \mathbf{v}_e \times \mathbf{B}) - \tau_{e\gamma}^{-1}\mathbf{v}_e + \tau_{ep}^{-1}(\mathbf{v}_p - \mathbf{v}_e). \tag{13}$$

Here $\mathbf{v}_e$ and $\mathbf{v}_p$ are the electron and proton velocities, $\mathbf{E}$ and $\mathbf{B}$ are the induced electric and magnetic fields, and

$$\tau_{e\gamma}^{-1} = 4\sigma_T \rho_\gamma c/3m, \qquad \tau_{ep}^{-1} \simeq 1.2 \ln\Lambda\, n_e (e^2/kT)^2 (kT/m)^{1/2},$$

are the characteristic frequencies of momentum exchange between the electrons and γ-quanta and between electrons and protons; $\sigma_T \simeq 6.65 \cdot 10^{-25}$ cm^2 is the Thompson scattering cross-section, $T \sim (1 + z)^2$ is the temperature of the matter, and $\ln \Lambda \simeq 30$ is the Coulomb logarithm. We have ignored the gradient of the gravitational potential, the inertial force, and the friction between the electrons and the neutral matter. It is easy to show that the inclusion of these factors makes no substantial difference.

Since the magnetic field is initially absent, the term $\mathbf{v}_l \times \mathbf{B}$ in Eq (13) may be omitted. On taking the curl and substituting into Maxwell equations, making an allowance for the Hubble expansion, we obtain the following induction equation for $\mathbf{B}$:

$$\frac{1}{a^2}\frac{\partial}{\partial t}\left(a^2 \frac{e}{m_p c}\mathbf{B}\right) = \frac{\omega}{\tau_{e\gamma}} + \frac{c}{4\pi n_e e \tau_{ep}}\nabla^2 \mathbf{B}. \tag{14}$$

The density ρ_γ and the temperature T in the post-recombination period are so small that $\tau_{ep} \ll \tau_{e\gamma}$ by a large factor. This justifies the replacement $\omega \to \omega_p = \omega$. Moreover, one may ignore for fields of galactic scale the last term, $(c^2/4\pi\sigma)\nabla^2\mathbf{B}$, in Eq. (17) which arises from the Coulomb interactions. It then follows that

$$\mathbf{B} = (mc^2/e)a^{-2} \int_{t_f}^{t} \tau_{e\gamma}^{-1} \omega a^2 \, dt.$$

To estimate the field strength, it is convenient to use, instead of t, the dimensionless redshift z,

$$dt = -\Omega^{1/2} \mathcal{H}_0^{-1} z^{-5/2} \, dz,$$

where $\Omega \simeq 0.03$ is the mean-to-critical density ratio of the matter (the critical density defines the geometry of the Universe); $\mathcal{H}_0 (\simeq 50 \text{ km/s Mpc})$ is the Hubble constant; $\tau_{e\gamma} = \tau_{e\gamma}(0)(1+z)^{-4}$; and, for the sake of simplicity, the angular velocity is assumed constant. We obtain

$$B \simeq 2 \frac{mc^2}{e} \frac{(1+z_f)^{5/2}}{\Omega^{1/2} \mathcal{H}_0 \tau_{e\gamma}(0)} \omega,$$

where z_f is the redshift at the time of the formation of the galaxies. Taking $1 + z_f = 10$, and $\omega = 10^{-15} \text{ s}^{-1}$ we obtain

$$B \simeq 10^{-21} \text{ Gauss.}$$

How negligibly small the protogalactic fields turn out to be! But we should remember that over cosmological periods ($\sim 10^{10}$ years) such fields are quite capable of growing exponentially (through dynamo action!) to the observed strength, $(2-3) \cdot 10^{-6}$ Gauss, of the Galactic field.

CHAPTER 16

Accretion on Black Holes

In this chapter we shall consider the properties of magnetic fields in the matter accreting onto compact gravitating objects which appear in the vicinity of stars, and possibly galactic nuclei, surrounded by gas. The great interest in accretion, born ten years ago and still flourishing today, was fueled by the opening of a new chapter in astrophysics, namely a relativistic astrophysics that provided an opportunity to explore general relativity theory (GRT) by astronomical observations.

The most exciting prediction of GRT in the field of stellar astronomy is that massive stars ($\geqslant 3M_\odot$) evolve naturally into black holes— objects which have so far collapsed due to their self-gravitation that they have "retired from the world", i.e. they have so completely disconnected themselves from our world that they can provide us with no information. Black holes do not possess magnetic fields (Section I), unlike other relativistic objects (for example pulsars) which have strong magnetic fields (and which will therefore be considered separately; see Chapter 17). We are here interested in the rôle that magnetic fields play in the observed manifestations of black holes. Only a few brief remarks will be made about accretion onto white dwarfs and neutron stars.

The natural, and so far the only, way to discover a black hole is by the radiation from the gas falling into it under the action of gravitational forces. In Section II we will show that the magnetic field can have an important effect on the emission by gas accreting onto a single black hole (Shvartzman, 1971; Pringle et al., 1973). If the accretion occurs in a binary stellar system (Section IV), with gas falling from the normal visible component onto the compact collapsed component, the magnetic field appears to be crucial for the outward transportation of the angular momentum of the gas (Shakura and Sunyaev, 1973). The field allows the gas to fall onto the black hole

306

o the accompaniment of a massive release of radiant energy. In the inal section (Section VI) we shall discuss a possible dynamo mechansm in accretion disks and mention some nonlinear effects, particularly ield buoyancy which may result in the formation of hot coronae adiating X-rays just like stellar coronae (Galeev *et al.*, 1979).

I Black Holes

n the fascinating book "Gravitation" (Misner, Thorne and Wheeler, 1973) it is explained, through an imaginary dialogue between the ceptic, Sagredus, and the erudite Salvatius, why a massive object, collapsing to its gravitational radius $r_g = 2GM/c^2$, should be called a 'black hole". "Black" is appropriate, because the luminosity of the ot surface of the object as measured by a remote observer decays exponentially with time. The e-folding time is very short, namely $3^{3/2}r_g/2c \simeq 2.6 \cdot 10^{-5}(M/M_\odot)$ seconds. The discreteness of energy mplies that the number of photons emitted is finite, i.e. even the exponential decrease of the luminosity will not persist indefinitely. 'Hole" is appropriate because neither matter nor radiation that falls nto the object can leave after crossing a certain critical surface. In he case of a non-rotating object this surface is spherical, and of radius r_g.

General relativity shows that asymptotically the external field of a black hole is completely defined by its mass M, charge Q and angular momentum J (Misner *et al.*, 1973). In fact, of all the quantities characterizing this isolated source of gravitational and electromagnetic fields, only the conserved quantities M, Q and J can be known to the external world. It may be recalled that these quantities can be represented, after using Gauss' theorem, in terms of fluxes over remote surfaces. By contrast, this is impossible for, say, the baryon number. To determine this it would be necessary to descend into the black hole and count the particles, but the results would be of no use to anybody outside.

Outside a black hole of charge Q there is a radial electric field $E_r = Q/r^2$. A rotating black hole creates, in addition, a dipolar magnetic field

$$H_r = (2QJ/M)r^{-3} \cos \theta, \qquad H_\theta = (QJ/M)r^{-3} \sin \theta,$$

where QJ/M is the magnetic moment of the black hole.

The case of a charged black hole is of no practical interest. It is clear that electrostatic forces, which greatly exceed gravitational forces, rapidly extract charges of opposite sign from the external medium to neutralize the charge, Q, of the black hole. For this reason we believe that an evolved black hole has no intrinsic magnetic moment.

It is important then to understand how the magnetic moment of an uncharged but initially magnetic star vanishes during gravitational collapse. This problem was first solved by Ginzburg (1964), [see also Ginzburg and Ozernoy (1964) for the nonrotating collapsing star]. In essence, one must generalize the law of magnetic field amplification under a contraction $(H \sim R^{-2})$ to the case of relativistic motion in the Schwarzschild metric

$$ds^2 = (1 - r^{-1}r_g)c^2 \, dt^2 - (1 - r^{-1}r_g)^{-1} \, dr^2 - r^2(d\theta^2 + \sin^2 \theta \, d\phi^2),$$

where c is the speed of light, (r, θ, ϕ) spherical coordinates, and

$$r_g = 2GM/c^2 \simeq 3(M/M_\odot) \text{ km},$$

is the Schwarzschild radius.

Let R_0 and μ_0 be the initial radius and magnetic moment of a gravitationally collapsing sphere. In the non-relativistic approximation the magnetic moment $\mu (\sim HR^3)$ will decrease as

$$\mu = \mu_0(R/R_0),$$

i.e. $\mu \to 0$ as $R \to 0$. In general relativity under the frozen-in condition, $H \sim R^{-2}$, also holds locally; globally, however, the influence of the curvature is important, and

$$\mu = \mu_0 r_g / \{3R_0 \ln [r_g/(R - r_g)]\}.$$

Now, $\mu \to 0$ as $R \to r_g$. Using the dependence of the radius on time (see Misner et al., 1973), one can write this formula as

$$\mu(t) = \mu_0 r_g^2 / 3R_0 ct.$$

Hence, to a remote external observer the magnetic moment of the collapsing star decays as t^{-1}. The magnetic field has the form

$$H_r = 2\mu \cos \theta \, r^{-3} f(r), \qquad H_\theta = \mu \sin \theta \, r^{-3} \psi(r),$$

where the functions $f(r)$ and $\psi(r) \to 1$ for $r \gg r_g$ and where, as $r \to r_g$,

$$f(r) \simeq -3 \ln (1 - r_g/r), \qquad \psi(r) \simeq 3(1 - r_g/r)^{-1/2}.$$

Thus, the magnetic field flattens itself against the critical surface of the collapsing object $(H_\theta \gg H_r, r \to r_g)$, though the field decays with time at any given radius.

II Accretion of Matter onto a Nonrotating Black Hole in the Presence of a Magnetic Field

Let us consider an idealized example. On the one hand, this example is to some extent of aesthetic interest only. It is an exact solution describing the nonstationary magnetic field of a conducting relativistic gas (i.e. a gas flowing with a relativistic velocity) moving in a strong gravitational field. On the other hand, the remarkable fact that a black hole cannot acquire a magnetic field due to accretion can be clearly seen.†

Suppose that a black hole of mass M does not rotate. The external gravitational field and the corresponding curved space are then completely described by the Schwarzschild metric. Let the pressure of the ambient gas be small so that the gas falls freely onto the black hole from infinity. Spherically symmetric gas flow has the four-velocity $u^i = (u_0, u^r, 0, 0)$, whose components follow from energy conservation $mc^2 g_{00} u^0 = $ constant and the relation $u_i u^i = 1$, $(i = 0, 1, 2, 3)$. We therefore have

$$u^0 = (1 - r_g/r)^{-1}, \qquad u^r = -(r_g/r)^{1/2}.$$

Consider an initially weak poloidal magnetic field $B^i = (B^0, B^r, B^\theta, 0)$,

---

† To avoid any misunderstanding, we should mention that the magnetic field created by an external source can penetrate a black hole.

embedded in such a flow which, at the outset, is parallel to the z-axis (Ginzburg and Ozernoy, 1964),

$$B^r(t=0) = B_0 \cos\theta \, (1-r_g/r_0)^{1/2},$$

$$r_0 B^\theta(t=0) = -B_0 \sin\theta \, (1-r_g/r_0)^{1/2}.$$

Frozen-in magnetic field evolves according to the four-equation

$$\partial[(-g)^{1/2}(B^i u^k - B^k u^i)]/\partial x_k = 0,$$

where g is the determinant of the metric tensor. This extension to the relativistic case of the induction equation for $\nu_m = 0$ was first derived by Lichnerowicz. Its solution in the Schwarzschild metric can be written in parametric form as (Bisnovatyi–Kogan and Ruzmaikin, 1974)

$$B_r = B_0 \cos\theta (r_0/r)^2 (1-r_g/r_0)^{1/2},$$

$$r B_\theta = -B_0 \sin\theta \, r_0^{1/2} r_g r^{-3/2} (1-r_g/r_0)^{1/2},$$

$$\frac{ct}{r_g} + \frac{2}{3}\left(\frac{r}{r_g}\right)^{3/2} + 2\left(\frac{r}{r_g}\right)^{1/2} + \ln\left[\frac{1-(r_g/r)^{1/2}}{1+(r_g/r)^{1/2}}\right]$$

$$= \frac{2}{3}\left(\frac{r_0}{r_g}\right)^{3/2} + 2\left(\frac{r_0}{r_g}\right)^{1/2} + \ln\left[\frac{1-(r_g/r_0)^{1/2}}{1+(r_g/r_0)^{1/2}}\right].$$

These covariant components of B^i represent the physical components of the magnetic field in a reference frame moving with the matter. It can readily be seen that, for any fixed radius, the magnetic field grows as the mass of the incident matter increases (i.e. r_0).

To put things more clearly, consider the explicit dependence of the field components on r and t in two limiting cases:

(i) in a non-relativistic, Newtonian, region $(r, r_0 \gg r_g)$

$$B^r \simeq (1 + \tfrac{3}{2}ctr_g^{1/2}r^{-3/2})^{4/3} B_0 \cos\theta,$$

$$r B^\theta \simeq -(1 + \tfrac{3}{2}ctr_g^{1/2}r^{-3/2})^{1/3} B_0 \sin\theta.$$

For any radius r, the field grows as a power of the time.

(ii) in a relativistic region in the vicinity of the Schwarzschild radius $(1-r_g/r \ll 1, 1-r_g/r_0 \ll 1)$

$$B^r \simeq [2(1-r_g^{1/2}/r^{1/2})]^{1/2} \, e^{ct/2r_g} B_0 \cos\theta,$$

$$r B^\theta \simeq -[2(1-r_g^{1/2}/r^{1/2})]^{1/2} \, e^{ct/2r_g} B_0 \sin\theta.$$

Here the field grows exponentially for any fixed radius r. The magnetic field vanishes, however, as $r \to r_g$. It should be noted that the field grows in the Lagrangian sense, but that fluid particles can reach $r = r_g$ only asymptotically ($t \to \infty$).

In the region $r \gg r_g$ the radial field grows faster than the θ-component, i.e. the field is asymptotically stretched in the radial direction as $t \to \infty$. The magnetic field strength at which growth terminates can easily be obtained for a spherically symmetric flow. For a fixed accretion rate $\dot{M}$, the gas density follows from the continuity equation

$$\dot{M} = 4\pi r^2 \rho v.$$

Hence,

$$B^2 \approx 4\pi \rho v^2 \approx \dot{M} c r^{-2} (r_g/r)^{1/2}.$$

The accreting gas is drawn from the interstellar medium and is therefore usually completely ionized. As was first pointed out by Shvarzman (1971), the majority of the radiated energy is then in the form of magnetic bremsstrahlung (the power radiated is proportional to the square of the magnetic field). This radiation comes predominantly from the inner part of the accreting region. We should emphasize that, in the absence of a magnetic field, a spherically accreting gas radiates only a tiny amount of energy (Zeldovich and Novikov, 1971). As a matter of fact, the steady flow is far from being spherically symmetric. The magnetic field does not prevent motion along the lines of force, but constrains the transverse motions. This results in the formation of a disk-like configuration at $\theta = \pi/2$, illustrated in Figure. 16.1 (Bisnovatyi–Kogan and Ruzmaikin, 1974; 1976).

The total luminosity of the disk is determined by the gravitational energy released by the gas falling onto the black hole and by its effective radius ($\sim 3r_g$ for a nonrotating hole and $\sim 1.5r_g$ for a rotating hole) at which the escape of photons is blocked due to the redshift and aberration, i.e.

$$L \simeq GM\dot{M}/1.5r_g,$$

where $G = 6.67 \cdot 10^{-8}$ cm^3 g^{-1} s^{-2} is the gravitational constant. Interstellar matter with $\rho_\infty = 10^{-24}$ g cm^{-3}, $T_\infty = 10^4$ K provides an energy flux of the order $\dot{M}c^2 = 10^{32}(M/M_\odot)^2$ erg s^{-1} (Zeldovich and Novikov,

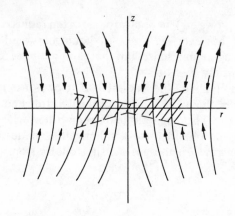

Figure 16.1   A schematic representation of the accretion of matter through a magnetic field and onto a black hole (in the center). The magnetic field is uniform at infinity. Short arrows indicate the direction of the gas flow.

1971). Thus, the total luminosity of a collapsed star with $M = M_\odot$ due to accretion is of the same order as the solar luminosity $(4 \cdot 10^{33} \text{ erg s}^{-1})$. It follows that the detection of such stellar black holes is, in practical, possible only in the immediate neighborhood of the Sun $(\leq 300 \text{ pc})$. If $M = 10^5 M_\odot$ the total luminosity is $L \approx 3 \cdot 10^{41} \text{ erg s}^{-1}$.

To estimate the spectrum of the radiation one usually assumes that every portion of the disk emits black-body radiation at its appropriate temperature (which depends on the radius r) and then integrates over the whole disk. The resultant spectrum is almost flat $(L \sim \omega^{1/3})$ in the low-frequency $(\leq 60 \text{ eV})$ range (infrared and optical), but cuts off exponentially at higher frequencies. The optical luminosity of the disk is 20 times smaller than its total luminosity (Bisnovatyi–Kogan and Ruzmaikin, 1976).

If the incoming material possesses angular momentum, then accretion also occurs via a disk surrounding the black hole. This is the case, for instance, when the black hole is embedded in a rotating gas cloud or moves through such a cloud [see the detailed review by Zeldovich and Novikov (1971) and the references therein]. The maximum energy output depends on the angular momentum of the black hole and is between 6 and 40% of Mc^2.

The observations of Popova and Vitrichenko (1978) have revealed a unique stellar object, KR Aur. The continuous spectrum of this object is practically flat in the optical range. The shape of the H_β hydrogen line suggests accretion with a mean velocity of 3200 km s^{-1}. The continuum radiation is polarized (the degree of polarization is 0.3%), implying the presence of a magnetic field. Taking into account the irregular variability of the radiation, the authors suggest that this object may be a lone black hole at a distance of about 200 pc. On the other hand, some close binary systems of dwarfs have similar characteristics so that conclusive identification is impossible.

Accretion also seems to be useful in explaining quasar phenomena. Quasars, discovered by M. Schmidt in 1963, have proved to be the most powerful (10^{46}–10^{47} erg s^{-1}) astrophysical sources of energy yet found. Moreover they are variable. One usually can discern both quasi-periodic brightness variations, with a typical time-scale of about a year, and strong (of about 60% of the maximum luminosity) non-periodic changes with a characteristic time of one week to a month. These strong variations imply that the linear dimensions of a quasar do not exceed one light-week ($\lesssim 10^{16}$–10^{17} cm). Recent observations have shown (Wyckoff et al., 1981) that quasars are surrounded by rather dim envelopes on the scale 70–100 kpc. Naturally, therefore, quasars are considered to be galactic nuclei at a certain stage of their evolution.

The first problem which faces theorists is to explain the enormous release of energy occurring in a very small volume. In 1969, Lynden–Bell showed that this problem could be solved by a model consisting of an accretion disk surrounding a massive ($\sim 10^7$–$10^9 M_\odot$) black hole. The magnetic field appears to be important for the generation of the observed non-thermal radiation (Takahara, 1979), and perhaps for the redistribution of angular momentum within the disk (Section IV).

It is tempting to invoke some of the insights gained from pulsar electrodynamics (Chapter 17) to explain the quasar phenomenon. For example, one may regard the jets associated with some quasars as products of particle acceleration in an electric field produced by the rotation of the black hole [see, for instance, Blandford and Znajek, (1977)]. We consider such an inductor in the next section. We should, however, note a drawback to this hypothesis; a pulsar has its own strong, frozen-in, "primordial" magnetic field; in contrast, a black hole cannot possess an internal magnetic field (the Ginzburg theorem, see Section I). Consequently, the necessary magnetic fields must be created by some external source. Electric currents flowing in the accretion disk may provide such a source.

III   The Rotating Black Hole

From the point of view of the Newtonian theory, it is immaterial to an orbiting gas flow whether the gravitating body rotates or not. As early as 1918, however, it was noted by Lense and Thirring that, in GRT, the rotation of the central body affects an orbiting particle. Roughly speaking, the effect of the rotation is to entrain the particles. The resulting angular velocity decreases with increasing radius as

$$\omega \approx a_c r_g / r^3,$$

where a_c is the angular momentum per unit mass.

The rotation of the gravitating center affects the majority of the accretion disk only slightly, the most pronounced influence being felt at the inner boundary of the disk. While the orbit of the innermost stable particle lies at a distance $3r_g$ for a non-rotating central mass, it moves to $1.5r_g$ in the case of the rotating black hole. A rotating gravitating body provides, however, a specific electromagnetic effect, which was pointed out by Blandford and Znajek (1977) and explored at length by Macdonald and Thorne (1981). The gravitational field of a rotating black hole is described by the Kerr metric

$$ds^2 = \Lambda c^2 \, dt^2 - R^2 (d\phi - \omega \, dt)^2 - (\rho^2 / \Delta) \, dr^2 - \rho^2 \, d\theta^2,$$

where

$$\Lambda = 1 - r_g r \rho^{-2} + \omega^2 R^2,$$

$$\rho^2 = r^2 + a^2 \cos^2 \theta,$$

$$\Delta = r^2 - r r_g^{-1} + a^2,$$

$$R^2 = (r^2 + a^2) \sin^2 \theta + a^2 r r_g \rho^{-2} \sin^4 \theta,$$

$$\omega = a r_g c r [\rho^2 (r^2 + a^2) + a^2 r_g r \sin^2 \theta]^{-1}.$$

The factor $\omega(r, \theta)$ can be thought of as the effective angular speed. This function is represented schematically in Figure 16.2 for $\theta = 0$

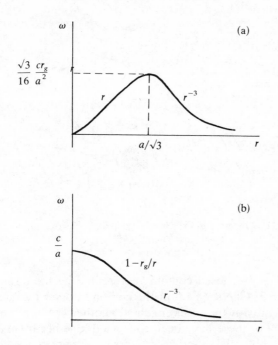

Figure 16.2 The effective angular velocity in the Kerr metric.

(near the pole) and $\theta = \pi/2$ (at the equator). It is this non-uniformity of "rotation" that creates the induction effect.

Suppose now that the black hole is within a gaseous disk which rotates with an angular velocity ω_d parallel to ω in a magnetic field created by an external source (say the currents flowing in nearby stars) or by dynamo-generated currents in the disk itself (Section V). We shall suppose, for the sake of simplicity, that a stationary axisymmetrical magnetic field is the result. The electromagnetic effect of the co-rotation of the gas is governed by the angular velocity difference, $\omega - \omega_f$, where ω_f is the angular velocity of the magnetic field. For a frozen-in field $\omega_f \approx \omega_d$, and therefore, in a non-rotating frame of reference, the electric field is

$$\mathbf{E} = -c^{-1}(\omega - \omega_f)R\,\mathbf{e}_\phi \times \mathbf{H},$$

according to the formula $\mathbf{E} = -\mathbf{v} \times \mathbf{H}/c$.

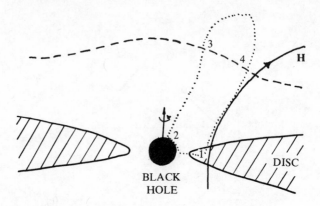

Figure 16.3   A diagram to explain the "unipolar" effect occurring near a rotating black hole (after Macdonald and Thorne, 1981).

Consider the closed contour represented by the dashed curve in Figure 16.3. One segment of the contour crosses the event horizon in the vicinity of the black hole and another lies in a remote region where $\omega \simeq 0$; these have been connected together by magnetic field lines. The potential differences across these segments are

$$V_{12} = \int_1^2 \mathbf{E} \cdot \mathbf{dl} = (\omega - \omega_f)\Phi/2\pi c,$$

$$V_{34} = \int_3^4 \mathbf{E} \cdot \mathbf{dl} \cong \omega_f \Phi/2\pi c,$$

where Φ is the magnetic flux through the surface swept out by the segment 12 rotating around the axis of the black hole. The potential differences along the lateral parts of the contour are zero, and therefore the e.m.f. is the sum of these components, i.e.

$$\text{e.m.f.} = V_{12} + V_{34} = \omega_0 \Phi/2\pi c,$$

where ω_0 is the value of ω on the event horizon, and which is non-zero only for the rotating (Kerr) black hole; it does not depend on the unknown quantity ω_f.

By noting that the boundary conditions at the horizon of the black hole are essentially the same as those holding at the end of a waveguide open to a vacuum, we can calculate by Kirchhoff's law the complex resistivity (impedance) of both the segment of the contour close to the horizon and the distant segment. These impedances depend significantly on ω_f (Blandford and Znajek, 1977; Macdonald and Thorne, 1981):

$$Z_{12} = \Phi/\pi c \rho H_\perp, \qquad Z_{34} = [\omega_f/(\omega_0 - \omega_f)]Z_{12}.$$

The energy outputs $I^2 Z$ from the segments are respectively

$$P_{12} = \frac{(\omega_0 - \omega_f)^2}{4\pi c} \rho^2 H_\perp \Phi, \qquad P_{34} = \frac{\omega_f(\omega_0 - \omega_f)}{4\pi c} \rho^2 H_\perp \Phi.$$

In the case of the extreme Kerr metric, $a = \frac{1}{2}r_g$, the power output reaches a maximum when $Z_{12} = Z_{34}$; it is

$$(P_{12})_{max} = (P_{34})_{max} \sim H^2 c r_g^2.$$

Substituting the estimate $M = 10^8 M_\odot$, we find that the observed quasar luminosity ($\sim 10^{46}$ erg s^{-1}) would require a very strong magnetic field, of the order of 10^3 G. Thus, in spite of the beauty of this picture, the trivial fall of matter onto the quasar is possibly a more important physical process.

IV Accretion Disks in Binary Stellar Systems

Accretion regions, evolving in the vicinity of lone black holes and stars, are objects of low luminosity because the material flux in the interstellar medium is rather small (typically $\dot{M} \simeq 10^{-15} M_\odot$ yr^{-1} for $M \simeq M_\odot$). In contrast, the accretion rate in close binary systems, can reach $10^{-5} M_\odot$ due to the abundant flow of material from one violently evolving star to the other. The rapid development of X-ray astronomy in the past decade has resulted in the discovery of a great variety of X-ray sources associated with close binary systems (Giacconi and Ruffini, 1978). These binaries include systems in which the visible star is of between 1 and 50 solar masses and its less massive companion is invisible but is thought to be a white dwarf, a neutron star, or a black hole, i.e. a star in the final stage of its evolution. The accretion of matter from the visible massive component onto the compact one, even at the rate $\dot{M} = 10^{-9} M_\odot \, \text{yr}^{-1}$, results in powerful X-ray emission from the vicinity of the invisible component. A typical luminosity of such an X-ray source is of the order $10^{37} \, \text{erg} \, \text{s}^{-1}$ and its effective temperature is 1–10 keV.

There are an enormous number of papers and reviews dealing with the observational and theoretical problems of compact X-ray sources [see, for instance, Giacconi and Ruffini (1978), Lightman *et al.* (1978), Pines (1980), Pringle (1981)]. Here we shall discuss briefly the significance of magnetic fields for accretion onto white dwarfs and neutron stars and at greater length, the rôle of magnetic fields in the accretion disks around black holes. The latter problem is, at present, very obscure but, at the same time, extremely important, for it raises the key question of the existence and detection of black holes in the Universe.

White dwarfs are low-mass ($\lesssim 1.4 M_\odot$) stars in which the gravitational forces are balanced by the pressure of a degenerate electron gas. A typical radius for such a star is about $5 \cdot 10^3 \, \text{km} = 5 \cdot 10^8 \, \text{cm} = 7 \cdot 10^{-3} R_\odot$. As we have noted in Chapter 2, magnetic fields of strength up to 10^8 G have been detected on many white dwarfs. The majority of the energy of the accreting gas is released near the surface of a white dwarf where a shock wave develops in which the temperature and density of the incoming material change drastically. The rôle of the strong magnetic field of the star is to channel the incoming gas towards the magnetic poles. Thus, only a minor part of the stellar surface, encounters the accreting gas (Masters *et al.*, 1977). The radiation generated depends on the accretion rate $\dot{M}$, the field strength, and the relevant fraction of the surface area.

Incoming electrons are relatively cold, so their cyclotron emission falls in the ultraviolet range of the spectrum (<10 eV). Harder UV and X-ray black-body radiation arises from the absorption and re-emission of the cyclotron radiation. Hard X-rays (with energies of up to 100 keV) can be generated in the hot region beyond the shock wave due to electron-ion bremsstrahlung, but their intensity is less than the UV intensity.

Still stronger magnetic fields (up to 10^{13} G) may be found on neutron stars (NS), where the gravitational forces are balanced for most of the volume by the pressure of degenerate neutrons. The mass of a NS cannot exceed $3 M_\odot$, its typical radius is about 10 km, i.e. $1.4 \cdot 10^{-5} R_\odot$. An isolated rotating NS manifests itself as a pulsar (see

Chapter 17). The source of its luminosity is the rotational energy of the star. If a rotating NS is found in a close binary system, where it can draw on gas from its companion, it may become an X-ray source (X-ray pulsar). The gas flowing from the visible component onto the NS possesses an angular momentum and therefore cannot fall directly onto the surface of the star: instead it spirals around it, so forming a disk. Far from the NS the gas moves in nearly Keplerian orbits. In the vicinity of the NS, however, a strong stellar magnetic field, which is usually thought of as being approximately dipolar in form, forces the matter to rotate with the angular velocity of the NS. The boundary between these limiting regions is the so-called Alfvén surface. Here the kinetic energy density of the falling gas attains the energy density of the magnetic field $(4\pi\rho v^2 \simeq H^2)$. The Alfvén surface is usually rather remote from the stellar surface, $r_A \sim 100R$, R being the radius of the NS. Near the Alfvén surface the accreting flow is reorganized and the gas is channeled by the magnetic field towards the magnetic poles of the NS (Figure 16.4). As a consequence, hot spots of small area, $(R/r_A)\pi R^2 \simeq 10^{-2}\pi R^2$, are formed near the poles and nearly all the gravitational energy of the incoming material is released in these spots. It is in their vicinity that is generated the highly anisotropic radiation, of temperature exceeding 6 keV, which is the outward manifestation of an X-ray pulsar (Gnedin and Sunyaev, 1973; Lamb et al., 1973; Basko and Sunyaev, 1974).

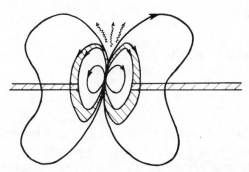

Figure 16.4 A schematic picture of accretion onto a rotating neutron star which possesses a strong magnetic field. The gas forms a disk far from the star and flows towards the poles, where the observed X-ray emission originates (indicated by wavy arrows).

The compact X-ray source Cyg X-1 (X-ray luminosity $4 \cdot 10^{37}$ erg s^{-1}) is commonly regarded as the most likely candidate for a black hole of stellar mass $M \simeq 10 M_\odot$ in a binary system [see, for instance the review by Lightman *et al.* (1978)]. This object emits radiation in a broad spectrum ranging from less than (about) 1 keV to over 100 keV, and showing remarkably strong (~25%), rapid (from 0.1 s to 1 ms) and chaotic fluctuations. The spectrum follows a power law with a spectral index of about -1.5, which suggests a non-thermal origin for the radiation.

According to observations, the visible component of the spectroscopic binary, Cygnus X-1, is the star HDE 226 868 (Henry Draper Catalogue), a supergiant of the class O 9.7, with an orbital period of $P = 5.60^d$. The mass of this star is about $25 M_\odot$. An analysis of some of the characteristics of this binary system and an investigation of the effect of the invisible component on the form (distortion from the spherical), and thus the luminosity, of the visible star established a lower limit of $6 M_\odot$ for the mass of the compact companion (Lyutyi *et al.*, 1973; Avni and Bachall, 1975). This lower limit, which exceeds the stability limit for both a white dwarf and a NS, taken in conjunction with the compactness of the X-ray source (inferred from the rapid variability of its radiation), are the bases for its nomination as a candidate for a black hole in a binary system (Lightman *et al.*, 1978).

In the following sections (V and VI) we will make an attempt to clarify the rôle of magnetic fields in disk accretion onto a black hole. As a check, we will refer all numerical estimates to Cyg X-1.

V   The Nature of the Angular Momentum Transport in the Accretion disk

As we have already noted in the case of a non-rotating black hole, the circular orbits of test particles with non-zero angular momentum are stable for radii larger than $3r_g$. For smaller radii, the particles fall radially into the black hole. The binding energy of a particle of mass m at the innermost stable orbit is $0.94 \, mc^2$. Thus, to reach such a radius a particle has to release $0.06 \, mc^2$ of its energy and rid itself of a corresponding angular momentum. If the black hole rotates, the numbers change (for an extremely fast rotation the last circular orbit has a radius $1.5r_g$ and the energy to be released is $0.42\ mc^2$) but qualitatively the situation is the same. Thus, for the matter to fall into the black hole and release energy, it is necessary that its angular momentum be transported outward. The loss of angular momentum due to both the molecular and the radiative viscosities is small. Indeed, radiation by itself is thought to be a consequence of accretion, and also the molecular magnetic viscosity is small; both the Reynolds number and the magnetic Reynolds number are large. For instance, according to Shakura and Sunyaev (1973), in the inner part of the disk where radiation dominates,

$$Re = hv/\nu_r \simeq 2 \cdot 10^2 (\dot{M}/3 \cdot 10^{-8} \dot{M}_\odot \,\mathrm{yr}^{-1})^{-2} (M/M_\odot)^{3/2} (r/3r_g)^3 \xi^{-1},$$

$$R_m = hv/\nu_m$$

$$\simeq 10^{14} (\dot{M}/3 \cdot 10^{-8} \dot{M}_\odot \,\mathrm{yr}^{-1})^{11/8} (M/M_\odot)^{19/8} (3r_g/r)^{33/16} \xi^{-1/4},$$

where ξ is the r.m.s. turbulent velocity divided by the speed of sound. It is worth noting that R_m is an increasing and Re a decreasing function of r. In the inner part of the disk the magnetic Reynolds number exceeds the hydrodynamic Reynolds number. The transport of angular momentum (i.e. accretion) in the disk is therefore possible only through turbulence and/or magnetic fields (Shakura and Sunyaev, 1973). The presence of magnetic fields in the matter flowing out from the visible component is hardly in doubt (every star has a magnetic field), but the presence of turbulence is more questionable. It is known that a medium in nonuniform rotation (Keplerian rotation in our case) in which the angular momentum density increases outward is stable against small perturbations. The well-known experiments of Taylor with two concentric rotating cylinders and the theoretical study by Zeldovich (1981) have, however, shown that for large Reynolds numbers turbulence due to nonlinear effects occurs, for any distribution of angular momentum.

The transport of angular momentum in a disk is associated predominantly with the $r\phi$-component of the stress tensor,

$$\sigma_{r\phi} = \rho(\nu_T r \, d\omega/dr - H_r H_\phi/4\pi\rho),$$

where ρ is the density of the accreting gas, ν_T the turbulent viscosity,

H_r and H_ϕ the components of the magnetic field in the disk and

$$\omega = (GM/r^3)^{1/2} = cr^{-1}(r_g/2r)^{1/2}$$

is the Keplerian angular velocity.

In the early papers on disk accretion it was usually assumed that $\sigma_{r\phi} = \xi^2 \rho v_s^2 \sim \rho v_s^2$ (v_s is the speed of sound, and $\xi = $ constant < 1), or some model was adopted for the turbulent and/or magnetic viscosity. This approach proved to be reasonable in the case of an optically thick disk with a sufficiently low luminosity (below the Eddington limit of $10^{38} M/M_\odot$ erg s^{-1}), when the spectrum of radiation depended but weakly on ξ or viscosity. The inner part of the accretion disk can, however, be optically thin—this is in fact the case for Cygnus X-I—and nonstationary phenomena and rapid variability are evident in the observed X-ray radiation. Detailed properties of the angular momentum transport are of great importance for the stability of the disk.

An attempt to calculate the stresses in a nonturbulent disk was made by Eardley and Lightman (1975). These authors envisaged that the magnetic fields which permeate the accreting gas would be intensified by the differential rotation and be dissipated due to the reconnection of the field lines. In the steady state a chaotic configuration of the magnetic field would be established, consisting of magnetic loops which reconnect with each other about once each Keplerian period. The structure and characteristics of the disk and the concomitant magnetic viscosity appeared similar to those of the early models based on less detailed forms for the viscosity. Here it is worth noting two points. First, the authors made the unwarranted assumption that the magnetic field in the disk would be two-dimensional in character. To be more precise, Eardley and Lightman first integrated the magnetic field over the "vertical" z-coordinate (orthogonal to the disk plane), and then analyzed the resulting two-dimensional picture. It would be more realistic to carry out the statistical averaging over the field loops first, and then to integrate over z those averaged quantities, which are quadratic in the field strength. Second, as we have noted above, there are forceful arguments in favor of turbulent models for accretion disks like Cygnus X-I. But the most important argument is, obviously, that under real astrophysical conditions a stable balance between the field amplification through differential rotation and dissipation by reconnections can hardly be expected. The reconnection process is usually considered in the presence of a strong field ($H^2 >$

$4\pi\rho v^2$) thus yielding the typical time $\tau_R \sim l/v_A$ for the reconnection of a loop of size l [here $v_A = H(4\pi\rho)^{-1/2}$]. The initial field which is carried by the gas flowing out of the normal companion is, however, relatively weak. As we show in the next section, the differential rotation of the disk alone cannot amplify the field to the required degree. Moreover, the field escapes from the disk relatively rapidly.

Finally, there is the interesting question of the outer boundary of the accretion disk. The angular momentum is concentrated here through its unremitting outward transfer. It is clear that in binary stellar systems the surplus angular momentum must somehow be transformed into the orbital angular momentum of the system. This may possibly occur through tidal action of the visible component on the outer parts of the disk (Papaloizou and Pringle, 1977). If a massive black hole has no companion, a gaseous cusp may be formed in the outer regions of the disk. Such a cusp would be unstable with respect to the clumping of matter and the formation of stars (Kolykhalov and Sunyaev, 1980).

VI Dynamos Operating in Accretion Disks

The nonuniform Keplerian rotation of the accretion disk is a powerful tool to amplify a magnetic field. We shall, however, now show that nonuniform rotation alone is not sufficient to maintain the field necessary to transport the angular momentum. Thus, dynamo action is a crucial ingredient of the theory of accretion disks.

The main contribution to the seed field is made by the normal star, and can originate from the outlying parts of its external dipole field as well as from the more intense fields frozen into the outflowing gas. Clearly, the seed field cannot exceed, say, 100 G. There now arises a fascinating question of field amplification concerning the gas flowing through the Lagrange point: the field lines are stretched in the longitudinal direction and pushed closer together in the transverse direction, just as in the flow considered in Chapter 5. In any case, the field drawn from the visible star is asymmetric with respect to the rotation axis of the disk. Fields that can be created in the disk itself by plasma processes are much weaker and will not be considered here.

The effect of a given nonuniform rotation on an asymmetric magnetic field has been considered at length, with a complete set of formal arguments, by Moffatt (1978) and Parker (1979). The main qualitative concepts are sufficient for our discussion.

Any open field line lying in the plane of the disk is stretched into a double spiral with oppositely directed neighboring coils. With every revolution, the typical scale of the field λ decreases as $\lambda = L/\omega t$, where L is the initial scale. The progressively decreasing scale implies a progressively increasing Ohmic dissipation. The characteristic dissipation time is

$$\tau_d \simeq \lambda^2/\nu_m = L^2/\nu_m(\omega t)^2.$$

On the other hand, a nonuniform rotation causes the azimuthal field to grow linearly with time: $H_\phi \simeq H_0\omega t$. The time at which the azimuthal field reaches its maximum,

$$(H_\phi)_{max} \simeq R_m^{1/3}H_0,$$

at which the dissipation and amplification rates become comparable, is

$$\tau_* \simeq R_m^{1/3}\omega^{-1},$$

where $R_m = L^2\omega/\nu_m$ is the magnetic Reynolds number of the initial scale. Afterwards the field decreases due to its decreasing scale. As a result, the field in the disk is negligible when $t \gg t^*$.

Applying these arguments to a real astrophysical situation, say to Cygnus X-I where $R_m \sim 10^{10}$–10^{12}, we conclude that t^* is much smaller than the typical time of radial displacements L/v_r. Moreover, the maximum field strength attainable in this way is rather low ($H_{max} \lesssim 10^4 H_0$), cf. Eardley and Lightman (1975).

The turbulent accretion disk is the ideal seat for hydromagnetic dynamo action. Mean helicity arises in the disk through the rotation and the "vertical" density gradient. The combined action of the nonuniform rotation and the helicity enables a magnetic field to be generated in precisely the same way as in the gaseous Galactic disk (see Chapters 12 and 13).

Accretion disks rotate far more rapidly than the Galactic disk (the angular velocity in Cygnus X-I is of order $10^4.s^{-1}$). The dimensionless

Rossby number $R_0 = v/\omega l$ does not exceed unity; in other words, the turbulence is strongly affected by the rotation. The dominance of the Keplerian flow suggests that the turbulence probably takes the form of an ensemble of random inertial waves. The characteristic time over which these waves interact nonlinearly replaces the characteristic eddy lifetime of the ordinary turbulence, and may be estimated to be $\tau_\omega \sim \omega^{-1} M^{-2}$, where $M = v/v_s < 1$ is the Mach number. Now we can readily estimate parameters essential for the dynamo process (Pudritz, 1981):

$$\nu_T \sim h^2/\tau_\omega \sim M^2 h^2 \omega, \qquad \alpha \sim h/\tau_\omega \sim M^2 h\omega,$$

where h is the half-thickness of the disk. It is interesting that the dynamo number

$$D = h^3 \alpha\omega/\nu_T^2 \sim M^{-2}$$

is determined exclusively by the Mach number and does not depend on the radial distance.

We shall not dwell on the large-scale distribution of the field, but turn our attention right away to the more important question of the small-scale magnetic fields $b \gg \langle H \rangle \equiv B$, which play a dominant rôle in the angular momentum transport.

The growth of the large scale field will cease when the Maxwell stress balances the Coriolis force:

$$B_\phi B_r \sim 4\pi\rho v\omega h.$$

The results of Section 8.IV suggest that the stresses produced in the small-scale fields will then be of order

$$\langle b_r b_\phi \rangle \sim R_m^\delta B_\phi B_r,$$

where the index δ is determined by the spectral index, q, of the turbulent energy distribution:

$$\delta = (5 - 3q)/(3 - q).$$

If the turbulence is essentially an ensemble of interacting Alfvén waves, then $q = \frac{3}{2}$ and $\delta = \frac{1}{3}$. At the present time it is unclear how different the spectral index will be in rotationally dominated turbulence, but we may expect $\delta < 1$ (since $\delta = 1$ in degenerate two-dimensional turbulence). It is clear, however, that the stresses associated with the small scale field exceed those of the large scale field by a very large factor. For $\delta = \frac{1}{3}$ and $R_m \sim 10^{12}$ this factor is of order 10^4. On the other hand, the Maxwell stresses are obviously of the same order of magnitude as the hydrodynamical Reynolds stresses $\rho \langle v_i v_j \rangle$. Hence, both contributions to the angular momentum transport in accretion disks should be calculated simultaneously.

The estimates we have just made can serve only as a hint for future detailed investigations of the angular momentum transport in the disk.

Another important problem is to calculate the time-spectrum of the fluctuations of the radiation. It is anticipated that, for a rotation-dominated turbulence, the low frequency cut-off in the spectrum will be of the same order as the angular velocity ω. The characteristic periods of the fluctuations in the X-ray radiation determined by the angular velocity were estimated and compared with the observational data by Sunyaev (1972). Variability would also be a consequence of thermal instability (Shakura and Sunyaev, 1975).

Proceeding from an analogy with stellar coronae, Galeev et al. (1979) suggested that the accretion disk may be surrounded by a hot corona which is created by loops of small-scale magnetic field "rising" from the disk (see also Section 11.II). The bremsstrahlung radiation, which originates from the transition zone between the magnetic loop and the disk (i.e. at the feet of the loops), could explain the observed hard X-ray emission from Cygnus X-I. Therefore, the buoyant rise of loops of the field generated in the disk is worthy of special attention.

CHAPTER 17

Strong Magnetic Fields

In some sense, the subject matter of this chapter is the most attractive of the book. It concerns the extremely strong magnetic fields that occur in Nature. These fields are, however, associated with unusual objects in poorly observed conditions, so the exposition is necessarily dominated by theoretical speculations.

We shall begin with a discussion of some of the uncommon features acquired by atoms and molecules in strong magnetic fields. Then we shall touch briefly on the influence of these fields on charged particles, quanta and a vacuum. These features, which until recently seemed highly esoteric, are now widely known to astrophysicists, through the need to portray the pulsar (Section III). The last section (Section IV) is deliberately provocative; it is devoted to the possible existence of magnetic monopoles.

I Atoms and Molecules in Strong Magnetic Fields

In the absence of a magnetic field, the energy and spatial structure of an atom are determined in the main by the electrostatic interaction between the electrons and the nucleus, as well as between the electrons and each other. The electron cloud has a spherically-symmetric form in the ground state. In the non-relativistic approximation, the energy levels of hydrogen and hydrogen-like atoms are determined by Bohr's well-known law

$$\mathcal{E}_n = -\tfrac{1}{2}Ze^2 a_B^{-1} n^{-2}, \qquad n = 1, 2, \ldots,$$

where $a_B = \hbar^2/mZe^2$ is the Bohr radius, m and e are the mass and charge of the electron, $2\pi\hbar$ is Planck's constant and Z is the atomic number (Ze is the nuclear charge). The Bohr radius of the hydrogen atom ($Z = 1$) is $0.5 \cdot 10^{-8}$ cm, and the energy of its ground state is 13.6 eV.

The behavior of a free electron in a (uniform) magnetic field has been well studied (see, for instance, Landau and Lifshitz, 1958). The electron moves freely along the field, the energy levels of its transverse motion (the Landau levels) being

$$\mathscr{E}_n = 2\pi\hbar\nu_H(n + \tfrac{1}{2} + \sigma), \qquad n = 0, 1, 2, \ldots,$$

where $2\pi\nu_H = eH/mc$ and $\sigma = \pm\tfrac{1}{2}$ is the projection of the electron spin on the magnetic field. The spatial distribution of the electrons over these states has a cylindrical form, the radius in the ground state being of order

$$r_0 = (\hbar/2\pi\nu_H m)^{1/2} \simeq 2.6 \cdot 10^{-4} H^{-1/2} \text{ cm}.$$

Let us now place an atom in a magnetic field. It is clear that for a sufficiently strong field the radius r_0 becomes smaller than the Bohr radius. In other words, the electron energy Ze^2/a_B becomes less than $\hbar\nu_H$. In this way, the notion of a critical magnetic field for the atom arises, i.e.

$$H_{ca} = m^2 e^3 c \hbar^{-3} Z^2 \simeq 2.3 \cdot 10^9 Z^2 \text{ G}. \qquad (1)$$

Essentially a strong magnetic field ($H > H_{ca}$) changes the structure and energy of the atom (Kadomsev, 1970; Cohen et al., 1970; Kadomsev and Kudryavtsev, 1971b). The effect of the magnetic field is then dominant, and the electrostatic Coulomb field can be considered as a perturbation. An electron moving on a line through the nucleus parallel to the field would have an infinite binding energy. The Larmor radius of the electron orbit is one factor which diminishes the Coulomb interaction. The atom acquires the shape of a needle, of thickness r_0 and length $a_B/\ln(a_B/r_0)$. The energy of the ground state for a hydrogen atom is

$$\mathscr{E} = -\tfrac{1}{2}e^2 a_B^{-1} \{\ln(H/H_{ca})\}^2,$$

and for a complex atom it is

$$\mathscr{E} = -(e^2 N/32 a_B)(4Z - N_e + 1)^2 \{\ln[H/(H_{ca}Z^3)]\}^2,$$

where N_e is the number of electrons in the atom. It may be noted that the field (1) hardly exceeds that at which classical electrodynamics becomes invalid, viz. $H_{cp} \sim 4 \cdot 10^{13}$ G [see Eq. (2) below].

The needle-like atoms have large electric quadrupole moments so they should be able to combine to form molecules. In this context, recall that, although the atoms interact electrostatically, they are electrically neutral and the mean value of their electric dipole moment vanishes due to parity invariance (i.e. invariance under the transformation $\mathbf{r} \to -\mathbf{r}$). The main contribution to the attraction of the needle-like atoms thus comes from the quadrupole–quadrupole interaction.

It is anticipated (Kadomsev and Kudryavtsev, 1971a) that molecules of an unusual form will be created. The common electron cloud forms a charged needle, the nuclei being spread along its axis. In matter composed of identical atoms, diatomic molecules should apparently be formed. If the matter is made up of heavy W and light $\mathscr{L}$ atoms, the molecules formed will be of the type $W_2 \mathscr{L}_p$, where $p \sim Z_W/Z_{\mathscr{L}}$. Such a molecule has a large binding energy $[\sim Z_W^2 Z_{\mathscr{L}} \ln (H/H_{ca})]$ which can (in apparent conflict with the common ideas about molecules) exceed the ionization energy. Hence the heated matter can turn into a plasma with molecular ions!

In a magnetic field exceeding the critical strength (1) and at sufficiently low temperatures ($\leq 10^6$ K, for $H \sim 10^{12}$ G), a crystallization of the atomic matter is possible. This takes the form of densely packed needles oriented along the field and weakly bound in the transverse direction. The crystallization strongly affects the density of the matter. For example, the density of iron at 10^{12} G is $2.7 \cdot 10^3$ g cm^{-3} (Baym and Pethick, 1979), while it is 7.8 g cm^{-3} when no magnetic field is present. It is believed that matter in a strong magnetic field behaves like a quasi-one-dimensional metal which conducts well in the direction of the field and poorly in the transverse direction. This concept is applied to the surface physics of sufficiently cooled neutron stars observed as pulsars (Section III). The surface of such a star, in contrast to the surfaces of all other stars, is not gaseous but solid!

II Particles, Photons and Vacuum in Strong Magnetic Fields

In this section we consider the influence of a super strong magnetic field on charged particles, say electrons, on quanta of an electromagnetic field and on a physical vacuum. The exposition will be short and fragmentary, but this simple approach seems to be sufficiently good for the present state of astrophysics. [For a more detailed review, see for example Pavlov and Gnedin (1983).] The physics of pulsars (Section III) already requires, however, a more sophisticated treatment of the effects presented below.

The natural quantum mechanical measure of magnetic field strength is

$$H_{cp} = m^2 c^3 / e\hbar \approx 4.4 \cdot 10^{13} \text{ G}, \tag{2}$$

at which the radius of the electron distribution in the magnetic field, i.e. the cyclotron radius $(\hbar/2\pi\nu_H m)^{1/2}$, is equal to the classical radius of the electron, e^2/mc^2. In the rest of this section, when we talk about a strong magnetic field we will mean a field comparable with (2), which will be considered as static.

In general, magnetically induced conversion processes, such as synchrotron radiation, pair production and photon splitting, depend not only on the field strengths but also on the energy of the particle. The transition probabilities are principally determined by the parameter (Erber, 1966)

$$\chi = (E/mc^2)(H/H_{cp}).$$

The probabilities are increasing functions of χ in the range $\chi \lesssim 1$, and hence significant conversion rates require large energies and very strong magnetic fields. It is important that χ can exceed unity even when $H < H_{cp}$.

The radiation emitted by an electron, or any other charged particle which moves non-parallel to the magnetic field, is known as "magnetic bremsstrahlung" or "synchrotron radiation". In astrophysics this radiation is usually considered in the classical approximation (Ginzburg and Syrovatsky, 1964). In magnetic fields comparable with H_{cp}, quantum aspects are more relevant [see the review by Erber (1966)]. It is convenient to idealize the situation by assuming that the process takes place in a uniform magnetic field. The electron states in this field are classified by the energy ($\mathscr{E}$) and the spin projection (σ) parallel to the magnetic field,

$$\mathscr{E}_n = \pm mc^2[1 + 2(n + \sigma + \tfrac{1}{2})H/H_{\mathrm{cp}}]^{1/2}, \qquad n = 0, 1, 2, \ldots; \qquad \sigma = \pm\tfrac{1}{2}.$$

Motion in a direction parallel to the magnetic field may be excluded through a transformation to the "center of drift" system. There is, of course, an uncertainty in the position of the center in the plane perpendicular to the magnetic field. The spectral distribution of the radiation emitted per unit distance has a peak at $3E\chi/(2+3\chi)$ and, in the range $\chi \gtrsim 1$ of interest, may be written as

$$I(\mathscr{E}, H, \nu) \simeq \frac{1}{2} \frac{e^2}{\hbar c} \frac{mc^2}{\lambda_c} \chi^{2/3} \left(\frac{2\pi\hbar\nu}{\mathscr{E}}\right)^{1/3} \left(1 - \tfrac{4}{3}\pi \frac{\hbar\nu}{\mathscr{E}}\right)$$

(Klepikov, 1954; Erber, 1966), where $\lambda_c = \hbar/mc$ is the Compton wavelength.

It is useful to compare the intensity of the magnetic bremsstrahlung with that in matter. The spectral intensity of the bremsstrahlung, per unit distance travelled by a high-energy electron (in matter characterized by density ρ, atomic weight A, and atomic number Z), is

$$I_m(\nu, z) \simeq (e^2/\hbar c)^3 (N\lambda_c^3/\lambda_c)g(Z),$$

where N is Avogadro's number, $N\lambda_c^3$ is the Compton volume and the function $g(Z)$ describes Coulomb corrections. The ratio I/I_m contains the large factor $(N\lambda_c^3)^{-1} \simeq 3 \cdot 10^9$, which highlights the significance of synchrotron losses as compared with bremsstrahlung from the matter in strong magnetic fields (Erber, 1966).

It is important to understand that the homogeneous magnetic field does not admit the full group of translations. It is clear, formally, that the field H_z corresponds to a vector potential depending on x and/or y. Therefore the state is invariant under a translation and dilation (Lorentz transformation) along the z axis. By Noether's theorem, the z-component of the momentum is conserved, but not the x and y components. This makes possible new processes which are strictly forbidden in the absence of a magnetic field. The first of these is the conversion of one real photon (with definite k and $\omega = |k|/c$) into an electron–positron pair, e^-e^+.

To estimate the energy threshold, let us consider a photon of energy $2\pi\hbar\nu$ moving at a small angle ψ to the direction of the magnetic field. The photon energy is minimal and is equal to $2\pi\hbar\nu\psi$ in the frame of reference where the photon moves across the magnetic field. In this frame of reference pair creation is possible if the photon energy exceeds $2mc^2$, therefore the process is kinematically possible when

$$2\pi\hbar\nu\psi \geq 2mc^2. \tag{3}$$

It is appropriate to express the probability that the photon is converted into an electron–positron pair in terms of a photon attenuation coefficient $k(\chi)$, which determines the number of actual pairs n_p created in a photon path length d in a magnetic field perpendicular to the direction of light propagation:

$$n_p = n_\gamma[1 - \exp(-k_p d)].$$

A direct calculation of the amplitude of the process

$$\gamma + H \rightarrow e^+ + e^- + H$$

results in a simple expression for the attenuation coefficient in the case of practical interest, in which $H \leq H_{cp}$ and the photons are ultrarelativistic:

$$k(\chi) = \frac{e^2}{2\hbar c\lambda_c} \frac{H_\perp}{H_{cp}} T(\chi), \qquad H_\perp = H\sin\psi \tag{4}$$

(Klepikov, 1954; Erber, 1966), where $T(\chi)$ is a function with a maximum at $\chi_{max} \simeq 12$,

$$T(\chi) = \begin{cases} 0.46\exp(-8/3\chi), & \chi \ll 1 \\ 0.73\chi^{-1/3}, & \chi \gg 1. \end{cases} \tag{5}$$

If H is considered fixed, it is easy to verify that $k(\chi)$ has a maximum at the photon energy

$$2\pi\hbar\nu_{max} \simeq 12mc^2 H_{cp}/H_\perp.$$

On the other hand, at fixed ν the attenuation coefficient is an increasing function of the magnetic field strength. This means that strong magnetic fields are efficient pair converters for photons of high energy. The mean free path of the photon for this process depends strongly on $\mathscr{E}_\gamma H_\perp$. It is interesting that, in the case where $\hbar\nu \gg mc^2$

and $H = 10^6$-10^{12} G relevant to pulsar magnetospheres, the attenuation coefficient is given by (Ozernoy and Usov, 1977; see also Section III)

$$k^{-1} \simeq 10^6 H^{-1} \exp{(10^{19.7}/2\pi\hbar\nu H_\perp)} \text{ cm.} \qquad (6)$$

In a strong magnetic field a process of photon splitting exists that has no threshold. This process may be thought of as $\gamma \to (e^+ e^-) \to 2\gamma$, i.e. a photon going via a virtual intermediate pair to two photons. The magnetic field is responsible for binding the pairs in the intermediate states. Suppose we consider a photon propagating through a uniform magnetic field in a direction normal to the field lines. Then, according to Adler (1971), the attenuation coefficient for this photon splitting is given by

$$k_\gamma \simeq 0.1 (2\pi\hbar\nu/mc^2)^5 (H_\perp/H_{cp})^6 \text{ cm}^{-1}.$$

Comparing this with (4) we see that this process is more effective at low energies than the threshold process. It is, however, insignificant when applied to pulsars because it is proportional to $(H_\perp/H_{cp})^6 \ll 1$.

For a discussion of some other interesting processes that can occur in a strong magnetic field, for instance magnetic Cherenkov radiation or trident cascades $e + H \to e + \gamma + H \to (e^+ e^-) + H$, see the review by Erber (1966).

Quantum electrodynamics tells us that, in a strong external static electric field, the creation of real electron–positron pairs is possible. Using, for the purposes of demonstration, a semi-classical jargon, we may describe this process as follows ["Magic without Magic", Zeldovich (1972)]. A pair is born at the distinct points r_1 and r_2. The difference between the electrostatic potential energies at these points must exceed the rest energy of the pair, i.e.

$$e \mathbf{E} \cdot (\mathbf{r}_1 - \mathbf{r}_2) \geq 2mc^2.$$

The surplus energy appears as kinetic energy in the particles created. In order to be in different positions, the electron and positron must penetrate each other's potential barriers. Hence the probability of creating a pair is evidently

$$w \sim \exp{(-\beta \Delta r/\lambda_c)},$$

where $\Delta r = 2mc^2/eE$ is the barrier width, and β is a numerical factor of order unity. It follows immediately that for this probability to be high, the electric field strength must exceed $m^2c^3/e\hbar$, i.e. the critical field (2).

In the absence of magnetic monopoles (see, however, Section IV), a magnetic field is not equivalent to an electric field. Quantum electrodynamics ensures the stability of a vacuum with respect to pair production in a static magnetic field.

III   Pulsars

The final products of stellar evolution are the trio of compact objects: white dwarfs, neutron stars and black holes. The extreme compactness of these stars and the relativistic effects involved are reasons to expect that their magnetic properties will be peculiar.

First of all let us estimate the strength of the magnetic field which can be attained through a simple contraction of an ordinary star, say of radius $R_0 = 10^6$ km, to the compact state which characterizes white dwarfs and neutron stars. We shall assume that the initial magnetic field is H_0 and that the contraction proceeds with magnetic flux conserved, that is

$$H = H_0(R_0/R)^2. \tag{7}$$

The magnetic energy density varies as R^{-4} so that the ratio of the magnetic and gravitational energies remains unchanged.

The minimal radius of a white dwarf is $R \sim 10^3$ km [see, for instance, Zeldovich and Novikov (1975)], therefore the magnetic field may be intensified by 10^6. For $H_0 = 10^2$ G, we have $H \simeq 10^8$ G so, apparently, this is the maximum field strength for a white dwarf. In the case of a neutron star ($R \sim 10$ km) the amplification factor is larger (10^{10}) so that the same initial field is compressed to 10^{12} G. It is just such strong fields that are today associated with pulsars, the cosmic objects which emit short recurring pulses of radiation (Hewish et al., 1968; Ruderman, 1974; Ginzburg and Zheleznyakov, 1975).

A pulsar is identified with a rotating neutron star (Gold, 1968) and it is commonly accepted that its rotation is a reservoir from which all its active phenomena draw their energy. The magnetic field is the "transmission belt" between the rotation and the accelerated particles emitting the observed radiation.

Based on this concept, it is easy to estimate the strength of a pulsar magnetic field. Observations of the frequency of incoming pulses determine the angular velocity ω of the pulsar, and the rate at which it is slowing down $\dot{\omega}/\omega$. The observed radiation is evidently generated in a plasma on or near the pulsar. Due to magnetic field stresses, the plasma rotates approximately with the same angular velocity as the pulsar up to distances from the axis of rotation of order c/ω, this being the radius of the so-called "light cylinder", c being the speed of light. The rate at which the star loses angular momentum is determined by the magnetic stresses on the light cylinder:

$$I\dot{\omega} \sim H_{lc}^2 (c/\omega)^3,$$

where I is the moment of inertia of the neutron star (g cm^2), and H_{lc} is the magnetic field strength at the light cylinder. Extrapolating this field to the surface of the pulsar assuming (say) a dipolar law, we obtain the desired estimate:

$$H \sim [(Ic^3/R^6\omega^2)(\dot{\omega}/\omega)]^{1/2}.$$

This is, apparently, a lower limit on the magnetic field strength on the surface of the pulsar, because the field may decrease more rapidly than for a dipole.

The moment of inertia of neutron stars is known to be in the range 10^{44}–10^{45} g cm^2, the observed periods, $2\pi/\omega$, being between 0.03 and 4.3 s. Taking as an example the pulsar in the Crab Nebula, the rate of slowing down is $\dot{\omega}/\omega \sim 10^{-3}$ (year)$^{-1} \sim 3 \cdot 10^{-11}$ s^{-1}. This gives a crude estimate for the magnetic field strength on the surface of the pulsar, as $H \sim 10^{12}$ G. The agreement of this estimate with the previous one, obtained by speculating about the contraction of the star, is a central argument in favor of an evolutionary origin for the magnetic field of the pulsar. The scale height kTR^2/GmM of the atmosphere of a non-magnetic neutron star with $M \sim M_\odot$ and $T \lesssim 10^6$ K is negligible ($\lesssim 1$ cm), so the star could not, in practice, have an atmosphere. The magnetic field and rotation change this situation decisively (Goldreich and Julian, 1969). The exact solution of the problem of the field outside a conducting, rotating sphere, in a vacuum, which possesses a dipole-type field, reveals that an electric quadrupole field is induced outside the star. The electric field in the direction of the magnetic field is of order

$$E_{\parallel} \simeq (\omega R/c)H.$$

For the Crab pulsar ($\omega \sim 200 \text{ s}^{-1}$, $R \sim 10 \text{ km}$, $H \sim 10^{12} \text{ G}$), $E_{\parallel}$ is about 10^{10} v/cm at the surface of the pulsar. The corresponding electric force significantly exceeds the gravitational force acting on a charged particle (electron or proton) on the surface. If the surface of the star were gaseous it would be an easy matter to estimate the density of electric charges in the resulting magnetosphere. Indeed, in the frame of reference rotating with the star, the electric field created by the outflowing matter is greatly reduced compared to the vacuum field, $E_{\parallel}$. So, for an observer, rotating with an angular velocity ω relative to the star, there is an electric field $-H\omega r/c$ and a corresponding charge density of

$$n_e = (4\pi e)^{-1}\boldsymbol{\nabla} \cdot \mathbf{E} = \omega H/2\pi ce$$
$$\simeq 10^{10}(H/10^{12} \text{ G})(\omega/1 \text{ s}^{-1}) \text{ cm}^{-3}.$$

The surface of the pulsar, in contrast to the surfaces of other stars, is however not gaseous (Ruderman, 1974). As we noted in Section I, when the magnetic field exceeds $10^9 Z$ Gauss (Z is the atomic number), the atomic electron shells are reshaped. The resulting needle-like atoms have large electric quadrupole moments and, hence, effectively attract each other. If the surface temperature is small compared with the binding energy of the atoms, then the surface is similar to a quasi-one-dimensional metal, conducting well in the direction of the field and acting as an insulator in transverse directions. According to both the observations of soft X-rays and ultraviolet radiation and to theoretical estimates, the surfaces of all pulsars so far observed are solid (Usov, 1981).

Under the action of the component, $E_{\parallel}$, of electric field parallel to the magnetic field, the cold emission of electrons from the solid surface is possible (Ruderman, 1974). This field is, however, too weak (for $H \sim 10^{12}$ G) to extract ions (Ginzburg and Usov, 1972).

The primary electrons pulled out from the surface are accelerated by $E_{\parallel}$ to ultra-relativistic energies ($\sim 10^8 mc^2$). This acceleration is limited by a "curvature radiation"—the radiation of particles forced to move along the curved magnetic field lines. To estimate the energy lost in curvature radiation one can use the usual formulae of synchrotron radiation theory with a replacement of the orbital radius of the particle, $\mathscr{E}/eH_{\perp}$, by the curvature radius R_c of a magnetic field line, which is determined only by the geometry of the magnetic field. As a result, in the classical limit $2\pi\hbar\nu \ll \mathscr{E}$, we have

$$d\mathscr{E}/dt = -(2e^2c/3R_c^2)\Gamma^4,$$

$$I(\nu, \Gamma) = (3^{1/2}e^2/2\pi R_c)\Gamma\nu_c^{-1}\nu \int_{\nu/\nu_c}^{\infty} K_{5/3}(x)\, dx \sim \nu^{1/3}\, e^{-\nu/\nu_c},$$

$$\nu_c = (3c/4\pi R_c)\Gamma^3,$$

where $I(\nu, \Gamma)$ and ν_c are the spectral intensity and characteristic frequency of the curvature radiation and $\Gamma = (1 - v^2/c^2)^{-1/2}$ is the Lorentz-factor of the particles; for the high energy particles, this can actually reach values between 10^6 and 10^7. Ultra-relativistic particles in pulsar magnetospheres, with $R_c \sim 10^7$–10^8 cm and such Γ, emit γ-quanta with energies of up to 10^{13} eV. The electrons near the surface of the pulsar move almost precisely along the magnetic field lines (the so-called "pitch-angle" is close to zero). This is because they are at the lowest Landau level due to the very short time-scale of the synchrotron radiative losses ($\sim 10^{-14}$ s). The energy barrier ($2\pi\hbar\nu_c \sim 10$ keV at $H = 10^{12}$ G) greatly impedes the transition to the next Landau level (see Section I).

The γ-quanta are also emitted along the magnetic field lines. The angle between their direction of propagation and the magnetic field subsequently increases, and ultimately the inequality

$$2\pi\hbar\nu\psi \geq 2mc^2$$

is satisfied. The absorption of the quantum, followed by the creation of an electron–positron pair, can then occur:

$$\gamma + H \to e^+ e^- + H.$$

In order for the γ-quantum to be absorbed into the pulsar magnetosphere, its mean free path for this process should be less than the radius of the light cylinder, c/ω. Using the formulae (4)–(6) of Section II we obtain (Tademaru, 1973; Ozernoi and Usov, 1977)

$$H_{\perp}\mathscr{E}_{\gamma} \gtrsim 10^{18}\, \text{G} \cdot \text{eV}.$$

It follows that the γ-quanta of curvature radiation, which is generated

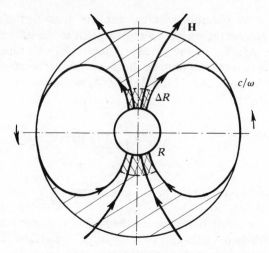

Figure 17.1 Sketch of the pulsar magnetosphere. The rotation axis is perpendicular to the plane of the figure. The near zone $r < c/\omega$ (shaded) is filled with plasma flowing out from the pulsar. The region $\mathbf{E} \cdot \mathbf{H} \neq 0$ with the dimensions $\Delta R(\omega R/c)^{1/2}R$ is cross-hatched. In the remaining regions $\mathbf{E} \cdot \mathbf{H} = 0$.

in the vicinity of the surface of the pulsar and, which has energy exceeding several GeV, is absorbed with the production of electron–positron pairs (Sturrock, 1971). It must be emphasized that this pair production is possible only in the presence of a strong magnetic field.

The structure of the pulsar magnetosphere is illustrated qualitatively in Figure 17.1, the regions of plasma outflow being shaded. The greatest concentration of pair production occurs within a distance ΔR (as measured along the field lines) of the surface, and beyond this the electric field $E_{\parallel}$ is completely screened. The only part of the plasma in which $E_{\parallel} \neq 0$ is quite small, its lateral extent being only of order $(\omega R/c)^{1/2}R$ (Tademaru, 1973). Estimates of ΔR are model-dependent, but $\Delta R \leq R$ (in the Sturrock model $\Delta R = R$). A particle inside this region is accelerated by the electric field and radiates energy. Outside the region, curvature radiation decelerates the particle. The energy spectrum of the accelerating beam of primary particles in the region where $E_{\parallel} \neq 0$ is almost mono-energetic. The spectrum of the plasma streams it produces in the magnetosphere is determined by the curvature radiation spectrum (Tademaru, 1973; Usov, 1981).

Thus, the plasma in pulsar magnetospheres is composed mainly of electrons and positrons. The great majority of these form a low-energy component (Lorentz-factor $\Gamma \approx 10^2-10^3$) moving away from the pulsar surface. This plasma is pierced by a high-energy beam (or several beams) of particles with $\Gamma \sim 10^6-10^8$. For instance, in the case of the Crab pulsar NP 0531 − 21, the densities expected for the secondary low-energy plasma and the high-energy beam near the surface may be as large as 10^{20} cm^{-3} and 10^{17} cm^{-3} respectively (Usov, 1981). The particles created rapidly lose their momentum perpendicular to the magnetic field due to synchrotron radiation, so their distribution is close to being one-dimensional.

By studying in detail the various instabilities to which a one-dimensional relativistic electron–positron plasma pierced by beams of ultra-relativistic particles is prone, Lominadze, Mikhailovsky, Usov and their colleagues have reached an understanding of the mechanisms of high-energy (from optical to γ wavelengths) pulsar radiation (Lominadze et al., 1980; Usov, 1981). For instance, as was early predicted by Shklovsky (1970), the high energy radiation from the Crab pulsar is basically the synchrotron produced by relativistic particles near the light cylinder. The requisite pitch-angles arise through plasma instabilities in this region. The marked contributions to some spectral bands can be produced by other radiation mechanisms, for instance by the Compton scattering of Langmuir waves. The curvature radiation, which plays an essential rôle in the creation of the electron–positron plasma, apparently contributes in the main to the γ-radiation, which in the case of the Vela pulsar PSR 0833 − 45 is generated near the magnetic poles (Ozernoy and Usov, 1977).

Plasma instabilities near the light cylinder excitate electromagnetic oscillations, at approximately the cyclotron (radio) frequency. This may explain pulsar radio-emission (Usov, 1981).

IV Monopoles

In the last section of our book we shall discuss the problem of magnetic charges, the so-called "monopoles". Up to this point an absence of monopoles was implied.

In contrast to electric charges, there are no magnetic charges in classical Maxwell theory, which treats the electric and magnetic fields asymmetrically:

$$\nabla \cdot \mathbf{E} = 4\pi\rho_e, \qquad \nabla \cdot \mathbf{H} = 0.$$

Here ρ_e is the electric charge density. In the absence of a magnetic field, the electric field is rapidly neutralized in a plasma by the redistribution of charges; it vanishes in the characteristic time σ^{-1}, where σ is the plasma conductivity. So a plasma in astrophysics is electrically neutral. There is no such serious limitation with respect to magnetic fields. In the absence of monopoles, these are supported by electric currents, and decay over the characteristic time $4\pi\sigma L^2/c^2$, which in astrophysics is very large due to the huge scale, L, involved. It is for this reason that a large-scale magnetic field, if once created, can exist in "non-monopole" astrophysics.

History records that physicists have changed their minds a few times on the question of magnetic monopoles. Faith in their existence has ranged from a reticent optimism to a dogmatic pessimism. For Coulomb, to whom we are indebted for the first results in electricity, the magnetic charge was apparently as natural as the electric charge. In fact, his experiments demonstrated that the poles of two magnets interact inversely as the square of the distance between them. The difficulty arose because of the lack of isolated magnetic charges. Electrodynamics, which started with Ampère's hypothesis explaining magnetism as due to an electrical current, was completed by Maxwell's theory, which made no reference to any magnetic charges. The question of the magnetic charge appeared again in quantum theory. In Dirac's paper of 1931 it was still looked upon as a graceful toy. Dirac symmetrized Maxwell's equations in the electric and magnetic fields. He kept the concept of a vector potential, however, in spite of the fact that such a potential invokes the now rejected condition, $\nabla \cdot \mathbf{H} = 0$. As a result the vector potential became non-unique, and singular curves appeared. From the condition of phase invariance of the wave function of the electric charge (e) for a closed circuit surrounding such a curve, a quantization of charge followed:

$$g_n = (\hbar c/e^2)\tfrac{1}{2}ne, \qquad n = 0, \pm 1, \pm 2, \ldots.$$

The minimum, nonvanishing magnetic charge is $68.5e$, but all attempts to detect such magnetic charges have been unsuccessful.

Unexpectedly, the problem of monopoles is of crucial importance in modern quantum field theory, which unites the theory of electromagnetism with those governing strong and weak interactions, Polyakov (1974) and t'Hooft (1974) have demonstrated that, in a non-abelian gauge theory containing an electromagnetic interaction within a large compact covering group, there are monopoles of mass 5–10 Tev and magnetic charge $g = \hbar c/e$.

In accordance with these theories, monopoles will occur in cosmology during the early stages of the big-bang (hot) Universe, and estimates give them a high concentration at that time (Zeldovich and Khlopov, 1978). On the other hand, a great profusion of monopoles would contradict the existence of the present-day large-scale magnetic fields (Parker, 1979). The predictions of the unified theory (if it is correct) thus appear to be incompatible with astrophysical observation. With this puzzling but challenging paradox, we conclude this, the last chapter of our book.

References

Acuna, M. H., Ness, N. F. and Connerney, J. E., "The magnetic field of Saturn: Further studies of the Pioneer-11 observations," *J. Geophys. Res.* **85**, 5675–5678 (1980).

Adler, S. L., "Photon splitting and photon dispersion in a strong magnetic field," *Ann. Phys.* **67**, 599–647 (1971).

Aharonov, Y. and Bohm, D., "Significance of electromagnetic potentials in the quantum theory," *Phys. Rev.* **115**, 485–491 (1959).

Aleksandrov, E. V., "Optical manifestations of the interference of non degenerate atomic states," *Usp. Fiz. (SSSR)* **107**, 595–622 (1972), [*Soviet Phys. Usp.* **15**, 436–451 (1973)].

Alfvén, H., *On the origin of the solar system*, Univ. Press, Oxford (1954).

Alfvén, H., "Electric currents in cosmic plasmas," *Rev. Geophys. Space Phys.* **15**, 271–287 (1977).

Allan, D. W., "On the behaviour of systems of coupled dynamos," *Proc. Camb. Phil. Soc.* **58**, 671–693 (1962).

Altschuler, M. D., Trotter, D. E. and Newkirk, G. Jr., "The large-scale solar magnetic field," *Sol. Phys.* **39**, 3–17 (1974).

Andronov, A. A., "Les cycles limites de Poincaré et la théorie des oscillations auto-entretenues," *Comp. R. Acad. Sci. Paris*, **189**, 559–561 (1929).

Andronov, A. A. and Witt, A. A., "Zur Theorie des Mitnehmens von van der Pol," *Arch. Electrotech.* **24**, 99–110 (1930).

Angel, J. R. P., "Magnetic white dwarfs," *Annu. Rev. Astron. Astrophys.* **16**, 487–520 (1978).

Anufriev, A. P. and Fishman, V. M., "Magnetic field structure in two-dimensional motion of a highly conducting fluid, "*Geomagn. & Aeron.* **22**, 292–296 (1982).

Appenzeller, J., "Interaction between the Barnard loop nebula and the interstellar magnetic field," *Astron. & Astrophys.* **36**, 99–105 (1974).

Arnold, V. I., "Sur la géométrie différentielle des groupes de Lie de dimension infinie et ses applications à l'hydrodynamique des fluides parfaits," *Ann. Inst. Fourier*, **16**, 319–361 (1966).

Arnold, V. I., "Notes on the three-dimensional flow pattern of a perfect fluid in the presence of a small perturbation of the initial velocity field," *Prikl. Mat. & Mekh. (SSSR)* **36**, 255–262 (1972), [*Appl. Math. & Mech.* **36**, 236–242 (1972)].

Arnold, V. I., *Mathematical methods of classical mechanics*, Springer-Verlag, New York–Heidelberg–Berlin (1980).

Arnold, V. I., Zeldovich, Ya.B., Ruzmaikin, A. A. and Sokoloff, D. D., "Magnetic field in a stationary flow with stretching in a Riemannian space," *JETP.* **81**, 2052–2058 (1981), [*Sov. Phys. JETP* **56**, 1083–1086 (1981)].

Avni, Y. and Bachall, J., "Ellipsoidal light variations and masses of X-ray binaries," *Astrophys. J.* **197**, 675–688 (1975).

Babcock, H. W., "Zeeman effect in stellar spectra," *Astrophys. J.* **105**, 105–191 (1947).

Babcock, H. W., "The topology of the Sun's magnetic field and the 22-year cycle," *Astrophys. J.* **133**, 572–587 (1961).

Backus, G. E., "A class of self-sustaining dissipative spherical dynamos," *Ann. Phys.* **4**, 372–447 (1958).

Backus, G. E. and Chandrasekhar, S., "On Cowling's Theorem on the impossibility of self-maintained axisymmetric homogeneous dynamos," *Proc. Natl. Acad. Sci.* **42**, 105–109 (1956).

Baierlein, R., "The amplification of magnetic fields during the radiation era," *Mon. Not. R. Astron. Soc.* **184**, 843–870 (1978).

Baker, N. and Tamesvary, S., *Tables of convective stellar envelope models*, New York (1966).

Basko, M. M. and Sunyaev, R. A., "Radiative transfer in a strong magnetic field and accreting of X-ray pulsars," *Astron. & Astrophys.* **42**, 311–321 (1974).

Batchelor, G. K., "On the spontaneous magnetic field in a conducting fluid in turbulent motion," *Proc. R. Soc. Lond.* **A201**, 405–416 (1950).

Batchelor, G. K., *The theory of homogeneous turbulence*, Univ. Press, Cambridge (1953).

Batchelor, G. K., "Small-scale variation of convected quantities like temperature in turbulent fluid," *J. Fluid Mech.* **5**, 113–133 (1959).

Batchelor, G. K., Howells, I. D. and Townsend, A. A., "Small-scale variation of convected quantities like temperature in turbulent fluid. II. The case of large conductivity," *J. Fluid Mech.* **5**, 134–139 (1959).

Baym, G. and Pethick, C. J., "Physics of neutron stars," *Annu. Rev. Astron. Astrophys.* **17**, 415–444 (1979).

Beck, R., "Magnetic field in M31," *Astron. & Astrophys.* **106**, 121–132 (1982).

Beckers, J. M. and Schröter, E. H., "The intensity, velocity and magnetic structure of a sunspot region. I: Observational Technical Properties of Magnetic Knots," *Sol. Phys.* **4**, 142–164 (1968).

Berlin, A. B., Korolkov, D. B., Pariiski, Yu. N., Soboleva, N. S., and Timofeeva, G. M., "Occulation of polarized radioemission by the solar corona at the period of weak solar activity," *Pisma Astron. J. (SSSR)* **4**, 191–192 (1978), [*Sov. Astron. Lett.* **4**, 102–103 (1978)].

Bisnovatyi-Kogan, G. S. and Ruzmaikin, A. A., "The accretion of matter by a collapsing star in the presence of a magnetic field," *Astrophys. & Space Sci.* **28**, 45–59 (1974).

Bisnovatyi-Kogan, G. S. and Ruzmaikin, A. A., "The accretion of matter by a collapsing star in the presence of a magnetic field. II. Self-consistent stationary picture," *Astrophys. & Space Sci.* **42**, 401–424 (1976).

Bisnovatyi-Kogan, G. S., Ruzmaikin, A. A. and Sunyaev, R. A., "Condensation of stars and formation of a magnetic field in protogalaxies," *Astron. J. (SSSR)* **50**, 210–213 (1973), [*Sov. Astron.* **17**, 137–139 (1973)].

Black, D. C. and Scott, E. H., "A numerical study of the effects of ambipolar diffusion on the collapse of magnetic gas clouds," *Astrophys. J.* **263**, 696–715 (1982).

Blandford, R. D. and Znajek, R. L., "Electromagnetic extraction of energy from Kerr black holes," *Mon. Not. R. Astron. Soc.* **179**, 433–456 (1977).

Bolton, J. G. and Wild, J. P., "On the possibility of measuring interstellar magnetic fields by 21-cm Zeeman splitting," *Astrophys. J.* **125**, 296–297 (1957).

Bopp, B. W. and Evans, D. S., "The spotted flare stars BY Dra and CCER I: A model for the spots and some astrophysical implications," *Mon. Not. R. Astron. Soc.* **164**, 343–356 (1973).

Braginsky, S. I., "Self-excitation of a magnetic field during the motion of a highly conducting fluid," *JETP* **47**, 1084–1098 (1964), [*Sov. Phys. JETP* **20**, 726–735 (1964)]; "Theory of the hydromagnetic dynamo," *JETP* **47**, 2178–2193 (1964), [*Sov. Phys. JETP* **20**, 1462–1471 (1964)].

Braginsky, S. I., "Magnetic waves in the Earth's core," *Geomagn. & Aeron.* **7**, 1050–1060 [*Transl.* **7**, 851–859] (1967).

Bräuer, H., "The nonlinear dynamo problem: Small oscillatory solutions in a strongly simplified model," *Astron. Nachr.* **300**, 43–49 (1979).

Breit, G., "Quantum theory of dispersion (Continued). Parts VI and VII," *Rev. Mod. Phys.* **5**, 91–140 (1933).

Bullard, E. C., "The stability of a homopolar dynamo," *Proc. Camb. Phil. Soc.* **51**, 744–760 (1955).
Bullard, E. C. and Gellman, H., "Homogeneous dynamos and terrestrial magnetism," *Philos. Trans. R. Soc. Lond.* A**247**, 213–278 (1954).
Bumba, V. and Howard, R., "Large-scale distribution of solar magnetic fields," *Astrophys. J.* **141**, 1502–1512 (1965).
Burn, B. J., "On the depolarization of discrete radio sources by Faraday dispersion," *Mon. Not. R. Astron. Soc.* **133**, 68–87 (1966).
Burstein, P., Borken, R. J., Kraushaar, W. L. and Sanders, W. T., "Three-band observations of the soft X-ray background and some implications for thermal emission models," *Astrophys. J.* **213**, 405–420 (1977).
Busse, F. H., "Differential rotation in stellar convective zones II," *Astron. & Astrophys.* **28**, 27–37 (1972).
Busse, F. H., "Generation of planetary magnetism by convection," *Phys. Earth. & Planet. Inter.* **12**, 350–358 (1976).
Cameron, A. G. W., "The formation of the Sun and planets," *Icarus* **1**, 13–21 (1962).
Cassen, P. and Moosman, A., "On the formation of protostellar disks," *Icarus* **48**, 353–376 (1981).
Chaisson, E. J. & Vrba, F. J., "Magnetic field structures and strengths in dark clouds," in *Protostars and Planets*, pp. 180–208 (ed. T. Gehrels), Univ. Arizona Press, Tucson (1978).
Chanan, G., Ku, W. H.-M., Simon, M. and Charles, P., "Further Einstein observations of the Orion Nebula," *Bull. Am. Astron. Soc.* **11**, 623 (1979).
Chandrasekhar, S. and Fermi, E., "Magnetic fields in spiral arms," *Astrophys. J.* **118**, 113–115 (1953a).
Chandrasekhar, S. and Fermi, E., "Problems of gravitational stability in the presence of a magnetic field," *Astrophys. J.* **118**, 116–141 (1953b).
Charvin, P., "Étude de la polarization des raies interdites de la couronne solaire. Application au cas de la raie verte 5303," *Ann. Astrophys.* **28**, 877–934 (1965).
Chibisov, G. V. and Ptuskin, V. S., "Angular variations of nonthermal radio emission from the Galaxy relevant to the structure of interstellar magnetic field," in *17ème Conférence Internationale sur le Rayonnement Cosmique*, **2**, pp. 233–235, Paris (1981).
Childress, S., "Alpha-effect in flux ropes and sheets," *Phys. Earth & Planet. Inter.* **20**, 172–180 (1979).
Chugainov, P. F., "New microvariable HD 117555," *Commiss. 27 IAU Inf. Bull. Var. Stars.* No. **172**, pp. 1–2 (1966).
Clark, D. E. and Stephenson, F. R., "An interpretation of the pretelescopic sunspot records from the Orient," *Q. J. R. Astron. Soc.* **19**, 387–410 (1978).
Clarke, A. Jr., "Some exact solutions in magnetohydrodynamics with astrophysical applications," *Phys. Fluids* **8**, 644–649 (1965).
Cohen, R., Lodenquai, J. and Ruderman, M., "Atoms in super-strong magnetic fields," *Phys. Rev. Lett.* **25**, 467–469 (1970).
Consolmagno, G. J. and Jokipii, J. R., "^{26}Al and partial ionization in the solar nebula," *Moon & Planets* **19**, 253–259 (1978).
Cook, A. E. and Roberts, P. H., "The Rikitake two-disc dynamo system," *Proc. Camb. Phil. Soc.* **68**, 547–569 (1970).
Cowling, T. G., "Magnetic fields of sunspots," *Mon. Not. R. Astron. Soc.* **94**, 39–48 (1934).
Cowling, T. G., *Magnetohydrodynamics*, Interscience Publishers, New York–London (1957).
Cox, A., "Geomagnetic reversals," *Science* **163**, 237–245 (1969).

Cox, D. P. and Smith, B. W., "Large-scale effects of supernova remnants on the Galaxy: generation and maintenance of a hot network of tunnels," *Astrophys. J. Lett.* **189**, L105–L108 (1974).

Cram, L. E., "Nonthermal structure of stellar atmospheres," *Comm. Astrophys.* **9**, 25–49 (1980).

Crowell, R. H. and Fox, R. H., *Introduction to knot theory*, Ginn & Co., Boston–New York (1963).

Crutcher, R. M., Evans, N. J., II, Troland, T. and Heiles, C., "OH Zeeman observations of interstellar dust clouds," *Astrophys. J.* **198**, 91–93 (1975).

Davis, L. and Greenstein, J. L., "The polarization of starlight by alignment of dust grains," *Astrophys. J.* **114**, 209–240 (1950).

Deinzer, W. and Stix, M., "On the eigenvalues of Krause-Steenbeck's solar dynamo," *Astron. & Astrophys.* **12**, 111–119 (1971).

Deubner, F. L., Ulrich, R. K. and Rhodes, E. J., "Solar p-mode oscillations as a tracer of radial differential rotation," *Astron. & Astrophys.* **72**, 177–185 (1979).

Dicke, R. H., "The 5-minute oscillations of the Sun are incompatible with a rapidly-rotating core," *Nature* **300**, 693–697 (1983).

Dirac, P. A. M., "Quantized singularities in electromagnetic fields," *Proc. R. Soc. Lond. A***133**, 60–78 (1931).

Dogel, V. A. and Syrovatsky, S. I., "On the possible nature of the Maunder minimum," *Izv. Akad. Nauk SSSR, Ser. Fiz.* **43**, 716–723 (1979), [*Bull. Acad. Sci. USSR (Phys. Ser.*), **43**(4), 35–41 (1979)].

Dolginov, A. Z., Gnedin, Yu. N. and Silantev, N. A., *Propagation and polarization of radiation in the cosmic medium* (in Russian) Nauka, Moscow (1979).

Dolginov, A. Z. and Urpin, V. A., "The inductive generation of the magnetic field in binary systems," *Astron. & Astrophys.* **79**, 60–69 (1979).

Dolginov, Sh. Sh., Yeroshenko, Ye. G. and Zhuzgov, L. N., "Magnetic fields in the very close neighbourhood of Mars according to data from the Mars-2 and Mars-3 spacecrafts," *J. Geophys. Res.* **78**, 4779–4786 (1973).

Dorfi, E., "3D models for self-gravitating rotating magnetic interstellar clouds," *Astron. & Astrophys.*, **114**, 151–164 (1982).

Doroshkevich, A. G., "Model of a universe with a uniform magnetic field," *Astrofizika* **1**, 255–266 [Transl. **1**, 138–142] (1965).

Doroshkevich, A. G., "The origin of rotation of galaxies," *Astrophys. Lett.* **14**, 11–13 (1973).

Drobyshevski, E. M., "Magnetic field transfer by two-dimensional convection and solar 'semi-dynamo'," *Astrophys. & Space Sci.* **46**, 41–49 (1977).

Drobyshevski, E. M. and Yuferev, V. S., "Topological pumping of magnetic flux by three-dimensional convection," *J. Fluid Mech.* **65**, 33–44 (1974).

Durney, B. R., "On theories of solar rotation," in Proc. IAU Symp. No. 71, *Basic Mechanisms of Solar Activity*, pp. 243–296 (eds. V. Bumba and J. Kleczek,), Reidel, Dordrecht (1976).

Eardley, D. M. and Lightman, A. P., "Magnetic viscosity in relativistic accretion disks," *Astrophys. J.* **200**, 187–203 (1975).

Ebert, R., von Hoerner, S. and Tamesvary, S., *Die Entstehung von Sterner durch Kondensation diffuser Materie*, Springer-Verlag, Berlin (1960).

Eddy, S. A., "The Maunder minimum," *Science* **192**, 1189–1202 (1976).

Eddy, S. A., "The case of the missing sunspots," *Sci. Am.* **236**, (5), 80–95 (1977).

Eddy, S. A., Gilman, P. A. and Trotter, D. E., "Anomalous solar rotation in the early 17th century," *Science* **198**, 824–829 (1977).

Efimov, N. V., "Appearance of singularities on surfaces of negative curvature," *Matem. Sb.* **64**, 286–320 (1964).

346 YA. B. ZELDOVICH, A. A. RUZMAIKIN AND D. D. SOKOLOFF

Eichendorf, W. and Reinhardt, M., "Polarization properties of extragalactic radio sources," *Zesz. Nauk UJ.*, No. 570, 7–48 (1980), [*Acta Cosmologica* **9**, 7–48 (1980)].
Elsasser, W. M., "Induction effects in terrestrial magnetism," *Phys. Rev.* **69**, 106–116 (1946).
Erber, T., "High-energy electromagnetic conversion processes in intense magnetic fields," *Rev. Mod. Phys.* **38**, 626–668 (1966).
Field, G. B., "Thermal instability," *Astrophys. J.* **142**, 531–567 (1965).
Frank-Kamenetsky, M. D. and Vologodsky, A. V., "Topological aspects of the physics of polymers: the theory and its biophysical applications," *Usp. Fiz.*, (*SSSR*) **134**, 641–673 (1981), [*Sov. Phys. Usp.* **24**, 679–696 (1981)].
Friedmann, A. M. and Polyachenko, V. L., *Equilibrium and stability of gravitating systems* (in Russian) Nauka, Moscow (1976).
Frisch, U., Pouquet, A., Léorat, J. and Mazure, A., "Possibility of an inverse cascade of magnetic helicity in magnetohydrodynamical turbulence," *J. Fluid Mech.* **68**, 769–778 (1975).
Gailitis, A. K., "Theory of the Herzenberg dynamo," *Magn. Gidrodin.* 12–16 (1973), [*Magnetohydrodynam.* **9**, 445–449 (1975)].
Gailitis, A. K. and Freiberg, Ya. G., "Theory of a helical MHD dynamo," *Magn. Gidrodin.* 3–6 (1976), [*Magnetohydrodynam.* **12**, 127–134 (1977)].
Gailitis, A. K., Lielausis, O. and Freiberg, Ya. G., *On possibilities to observe the magnetic field generation in the fluid sodium*, Preprint No. 1, Institute of Physics, Latvian Academy of Sciences, Riga (1977).
Galeev, A. A., Rosner, R. and Vaiana, G. S., "Structured coronae of accretion discs," *Astrophys. J.* **229**, 318–326 (1979).
Galloway, D. J. and Proctor, M. R. E., *Magnetic flux expulsion in hexagons*, preprint of Max-Plank Institut für Physik and Astrophysik, MPI-PAE/Astro 250 (1981). See also "The kinematics of hexagonal magnetoconvection," *Geophys. & Astrophys. Fluid Dyn.* **24**, 109–136 (1983).
Gamov, G., "The role of turbulence in the evolution of the Universe," *Phys. Rev.* **86**, 251 (1952).
Gaponov-Grekhov, A. V. and Rabinovich, M. I., "L. J. Mandelshtam and the modern theory of non-linear oscillations and waves," *Usp. Fiz.* (*SSSR*), **128**, 579–624 (1979), [*Sov. Phys. Usp.* **22**, 590–614 (1979)].
Gardner, F. and Whiteoak, J. B., "Polarization of radio sources and Faraday rotation effects in the Galaxy," *Nature* **197**, 1162–1164 (1963).
Gardner, F., Morris, D. and Whiteoak, J. B., "The linear polarization of radio sources between 11 and 20 cm wavelength. III. Influence of the Galaxy on source depolarization and Faraday rotation," *Aust. J. Phys.* **22**, 813–838 (1969).
Gehrels, T. (ed.), *Jupiter: Studies of the interior, atmosphere, magnetosphere and satellites*, Univ. of Arizona Press, Tucson (1976).
Gershberg, R. E., "Flares of red dwarf stars and Solar activity," (Review), in Proc. IAU Symp. No. 67, *Variable stars and stellar evolution* (eds. V. E. Sherwood and L. Plaut), D. Reidel, Dordrecht, pp. 47–64 (1975).
Giacconi, R., "The Einstein X-ray observatory," *Sci. Am.* **242**, 70–82 (1980).
Giacconi, R. and Ruffini, R. (eds.), *Physics and astrophysics of black holes and neutron stars*, North-Holland, Amsterdam (1978).
Gibson, R. D., "The Herzenberg dynamo," *Q. J. Mech. & Appl. Math.* **21**, 243–255, 257–267 (1968).
Gillis, J., Mestel, L. and Paris, R. B., "Magnetic breaking during star formation. II," *Mon. Not. R. Astron. Soc.* **187**, 311–335 (1979).
Gilman, P. A., "Theory of convection in a deep rotating spherical shell, and its application to the Sun," in Proc. IAU Symp. No. 71, *Basic Mechanisms of Solar Activity* (eds. V. Bumba and J. Klechek), pp. 207–228. D. Reidel, Dordrecht (1976).

Ginzburg, V. L., "The origin of cosmic rays and radioastronomy," *Usp. Fiz. (SSSR)* **51**, 343–392 (1953).

Ginzburg, V. L., "Magnetic fields of collapsing masses and the nature of superstars," *Dokl. Akad. Nauk SSSR*, **156**, 43–46 [*Sov. Phys. Dokl.* **9**, 329–332] (1964).

Ginzburg, V. L., *The propagation of electromagnetic waves in plasmas*, Pergamon Press (1970).

Ginzburg, V. L. and Ozernoy, L. M., "On gravitational collapse of magnetic stars," *JETP* **47**, 1030–1040 (1964), [*Sov. Phys. JETP*, **20**, 689–698 (1965)].

Ginzburg, V. L. and Syrovatsky, S. I., *The origin of cosmic rays*, Moscow (1963). English translation: Pergamon Press (1964).

Ginzburg, V. L. and Usov, V. V., "Concerning the atmosphere of neutron stars (pulsars)," *Pisma JETP* **15**, 280–282 (1972) [*Sov. Phys. JETP Lett.* **15**, 196–198 (1972)].

Ginzburg, V. L. and Zheleznyakov, V. V., "On pulsar emission mechanisms," *Annu. Rev. Astron. Astrophys.* **13**, 511–535 (1975).

Gleisberg, W., "The eighty-year sunspot cycle," *J. Brit. Astron. Assoc.* **68**, 148–152 (1958).

Gleisberg, W. and Damboldt, T., "Reflections of the Maunder minimum of Sunspots," *J. Brit. Astron. Assoc.* **89**, 440–449 (1979).

Gnedin, Yu. N. and Sunyaev, R. A., "The beaming of radiation from an accreting magnetic neutron star and X-ray pulsars," *Astron. & Astrophys.* **25**, 233–239 (1973).

Gnedin, Yu. N. and Sunyaev, R. A., "Polarization of optical and X-radiation from compact thermal sources with magnetic field," *Astron. & Astrophys.* **36**, 379–394 (1974).

Gold, T., "Rotating neutron stars as the origin of the pulsating radio sources," *Nature* **218**, 731–732 (1968).

Goldreich, P. and Julian, W. H., "Pulsar electrodynamics," *Astrophys. J.* **157**, 869–880 (1969).

Golitsyn, G. S., "Fluctuations of the magnetic field and current density in a turbulent flow of a weakly conducting fluid," *Dokl. Akad. Nauk. SSR*, **132**, 315–318, [*Sov. Phys. Dokl.*, **5**, 536–539] (1960).

Golitsyn, G. S., *An introduction to the dynamics of planetary atmospheres*, Hydrometeor. Publ. House, Leningrad (1963). English translation: NASA TT F-15, 627, NASA, Washington, D.C. (1974).

Golitsyn, G. S., "Structure of convection during fast rotation," *Dokl. Akad. Nauk SSSR*, **216**, 317–320 (1981).

Gough, D. O., "Internal rotation and gravitational quadrupole moment of the Sun," *Nature* **298**, 334–339 (1982).

Greenberg, J. M., "Interstellar grains," Ch. 6 of *Stars and stellar systems* **6** (Ed. G. P. Kuiper), Univ. Press, Chicago (1968).

Gudzenko, L. I. and Chertoprud, V. E., "Some dynamic properties of cyclic solar activity," *Astron. J. (SSSR)*, **41**, 697–706 (1964), [*Sov. Astron.* **8**, 555–562 (1964)].

Gurevich, L. J. and Chernin, A. D., *Introduction to cosmology*, Nauka, Moscow (1978).

Guskova, E. G. and Pochtarev, V. I., "Magnetic fields in space according to a study of the magnetic properties of meteorites," *Geomagn. & Aeronom.* **7**, 310–316 [Transl. **7**, 245–250] (1967).

Hale, G. E., "On the probable existence of a magnetic field in sunspots," *Astrophys. J.* **28**, 315–343 (1908).

Hanle, W., "Über magnetische Beeinflussung der Polarization der Resonanzfluorezenz," *Z. Phys.* **30**, 93–105 (1924).

Harrison, E. R., "Generation of magnetic fields in the radiation era," *Mon. Not. R. Astron. Soc.* **147**, 279–286 (1970).

Harrison, E. R., "Standard Model of the early Universe," *Annu. Rev. Astron. Astrophys.* **11**, 155–186 (1973).

Haves, P., "Polarisation parameters of 183 extragalactic radio sources," *Mon. Not. R. Astron. Soc.* **173**, 553–568 (1975).

Heiles, C., "The interstellar magnetic field," *Annu. Rev. Astron. Astrophys.* **14**, 1–22 (1976).

Henon, M. M., "Sur la topologie des lignes de courant dans un cas particulier," *C.R. Acad. Sci. Paris* **262**, 312–314 (1966).

Herzenberg, A., "Geomagnetic dynamos," *Philos. Trans. R. Soc. Lond. A* **250**, 543–585 (1958).

Hewish, A., Bell, S. I., Pilkington, I. D., Scott, P. F. and Collins, R. A., "Observations of rapidly rotating radio sources," *Nature*, **217**, 709–713 (1968).

Hide, R., "Jupiter and Saturn: giant magnetic rotating fluid planets," *Observatory* **100**, 182–192 (1980).

Hide, R and Palmer, T. N., "Generalisation of Cowling's theorem," *Geophys. & Astrophys. Fluid Dyn.* **19**, 301–309 (1982).

Hill, H. A., Bos, R. J. and Goode, P. R., "Preliminary determination of the Sun's quadrupole moment from rotational splitting of the global oscillations and its relevance to tests of general relativity," *Phys. Rev. Lett.* **49**, 1794–1797 (1982).

t'Hooft, G., "Magnetic monopoles in unified gauge theories," *Nucl. Phys. B* **79**, 276–284 (1974).

Hopper, R. B. and Disney, M. J., "The alignment of interstellar dust clouds," *Mon. Not. R. Astron. Soc.* **168**, 639–650 (1974).

Howard, R., "The rotation of the Sun," *Rev. Geophys. & Space Phys.* **16**, 721–732 (1978).

Howard, R. and Harvey, J. W., "Spectroscopic determination of solar rotation," *Sol. Phys.* **12**, 23–51 (1970).

Hoyle, F., "The steady state theory," in *La structure et l'évolution de l'Univers: XI Conseil de physique*, Bruxelle, pp. 53–73 (1958).

Hoyle, F., "On the origin of the solar nebula," *Q. J. R. Astron. Soc.* **1**, 28–55 (1960).

Hoyle, F., "Magnetic field and highly condensed objects," *Nature* **223**, 936 (1969).

Hyder, C. L., "Magnetic fields in the loop prominence of March 16," *Astrophys. J.* **140**, 817–818 (1964).

Iben, I. Jr., "Stellar evolution. I. The approach to the main sequence," *Astrophys. J.* **141**, 999–1018 (1965).

Iroshnikov, P. S., "Turbulence of a conducting fluid in a strong magnetic field," *Astron. J. (SSSR)* **40**, 742–750 (1963), [*Sov. Astron.* **7**, 566–571 (1964)].

Isakov, R. V., Ruzmaikin, A. A., Sokoloff, D. D. and Faminskaya, M. V., "The asymptotic properties of the disc dynamo," *Astrophys. & Space Sci.* **80**, 145–155 (1981).

Ivanova, T. S. and Ruzmaikin, A. A., "Magnetohydrodynamic dynamo-model of the solar cycle," *Astron. J. (SSSR)* **53**, 398–410 (1976), [*Sov. Astron.* **20**, 227–234 (1976)].

Ivanova, T. S. and Ruzmaikin, A. A., "Non-linear magnetohydrodynamic model of the solar dynamo," *Astron. J. (SSSR)* **54**, 846–858 (1977), [*Sov. Astron.* **21**, 479–485 (1977)].

Ivanova, T. S. and Ruzmaikin, A. A., "The role of differential rotation in the solar dynamo," *Astron. J. (SSSR)* **57**, 127–130 (1980), [*Sov. Astron.* **24**, 75–77 (1980)].

Jenkins, E. B. and Meloy, D., "A Survey with Copernicus of interstellar OVI absorption," *Astrophys. J. Lett.* **193**, L121–L125 (1974).

Jepps, S. A., "Numerical models of hydromagnetic dynamos," *J. Fluid Mech.* **67**, 629–646 (1975).

Jokipii, J. R. and Lerche, I., "Faraday rotation dispersion in pulsar signals and the turbulent structure of the Galaxy," *Astrophys. J.* **157**, 1137–1145 (1969).

Kadomsev, B. B., "Heavy atom in an ultrastrong magnetic field," *JETP*, **58**, 1765–1769 (1970), [*Sov. Phys. JETP*, **31**, 945–947 (1970)].

Kadomsev, B. B. and Kudryavtsev, V. S., "Molecules in an ultra-strong magnetic field," *Pisma JETP* **13**, 15–19 (1971), [*Sov. Phys. JETP Lett.* **13**, 9–12 (1971)].

Kadomsev, B. B. and Kudryavtsev, V. S., "Molecules in an ultra-strong magnetic field," *Pisma JETP* **13**, 15–19 (1971), [*Sov. Phys. JETP Lett.* **13**, 9–12 (1971)].

Kadomsev, B. B. and Kudryavtsev, V. S., "Atoms in a superstrong magnetic field," *Pisma JETP* **13**, 61–64 (1971), [*Sov. Phys. JETP Lett.* **13**, 42–44 (1971)].

van Kampen, N. G., "A cumulant expansion for stochastic linear differential equations," *Physica* **74**, 215–247 (1974).

Kaplan, S. A., *Interstellar gas dynamics*, Pergamon Press, Oxford (1966).

Kaplan, S. A. and Pikelner, S. B., *Interstellar medium*, Harvard Univ. Press, Cambridge, Mass. (1970), *Fizika mezhzvezdnosredy*, Nauka, Moscow (1979).

Kaplan, S. A. and Pikelner, S. B., "Large-scale dynamics of the interstellar medium," *Ann. Rev. Astron. Astrophys.* **12**, 113–134 (1974).

Kármán, Th. von, "Über laminare und turbulente Reibung," *Z. Ang. Math. Mech.* **1**, 233–251 (1921).

Kaverin, N. S., Kobrin, M. M., Korshunov, A. I. and Shushunov, V. V., "Fine structure in the 5–12 GHz radio spectrum of local sources on the Sun and the current sheets of active regions," *Astron. J.* (*SSSR*), **57**, 767–770 (1980), [*Sov. Astron.* **24**, 442–443 (1980)].

Kawabata, K., Fujimoto, M., Soful, J., Fukui, M., "A large-scale metagalactic magnetic field and Faraday rotation for extragalactic ratio sources," *Publ. Astron. Soc. Japan* **21**, 293–306 (1969).

Kazantsev, A. P., "Enhancement of a magnetic field by a conducting fluid," *JETP* **53**, 1806–1813 (1967) [*Sov. Phys. JETP* **26**, 1031–1039 (1968)].

Kemp, J. C., "Circular polarization of thermal radiation in a magnetic field," *Astrophys. J.* **162**, 169–179 (1970).

Kippenhahn, R., "Differential rotation in stars with convective envelopes," *Astrophys. J.* **137**, 664–678 (1963).

Kleeorin, N. I. and Ruzmaikin, A. A., "Properties of a nonlinear solar dynamo model," *Geophys. & Astrophys. Fluid Dyn.* **17**, 281–296 (1981).

Kleeorin, N. I. and Ruzmaikin, A. A., "Dynamics of the average turbulent helicity in a magnetic field," *Magn. Gidrodin.* 17–24 (1982), [*Magnetohydrodynam.* **18**, 116–122 (1982)].

Klepikov, N. P., "Photon and electron–positron pair emission in a magnetic field," *JETP* **26**, 19–34 (1954).

Knobloch, E., "The diffusion of scalar and vector fields by homogeneous stationary turbulence," *J. Fluid Mech.* **83**, 129–146 (1977).

Knobloch, E., "Turbulent diffusion of magnetic fields," *Astrophys. J.* **225**, 1050–1057 (1978).

Kobrik, M. and Kaula, W. M., "A tidal theory for the origin of the solar nebula," *Moon & Planets* **20**, 61–101 (1978).

Kolobov, V. S., Reinhardt, M. and Sazonov, V. N., "A test of the isotropy of the Universe," *Astrophys. Lett.* **17**, 183–185 (1976).

Kolykhalov, P. I. and Sunyaev, R. A., "Outer parts of accreting disks around supermassive black holes," *Pisma Astron. J.* (*SSSR*) **6**, 680–686 (1980), [*Sov. Astron. Lett.* **6**, 357–361 (1980)].

Komberg, B. V., Ruzmaikin, A. A., and Sokoloff, D. D., "Faraday rotation in galaxies and absorption lines in quasars," *Pisma Astron. J.* (*SSSR*) **5**, 73–76 (1979), [*Sov. Astron. Lett.* **5**, 40–42 (1979)].

Kraft, R. P., "Studies of stellar rotation, V. The dependence of rotation on age among solar-type stars," *Astrophys. J.* **150**, 551–570 (1967).

Kraichnan, R. H., "Inertial-range spectrum of hydromagnetic turbulence," *Phys. Fluids* **8**, 1385–1387 (1965).

Kraichnan, R. H., "Diffusion of weak magnetic fields by isotropic turbulence," *J. Fluid Mech.* **75**, 657–676 (1975).

Kraichnan, R. H., "Diffusion of passive-scalar and magnetic fields by helical turbulence," *J. Fluid Mech.* **77**, 753–768 (1976).

Kraichnan, R. H., "Consistency of the α-effect turbulent dynamo," *Phys. Rev. Lett.* **42**, 1677–1680 (1979).

Kraichnan, R. H. and Nagarajan, S., "Growth of turbulent magnetic fields," *Phys. Fluids* **10**, 859–870 (1967).

Krajcheva, Z. T., Popova, E. T., Tutukov, A. V. and Jungelson, L. E., "Some properties of spectroscopic binary stars," *Astron. J.* (*SSSR*) **55**, 1176–1189 (1978), [*Sov. Astron.* **22**, 670–677 (1978)].

Krause, F., "Zur Dynamotheorie magnetischer Sterne: der 'Symmetrische Rotator' als Alternative zum schiefen Rotator," *Astron. Nachr.* **293**, 187–193 (1971).

Krause, F. and Rädler, K.-H., *Mean-field magnetohydrodynamics and dynamo theory*, Pergamon Press (1980).

Kronberg, P. P., Reinhardt, M. and Simard-Normandin, M., "On the intergalactic contribution to the rotation measures of QSO's," *Astron. & Astrophys.* **61**, 771–776 (1977).

Kronberg, P. P. and Perry, J. J., "Absorption lines, Faraday rotation and magnetic field estimates for QSO absorption line clouds," *Astrophys. J.* **263**, 518–532 (1982).

Kuznetsova, I. P., "On the Faraday rotation in the intergalactic medium," *Astron. J.* (*SSSR*) **53**, 475–484 (1976), [*Sov. Astron.* **20**, 269–274 (1976)].

Lamb, F. K., Pethick, C. J. and Pines, D. A., "Model for compact X-ray sources: accretion by rotating magnetic stars," *Astrophys. J.* **184**, 271–289 (1973).

Landau, L. D. and Lifshitz, E. M., *Quantum mechanics, non-relativistic theory*, Addison-Wesley, Reading, Mass. (1958).

Landau, L. D. and Lifshitz, E. M., *Fluid mechanics*, Addison-Wesley, Reading, Mass. (1959).

Landau, L. D. and Lifshitz, E. M., *The classical theory of fields* (2nd edition), Addison-Wesley, Reading, Mass. (1962).

Larmor, J., "How could a rotating body such as the Sun become magnetic?," *Rep. Brit. Assoc. Adv. Sci.* 159–160 (1919).

Larson, R. B., "Numerical calculations of the dynamics of a collapsing proto-star," *Mon. Not. R. Astron. Soc.* **145**, 271–296 (1969).

Larson, R. B., "The stellar state: formation of the solar-type stars," in *Protostars and planets*, pp. 43–57 (ed. T. Gehrels), Univ. of Arizona Press, Tucson (1978).

Laue, M. von, *Geschichte der Physik*, Athenaum-Verlag, Bonn (1950).

Ledoux, P. and Renson, P., "Magnetic stars," *Ann. Rev. Astron. Astrophys.* **4**, 293–352 (1966).

Lehnert, B., "An experiment on axisymmetric flow of fluid sodium in a magnetic field," *Ark. Fysik.* **13**, 109–116 (1958).

Leighton, R. B., "A magneto-kinematic model of the solar cycle," *Astrophys. J.* **156**, 1–26 (1969).

Levy, E. H. and Sonett, C. P., "Meteorite magnetism and early solar system magnetic fields," in *Protostars and planets*, pp. 516–532 (ed. T. Gehrels), Univ. Arizona Press, Tucson (1978).

Lightman, A. P., Sunyaev, R. A., Shakura, N. I., Shapiro, S. L. and Eardley, D. M., "Status report on Cygnus X-1," *Usp. Fiz. (SSSR)* **126**, 515–526 (1978), [*Comments Astrophys.* C**7**, 151–160 (1978)].

Lin, C. C., "The dynamics of disk-shaped galaxies," *Annu. Rev. Astron. & Astrophys.* **5**, 453–464 (1967).

Linds, B. T., "The distribution of dark nebulae in late-type spirals," in Proc. IAU Symp. No. 38, *The spiral structure of our Galaxy*, pp. 26–34 (eds. W. Beckers and G. Contopoulos), D. Reidel, Dordrecht (1970).

Linsky, J. L., Ch. 7 of *Solar output and its variation*, (Ed. O. R. White) Colorado Associated Univ. Press, Boulder (1977).

Lo, K. Y., Walker, R. C., Burke, B. F., Moran, J. M., Johnston, K. J. and Ewing, M. S., "Evidence for Zeeman splitting in 1720-MHz OH line emission," *Astrophys. J.* **202**, 650–654 (1975).

Lominadze, D. G., Machabeli, G. Z., Michailowsky, A. B., Ochelkov, Yu. P. and Usov, V. V., "The nature of high-frequency radiation of pulsars and the activity of supernova remnants," *Usp. Fiz. (SSSR)* **131**, 516–518 (1980), [*Sov. Phys. Usp.* **23**, 422–424 (1980)].

Longair, M. and Sunyaev, R. A., "Electromagnetic radiation in the Universe," *Usp. Fiz. (SSSR)* **105**, 41–95 (1971), [*Sov. Phys. Usp.* **14**, 569–599 (1972)].

Lorenz, E. N., "Deterministic nonperiodic flow," *J. Atmos. Sci.* **20**, 130–141 (1963).

Lortz, D., "Impossibility of steady dynamos with certain symmetries," *Phys. Fluids* **11**, 913–915 (1968a).

Lortz, D., "Exact solutions of the hydromagnetic dynamo problem," *Plasma Phys.* **10**, 967–972 (1968b).

Lowes, F. J. and Wilkinson, I., "Geomagnetic dynamo: a laboratory model," *Nature* **198**, 1158–1160 (1963);

Lowes, F. J. and Wilkinson, I., "Geomagnetic dynamo: an improved laboratory model," *Nature* **219**, 717–718 (1968).

Lozinskaya, T. A., "Kinematics of old supernova remnants," *Astron. & Astrophys.* **84**, 26–35 (1980).

Lozinskaya, T. A. and Kardashev, N. S., "The thickness of the gas disk of the Galaxy from 21 cm line observations," *Astron. J. (SSSR)* **40**, 209–215 (1963), [*Sov. Astron.* **7**, 161–166 (1963)].

Lücke, M., "Statistical dynamics of the Lorenz model," *J. Stat. Phys.* **15**, 455–475 (1976).

Lynden-Bell, D., "Galactic nuclei as collapsed old quasars," *Nature* **223**, 690–694 (1969).

Lynden-Bell, D. and Pringle, J. E., "The evolution of viscous disks and the origin of the nebular variables," *Mon. Not. R. Astron. Soc.* **168**, 603–637 (1974).

Lyutyi, V. V., Sunyaev, R. A. and Cherepashchuk, A. M., "On the nature of optical variability of HZ Her = Her X1 and BD + 34°3815 = Cyg X1," *Astron. J. (SSSR)* **50**, 3–11 (1973), [*Sov. Astron.* **17**, 1–6 (1973)].

Macdonald, D. and Thorne, K. S., *Black-hole electrodynamics: an absolute-space universal-time formulation*, Preprint OAP-616 (1981).

Manchester, R. N., "Structure of the local magnetic field. Part I," *Astrophys. J.* **172**, 43–52 (1972).

Manchester, R. N., "Structure of the local magnetic field. Part II," *Astrophys. J.* **188**, 637–644 (1974).

Masters, R. A., Fabian, A. C., Pringle, J. E. and Rees, M. J., "Cyclotron emission from accreting magnetic white dwarfs," *Mon. Not. R. Astron. Soc.* **178**, 501–504 (1977).

Mathewson, D. S. and Ford, V. L., "Polarization observations of 1800 stars," *Mem. R. Astron. Soc.* **74**, 139–182 (1970).

Mathewson, D. S., van der Kruit, P. G. and Brown, W. N., "A high resolution continuum survey of M51 and NGC 5195 at 1415 MHz," *Astron. & Astrophys.* **17**, 468–486 (1972).

Mazets, E. P., Golenetskij, S. V., Aptekar, R. L., Guryanov, Yu. A. and Ilinskij, V. N., "Lines in gamma-burst energy spectra," *Pisma Astron. J. (SSSR)* **6**, 706–711 (1980), [*Sov. Astron. Lett.* **6**, 372–375 (1980)].

McKee, C. F. and Ostriker, J. P., "A theory of the interstellar medium: three components regulated by supernova explosions in an inhomogeneous substrate," *Astrophys. J.* **218**, 148–169 (1977).

McLaughlin, J. B. and Martin, O. C., "Transition to turbulence in a statically stressed fluid system," *Phys. Rev.* **A12**, 189–203 (1975).

McNally, D., "The distribution of angular momentum amomg main sequence stars," *The Observatory* **85**, 166–169 (1965).

Meneguzzi, M., Frisch, U. and Pouquet, A., "Helical and non-helical turbulent dynamics," *Phys. Rev. Lett.* **47**, 1060–1064 (1981).

Mestel, L., "Problems of star formation," *Q. J. R. Astron. Soc.* **6**, 265–300 (1965).

Mestel, L., "Magnetic braking by a stellar wind. I," *Mon. Not. R. Astron. Soc.* **138**, 359–391 (1968).

Mestel, L., "Theoretical processes in star formation," in *Star formation*, pp. 213–232 (eds. T. de Jong and A. Maeder), D. Reidel, Boston (1977).

Mestel, L. and Paris, R. B., "Magnetic braking during star formation. III," *Mon. Not. R. Astron. Soc.* **187**, 337–356 (1979).

Mirabelle, A. P. and Monin, A. S., "Two-dimensional turbulence," *Adv. Mech.* **2**, 47–95 (1979).

Mishurov, Yu. N., Pavlovskaya, E. D. and Suchkov, A. A., "Galactic spiral structure parameters derived from stellar kinematics," *Astron. J. (SSSR)* **56**, 268–278 (1979), [*Sov. Astron.* **23**, 147–152 (1979)].

Mishustin, I. N. and Ruzmaikin, A. A., "Occurrence of 'priming' magnetic fields during the formation of protogalaxies," *JETP* **61**, 441–444 (1971), [*Sov. Phys. JETP* **34**, 233–235 (1972)].

Misner, C. W., Thorne, K. S. and Wheeler, J. A., *Gravitation*, W. H. Freeman and Company, San Francisco (1973).

Mitton, S., "The polarization properties of 65 extragalactic sources in the 3C catalogue," *Mon. Not. R. Astron. Soc.* **155**, 373–381 (1972).

Moffatt, H. K., "Magnetic eddies in an incompressible viscous fluid of high electrical conductivity," *J. Fluid Mech.* **17**, 225–239 (1963).

Moffatt, H. K., "The degree of knottedness of tangled vortex lines," *J. Fluid Mech.* **35**, 117–129 (1969).

Moffatt, H. K. *Magnetic field generation in electrically conducting fluids*, Univ. Press, Cambridge (1978).

Mogilevsky, E. I., Jospha, B. A. and Obridko, V. N., "On the polarization of the solar coronal emission lines," *Sol. Phys.* **33**, 169–175 (1973).

Molchanov, S. A., Ruzmaikin, A. A. and Sokoloff, D. D., "Magnetic fields in random flows," *Proc. 4th USSR–Japanese symp. on prob. theory & math. stat.*, **2**, 113–115 (1982).

Monin, A. S., "On the nature of turbulence," *Usp. Fiz. (SSSR)* **125**, 97–122 [*Sov. Phys. Usp.* **21**, 429–442] (1978).

Monin, A. S., "The global hydrodynamics of the sun," *Usp. Fiz. (SSSR)* **132**, 123–167 [*Sov. Phys. Usp.* **23**, 594–619] (1980).

Monin, A. S. and Yaglom, A. M., *Statisticheskaya gidromekhanika part 2.* Nauka, Moscow (1967), [*Statistical Fluid Mechanics*, vol. 2, MIT Press (1971)].

Morris, D. and Berge, G., "Direction of the Galactic magnetic field in the vicinity of the sun," *Astrophys. J.* **139**, 1388-1392 (1964).

Morris, D. and Tabara, M., "A study of the depolarization and luminosity of radio sources," *Publ. Astron. Soc. Japan* **25**, 295-316 (1963).

Mouschovias, T. Ch., "Star formation in magnetic clouds," in *Protostars and planets*, pp. 209-242 (ed. T. Gehrels), Univ. Arizona Press, Tucson (1978).

Mouschovias, T. Ch. and Paleologou, E. V., "Magnetic braking of an aligned rotator during star formation," *Astrophys. J.* **237**, 877-902 (1980).

Nakano, T. and Tademaru, E., "Decoupling of magnetic fields in dense clouds with angular momentum," *Astrophys. J.* **173**, 87-108 (1972).

Ness, N. F., Behannon, K. W., Lepping, R. P. and Whang, Y. C., "The magnetic field of Mercury," *J. Geophys. Res.* **80**, 2708-2716 (1975).

Ness, N. F., Acuna, M. H., Lepping, R. P., Connerney, E. P., Behannon, K. W. and Burlaga, L. F., "Magnetic field studies by Voyager 1: Preliminary results at Saturn," *Science* **212**, 211-217 (1981).

Newton, H. W. and Nunn, M. L., "The Sun's rotation derived from sunspots 1934-1944 and additional results," *Mon. Not. R. Astron. Soc.* **111**, 413-421 (1951).

Novikov, E. A., "Energy spectrum of an incompressible fluid in turbulent flow," *Dokl. Akad. Nauk SSSR*, **139**, 331-334 (1961), [*Sov. Phys. Dokl.* **6**, 571-573 (1961)].

Oetken, L., "An equatorially symmetric rotator model for magnetic stars," *Astron. Nachr.* **298**, 197-208 (1977).

Oort, J. H., "The Galactic center," *Annu. Rev. Astron. Astrophys.* **15**, 295-362 (1977).

Orszag, S. A., "Analytical theories of turbulence," *J. Fluid Mech.* **41**, 363-386 (1970).

Ozernoy, L. M. and Chernin, A. D., "The fragmentation of matter in a turbulent metagalactic medium," *Astron. J.* (*SSSR*), **44**, 1131-1138 (1967), [*Sov. Astron.* **11**, 907-913 (1968)].

Ozernoy, L. M. and Chibisov, G. V., "Dynamical parameters of galaxies as a consequence of cosmological turbulence," *Astron. J.* (*SSSR*), **47**, 769-783 (1970), [*Sov. Astron.* **14**, 615-626 (1971)].

Ozernoy, L. M. and Usov, V. V., "The nature of pulsar gamma-rays," *Astron. J.* (*SSSR*) **54**, 753-765 (1977), [*Sov. Astron.* **21**, 425-431 (1979)].

Papaloizou, J. and Pringle, J. E., "Tidal torques on accretion disks in close binary systems," *Mon. Not. R. Astron. Soc.* **181**, 441-454 (1977).

Parker, E. N., "Hydromagnetic dynamo models," *Astrophys. J.* **122**, 293-314 (1955).

Parker, E. N., "The dynamical state of the interstellar gas and fluids," *Astrophys. J.* **145**, 811-833 (1966).

Parker, E. N., "The generation of magnetic fields in astrophysical bodies. II. The Galactic field," *Astrophys. J.* **163**, 252-278 (1971).

Parker, E. N., "The enigma of solar activity," in Proc. IAU Symp. No. 71 *Basic Mechanisms of solar activity*, pp. 3-15 (eds. V. Bumba and J. Kleczek), D. Reidel, Dordrecht (1976).

Parker, E. N., *Cosmical magnetic fields* (*their origin and activity*), Clarendon Press, Oxford (1979).

Pavlov, G. G. and Gnedin, Yu. N., "Vacuum polarization by magnetic fields and its astrophysical occurrences," in *Astrophysics and Space Physics*, **3**, Sov. Sci. Rev. Ser. E. Ed., R. A. Sunyaev, Harwood Acad. Publ., New York (1983).

Peebles, P. J. E., "Origin of the angular momentum of galaxies," *Astrophys. J.* **155**, 393-402 (1969).

Piddington, J. H., "The origin and form of the Galactic magnetic field," *Cosmical Electrodynamics* **3**, 60-70 (1972).

Pikelner, S. B., *Principles of the cosmical electrodynamics* (in Russian), Nauka, Moscow (1966).

Pikelner, S. B. and Khokhlova, V. L., "Magnetic stars," *Usp. Fiz. (SSSR)* **107**, 389-404 (1972) [*Sov. Phys. Usp.* **15**, 395-403 (1973)].

Pines, D., "Accreting neutron stars, black holes, and degenerate dwarf stars," *Science* **207**, 597-606 (1980).

Polovin, R. V., "Shock waves in magnetohydrodynamics," *Usp. Fiz. (SSSR)* **72**, 33-52 (1960), [*Sov. Phys. Usp.* **3**, 677-688 (1961)].

Polyakov, A. M., "Particle spectrum in quantum field theory," *Pisma JETP* **20**, 430-433 (1974), [*Sov. Phys. JETP Lett.* **20**, 194-195 (1974)].

Ponomarenko, Yu. B., "On the theory of hydromagnetic dynamo," *Zh. Prikl. Melch. Telch. Fiz. (USSR)* **6**, 47-51 (1973). [Trans. pp. 775-778.]

Popova, M. D. and Vetrichenko, E. A., "The unique object KR Aur," *Astron. J. (SSSR)* **55**, 765-775 (1978), [*Sov. Astron.* **22**, 438-444 (1978)].

Pouquet, A., "On two-dimensional magnetohydrodynamic turbulence," *J. Fluid Mech.* **88**, 1-16 (1978).

Pouquet, A., Frisch, U. and Léorat, J., "Strong MHD helical turbulence and the nonlinear dynamo effect," *J. Fluid Mech.* **77**, 321-354 (1976).

Pringle, J. E., "Accretion discs in astrophysics," *Annu. Rev. Astron. Astrophys.* **19**, 137-162 (1981).

Pringle, J. E., Rees, M. J. and Pacholczyk, A. G., "Accretion onto massive black holes," *Astron. & Astrophys.* **29**, 179-184 (1973).

Pudritz, R., "Dynamo action in turbulent accretion disks around black holes," *Mon. Not. R. Astron. Soc.* **195**, 881-914 (1981).

Rabinovich, M. I., "Stochastic self-oscillations and turbulence," *Usp. Fiz. (SSSR)* **125**, 123-168 (1978), [*Sov. Phys. Usp.* **21**, 443-469 (1978)].

Rädler, K.-H., "Zur Electrodynamik turbulent bewegter leitender Medien. II. Turbulenzbedingte Leitfähigkeits- und Permeabilitätsänderungen," *Z. Naturforsch.* **23**, 1851-1860 (1968).

Rayleigh, J. W. S., "Polarization of the light scattering by mercury vapour near the resonance periodicity," *Proc. R. Soc. Lond.* A**102**, 190-199 (1922).

Razin, V. A., "Polarization of cosmic radio radiation at 1.45 m and 3.3 m," *Astron. J. (SSSR)* **35**, 241-252 (1958), [*Sov. Astron.* **2**, 216-225 (1958)].

Reinhardt, M., "Interpretation of rotation measures of radio sources," *Astron. & Astrophys.* **19**, 104-108 (1972).

Rikitake, T., "Oscillation of a system of disk dynamos," *Proc. Camb. Phil. Soc.* **54**, 89-105 (1958).

Robbins, K. A., "A new approach to subcritical instability and turbulent transitions in a simple dynamo," *Math. Proc. Camb. Phil. Soc.* **82**, 309-325 (1977).

Roberts, G. O., "Spatially periodic dynamos," *Philos. Trans. R. Soc. Lond.* A**266**, 535-558 (1970).

Roberts, P. H., "Dynamo theory," in *Mathematical problems in the geophysical sciences*, Lectures in applied mathematics, vol. 14 (ed. W. H. Reid), Providence, Rhode Island, pp. 129-206 (1971).

Roberts, P. H. and Soward, A. M., "A unified approach to mean field electrodynamics," *Astron. Nachr.* **296**, 49-64 (1975).

Roberts, W. W., "Large-scale shock formation in spiral galaxies and its implication on star formation," *Astrophys. J.* **158**, 123-143 (1969).

Roberts, W. W. and Shu, F. H., "The role of gaseous dissipation in density waves of finite amplitude," *Astrophys. Lett.* **12**, 49-52 (1972).

Roberts, W. W. and Yuan, C., "Application of the density-wave theory to the spiral structure of the Milky Way system. III. Magnetic field: large-scale hydromagnetic shock formation," *Astrophys. J.* **161**, 887-902 (1970).

Ross, W. D., "Aristotle's Physics: a revised text with introduction and commentary," Clarendon Press, Oxford, England (1936).

Rougoor, G. W. and Oort, J. H., "Distribution and motion of interstellar hydrogen in the Galactic system with particular reference to the region within 3 kpc of the center," *Proc. Natl. Acad. Sci.* **46**, 1–12 (1960).

Ruderman, M., "Pulsars: structure and dynamics," *Annu. Rev. Astron. & Astrophys.* **10**, 427–476 (1972).

Rüdiger, G., "A theory of differential rotation based on the discussion of turbulent transport of angular momentum," *Astron. Nachr.* **298**, 245–252 (1977).

Ruelle, D. and Takens, F., "On the nature of turbulence," *Comm. Math. Phys.* **20**, 167–192 (1971).

Ruzmaikin, A. A., "Possible evaluation of the magnetic field near pulsars from the beat frequency of atomic transitions," *Astron. J. (SSSR)* **52**, 1173–1177 (1975), [*Sov. Astron.* **19**, 702–705 (1975)].

Ruzmaikin, A. A., "On the determination of the magnetic field in astrophysical conditions due to the polarization of atomic fluorescence," *Astron. J. (SSSR)* **53**, 550–557 (1976), [*Sov. Astron.* **20**, 311–315 (1976)].

Ruzmaikin, A. A., "The solar cycle as a strange attractor," *Comm. Astrophys.* **9**, 85–96 (1981).

Ruzmaikin, A. A. and Shukurov, A. M., "Magnetic field generation in the Galactic disc," *Astron. J. (SSSR)* **58**, 969–978 (1981), [*Sov. Astron.* **25**, 553–558 (1981)].

Ruzmaikin, A. A. and Shukurov, A. M., "Spectrum of the Galactic magnetic field," *Astrophys. Space Sci.* **82**, 397–407 (1982).

Ruzmaikin, A. A. and Sokoloff, D. D., "The scale and strength of the Galactic magnetic field according to the pulsar data," *Astrophys. & Space Sci.* **52**, 365–376 (1977a).

Ruzmaikin, A. A. and Sokoloff, D. D., "The interpretation of rotation measures of extragalactic radio sources," *Astron. & Astrophys.* **58**, 247–253 (1977b).

Ruzmaikin, A. A. and Sokoloff, D. D., "A method for determination of the large-scale magnetic field in the solar corona," *Pisma Astron. J. (SSRR)* **4**, 23–26 (1978), [*Sov. Astron. Lett.* **4**, 12–14 (1978)].

Ruzmaikin, A. A. and Sokoloff, D. D., "The calculation of Faraday rotation measures of cosmic radio sources," *Astron. & Astrophys.* **78**, 1–6 (1979).

Ruzmaikin, A. A. and Sokoloff, D. D., "Helicity, linkage and dynamo action," *Geophys. & Astrophys. Fluid Dyn.* **16**, 73–82 (1980).

Ruzmaikin, A. A., Sokoloff, D. D. and Kovalenko, A. V., "Galactic magnetic field parameters determined from Faraday rotation of radio sources," *Astron. J. (SSSR)* **55**, 692–701 (1978), [*Sov. Astron.* **22**, 395–401 (1978)].

Ruzmaikin, A. A., Sokoloff, D. D. and Shukurov, A. M., "Disk dynamo with concentrated helicity," *Magn. Gidrodin.* 20–26 (1980), [Magnetohydrodynam. **16**, 15–20 (1980)].

Ruzmaikin, A. A., Sokoloff, D. D. and Turchaninoff, V. I., "The turbulent dynamo in a disk," *Astron. J. (SSSR)* **57**, 311–320 (1980), [*Sov. Astron.* **24**, 182–187 (1980)].

Ruzmaikin, A. A., Sokoloff, D. D., Turchaninoff, V. I. and Zeldovich, Ya. B., "The disk dynamo," *Astrophys. Space Sci.* **66**, 369–384 (1979).

Ruzmaikin, A. A. and Vainshtein, S. I., "The magnetic field transfer in the solar convection zone," *Astrophys. & Space Sci.* **57**, 195–202 (1978).

Ruzmaikina, T. V., "On the role of the magnetic field and turbulence in the evolution of the pre-solar nebula," *Adv. Space Res.* **1**, 49–53, COSPAR, Great Britain (1981a).

Ruzmaikina, T. V., "The angular momentum of protostars giving birth to proto-planetary discs," *Pisma Astron. J. (USSR)* **7**, 188–192 (1981b), [*Sov. Astron. Lett.* **7**, 104–107 (1981)].

Ruzmaikina, T. V., in "Diskussionforum: Ursprung des Sonnensystems," (Ed. H. J. Völk), *Mitteilungen der Astronomischen Gesellschaft* **57**, 49–54 (1982).
Ruzmaikina, T. V. and Ruzmaikin, A. A., "Gravitational stability of the expanding universe in the presence of magnetic field," *Astron. J. (SSSR)*, **47**, 1206–1210 (1970), [*Sov. Astron.* **14**, 963–966 (1971)].
Saffman, P. G., "On the fine-scale structure of vector fields convected by a turbulent fluid," *J. Fluid Mech.* **16**, 545–572 (1963).
Safronov, V. S., *Evolution of the protoplanetary cloud and formation of the Earth and planets*, Nauka, Moscow (1969). Transl. from Russian by Israel Program for Scientific Translation, Jerusalem (1972).
Safronov, V. S. and Ruzmaikina, T. V., "On the angular momentum transfer and the accumulation of solid bodies in the solar nebula," in *Protostars and planets*, pp. 545–564 (ed. T. Gehrels), Univ. Arizona Press, Tucson (1978).
Sagdeev, R. Z. and Galeev, A. A., *Nonlinear plasma theory* (eds. T. O'Neill and D. Book), Benjamin, New York–Amsterdam (1969).
Salpeter, E. E., "Planetary nebulae, supernova remnants, and interstellar medium," *Astrophys. J.* **206**, 673–678 (1976).
Schatzman, E., "A theory of the rôle of magnetic activity during star formation," *Ann. d'Astroph.*, **25**, 18–29 (1962).
Schüssler, M., "Axisymmetric α^2-dynamo in the Hayashi-phase," *Astron. & Astrophys.* **38**, 263–270 (1975).
Seifert, H. and Threlfall, W., *Lehrbuch der Topologie*, Teubner (1934).
Shajn, G. A., "Diffuse nebulae and the interstellar magnetic field," *Astron. J. (SSSR)* **32**, 110–117 (1955).
Shakura, N. I. and Sunyaev, R. A., "Black holes in binary systems. Observational appearance," *Astron. & Astrophys.* **24**, 337–355 (1973).
Shakura, N. I. and Sunyaev, R. A., "A theory of the instability of disk accretion onto black holes and the variability of binary X-ray sources, galactic nuclei and quasars," *Mon. Not. R. Astron. Soc.* **175**, 613–632 (1975).
Shandarin, S. F., Doroshkevich, A. G., and Zeldovich, Ya. B., "The large-scale structure of the Universe," *Usp. Fiz. (SSSR)*, **139**, 83–134 (1983).
Shklovsky, J. S., "On the nature of the Crab nebula emission," *Dokl. Akad. Nauk (SSSR)* **90**, 983–986 (1953).
Shklovsky, J. S., "Nature of emission from pulsar NP 0532," *Nature* **225**, 251–252 (1970).
Shklovsky, J. S., *Supernovae* (in Russian), Nauka, Moscow (1976).
Shklovsky, J. S. and Sheffer, E. K., "Galactic spurs as possible sources of soft X-radiation," *Nature* **231**, 173–174 (1971).
Shukurov, A. M., Sokoloff, D. D. and Ruzmaikin, A. A., "Explosive growth of magnetic energy in a turbulent medium," *Magn. Gidrodin.* No. 2 (1983).
Shvartzman, V. F., "Halos around black holes," *Astron. J. (SSSR)* **48**, 479–488 (1971), [*Sov. Astron.* **15**, 377–384 (1971)].
Simon, G. W. and Weiss, N. O., "Supergranules and the hydrogen convective zone," *Z. Astrophys.* **69**, 435–450 (1968).
Skumanich, A., "Time scale for CaII emission decay, rotational braking and lithium depletion," *Astrophys. J.* **171**, 565–567 (1972).
Smith, E., Davis, L., Jones, D. E., Colburn, D. S., Colemann, I. J., Dyal, P. and Sonnet, C. P., "Magnetic field of Jupiter and its interaction with the solar wind," *Science* **183**, 305–306 (1974).
Smith, E., Davis, L., Jones, D. E., Colburn, D. S., Colemann, I. J., Dyal, P. and Sonnet, C. P., "Saturn's magnetic field and magnetosphere," *J. Geophys. Res.* **85**, 5655–5658 (1980).
Soward, A. M., "A thin disc model of the Galactic dynamo," *Astron. Nachr.* **299**, 25–33 (1978).

Soward, A. M. and Roberts, P. H., "Recent developments in dynamo theory," *Magn. Gidrodin.* 3–51 (1976); errata 146–147 (1977), [*Magnetohydrodynam.* **12**, 1–36 (1977)].

Spitzer, L. Jr., *The physics of full ionized gases*, Interscience, New York–London (1956).

Spitzer, L. Jr., *Diffuse matter in space*, Interscience, New York (1968).

Spoelstra, T. A. Th., "A survey of linear polarization at 1415 MHz. IV. Discussion of the results for the Galactic spurs," *Astron. & Astrophys.* **21**, 61–84 (1972).

Spoelstra, T. A. Th., "The Galactic magnetic field," *Usp. Fiz. (SSSR)* **121**, 679–694 (1977), [*Sov. Phys. Usp.* **20**, 336–342 (1977)].

Stacey, F. D. and Lovering, J. F., "Natural magnetic moments of two chondritic meteorites," *Nature* **183**, 529–530 (1959).

Steenbeck, M. and Krause, F., "Zur Dynamotheorie stellarer und planetarer Magnetfelder. I. Berechnung sonnenähnlicher Wechselfeldgeneratoren," *Astron. Nachr.* **291**, 49–84 (1969).

Steenbeck, M., Krause, F. and Rädler, K.-H., "Berechnung der mittlerer Lorentz-FeldStärke ʋ × 𝕭 für ein elektrisch leitendes Medium in turbulenter, durch Coriolis-Kräfte beeinflusster Bewegung," *Z. Naturforsch.* **21**, 369–376 (1966).

Stelzried, C. T., Levy, G. S., Sato, T., Rusch, W. V. T., Ohlson, J. B., Schatten, K. H. and Wilcox, J. M., "The quasi-stationary coronal magnetic field and density as determined from a Faraday rotation experiment," *Sol. Phys.* **14**, 440–456 (1970).

Stenflo, J. O., "Small-scale solar magnetic fields," in Proc. IAU Symp. No. 71, *Basic mechanisms of solar activity* (eds. V. Bumba and J. Kleczek), D. Reidel, Dordrecht, pp. 69–99 (1976).

Stix, M., "Non-linear dynamo waves," *Astron. & Astrophys.* **20**, 9–12 (1972).

Stix, M., "The Galactic dynamo," *Astron. & Astrophys.* **42**, 85–89 (1975), Erratum **68**, 459 (1978).

Stix, M., "Differential rotation and the solar dynamo," *Astron. & Astrophys.* **47**, 243–254 (1976).

Strittmatter, P. A., "Gravitational collapse in the presence of magnetic field," *Mon. Not. R. Astron. Soc.* **132**, 359–379 (1966).

Sturrock, P. A., "A model of pulsars," *Astrophys. J.* **165**, 529–556 (1971).

Suchkov, A. A., "The observational foundations of density-wave models for spiral structure," *Astron. J. (SSSR)*, **55**, 972–982 (1978), [*Sov. Astron.* **22**, 555–561 (1978)].

Sunyaev, R. A., "On the variability of X-ray radiation from black holes with accretion disks," *Astron. J. (SSSR)* **49**, 1153–1157 (1972), [*Sov. Astron.* **16**, 941–944 (1972)].

Syrovatsky, S. I., in discussion following the report by Kadomtsev and Tsytovich, "Collective plasma phenomena and their medium," in Proc. IAU Symp. No. 39, *Interstellar gas dynamics* (ed. J. Habing), pp. 135–138, D. ʀeidel, Dordrecht (1970).

Syrovatsky, S. I. and Somov, B. V., "Physical driving forces and models of coronal responses," in Proc. IAU Symp. No. 91, *Solar and Interplanetary Dynamics*, pp. 425–441, D. Reidel, Holland (1980).

Tademaru, E., "On the energy spectrum of relativistic electrons in the Crab Nebula," *Astrophys. J.* **183**, 625–635 (1973).

Takahara, F., "Magnetic flare model of quasars and active galactic nuclei-magnetized accretion disk around a massive black hole," *Progr. Theor. Phys.* **62**, 629–643 (1979).

Terzian, Y. and Davidson, K., "Pulsars: observational parameters and a discussion on dispersion measures," *Astrophys. & Space Sci.* **44**, 479–500 (1976).

Troland, T. H. and Heiles, C., "The Zeeman effect in radio frequency recombination lines," *Astrophys. J.* **214**, 703–705 (1977).

Trotter, H. F., "Noninvertible knots exist," *Topology* **2**, 341–358 (1963).

Trümper, J., Pietsch, W., Repping, C. and Voges, W., "Evidence for strong cyclotron line emission in the hard X-ray spectrum of Hercules X-1," *Astrophys. J. Lett.* **219**, L105–L110 (1978).

Tscharnuter, W. M., "Collapse of the presolar nebula," *Moon and planets*, **19**, 229–236 (1978).

Tverskoy, B. A., "Theory of hydrodynamical self-excitation of regular magnetic fields," *Geomagn. & Aeron.* **5**, 11–18 (1965), [Transl. pp. 7–12].

Usov, V. V., "Pulsar theory," *Adv. Space Lec.* **1**, 125–139 (1981).

Vaiana, G. S., Cassinelli, J. P., Faviano, G., Giacconi, R., Golub, L., Gorenstein, P., Haisch, B. M., Hardner, F. R. Jnr., Johnson, H. M., Linsky, J. L., Maxson, C. M., Mewe, R., Rosner, R., Seward, F., Topka, K., and Zwaan, C., "Results from an extensive Einstein stellar survey," *Astrophys. J.*, **244**, 163–182 (1981).

Vaiana, G. S. and Rosner, R., "Recent advances in coronal physics," *Annu. Rev. Astron. Astrophys.* **16**, 393–428 (1978).

Vainstein, S. I., "A unified approach to the nonlinear theory of the turbulent dynamo," *Magn. Gidrodin.* 3–9 (1980a), [*Magnetohydrodynam.* **16**, 111–116 (1980)].

Vainstein, S. I., "The dynamo of small-scale magnetic field," *JETP* **79**, 2175–2183 (1980b), [*Sov. Phys. JETP* **52**, 1099–1107 (1980)].

Vainstein, S. I. and Ruzmaikin, A. A., "Generation of the large-scale Galactic magnetic field. I," *Astron. J. (SSSR)* **48**, 902–909 (1971), [*Sov. Astron.* **15**, 714–719 (1971)].

Vainstein, S. I. and Ruzmaikin, A. A., "Generation of the large-scale Galactic magnetic field. II," *Astron. J. (SSSR)* **49**, 449–452 (1972), [*Sov. Astron.* **16**, 365–367 (1972)].

Vainstein, S. I. and Zeldovich, Ya. B., "Origin of magnetic fields in astrophysics," *Usp. Fiz. (SSSR)* **106**, 431–457 [*Sov. Phys. Usp.* **15**, 159–172] (1972).

Vainstein, S. I., Zeldovich, Ya. B. and Ruzmaikin, A. A., *Turbulent dynamo in astrophysics* (in Russian), Nauka, Moscow (1980).

Valée, J. P., "Magnetic field in the intergalactic region," *Nature* **254**, 23–26 (1975).

Valée, J. P., "The rotation measures of radio sources and their data processing," *Astron. & Astrophys.* **86**, 251–253 (1980).

Valée, J. P. and Kronberg, P. P., "Magnetic field in the Galactic spiral arm," *Nature Phys. Sci.* **246**, 49–51 (1973).

Valée, J. P. and Kronberg, P. P., "The rotation measures of radio sources and their interpretation," *Astron. & Astrophys.* **43**, 233–242 (1975).

Vandakurov, Yu. V., *Convection in the Sun and the 11-year cycle* (in Russian), Nauka, Leningrad (1976).

Varshalovich, D. A., "Spin state of atoms and molecules in the cosmic medium," *Usp. Fiz. (SSSR)* **101**, 369–383 (1970), [*Sov. Phys. Usp.* **13**, 429–437 (1971)].

Varshalovich, D. A. and Burdjuzha, V. V., "Polarization of cosmic maser OH emission," *Astron. J. (SSSR)* **52**, 1178–1186 (1975), [*Sov. Astron.* **19**, 705–710 (1975)].

Vasiliev, V. A. and Dergachev, V. A., "Solar activity during the Maunder minimum," *Izv. Acad. Nauk SSSR, Ser. Fiz.* **44**, 2510–2527 (1980), [*Bull. Acad. Sci. USSR (Phys. Ser.)* **44**(12), 52–64 (1980)].

Vershuur, G. L., "Observational aspects of the Galactic magnetic field," in *Interstellar gas dynamics*, pp. 150–167 (ed. J. Habing), D. Reidel, Dordrecht (1970).

Vershuur, G. L., "The Galactic magnetic field," in *Galactic and extragalactic radio astronomy* (eds. G. L. Vershuur and K. I. Kellerman), Springer-Verlag (1974).

Vitinsky, Yu. I., *Cyclicity and prediction of solar activity*, in Russian, Leningrad (1973).

Vitinsky, Yu. I., "Comments on the so-called Maunder minimum," *Sol. Phys.* **57**, 475–478 (1979).

Vogel, S. N. and Kuhi, L. V., "Rotational velocities of pre-main sequence stars," *Astrophys. J.* **245**, 960–979 (1981).

Vrba, F. J., Strom, S. E. and Strom, K. M., "Magnetic field structure in the vicinity of five dark cloud complexes," *Astron. J.*, **81**, 958–969 (1976).

Waldmeier, M., *Ergebnisse und Probleme der Sonnenforschung*, 1st Edition, Zürich (1941); 2nd Edition, Leipzig (1949).

Wasson, J. T., *Meteorites-classification and properties*, Springer, Heidelberg (1974).

Waxman, A. M., "Boundary layer circulation in disk-halo galaxies," *Astrophys. J. Suppl.*, **41**, 635–667 (1979).

Weiss, J. E. and Weiss, N. O., "Andrew Marvell and the Maunder minimum," *Q. J. R. Astron. Soc.* **20**, 115–118 (1979).

Weiss, N. O., "The expulsion of magnetic flux by eddies," *Proc. R. Soc. Lond.* A**293**, 310–328 (1966).

Westerhout, G., Seeger, Ch. L., Brown, W. N. and Tinbergen, J., "Polarization of the Galactic 75 cm radiation," *Bull Acad. Ned.* **16**, 187–212 (1962).

White, M. P., "Numerical models of the Galactic dynamo," *Astron. Nachr.* **299**, 209–216 (1978).

Wielen, R., "Density-wave theory of the spiral structure of galaxies," *Publ. Astron. Soc. Pacific* **86**, 341–362 (1975).

Wilson, O. C., "Chromospheric variations in main-sequence stars," *Astrophys. J.* **226**, 379–396 (1978).

Woller, P., *The posthumous works of Robert Hooke*, London (1705).

Woltjer, L., "A theorem on force-free magnetic fields," *Proc. Natl. Acad. Sci.* **44**, 489–491 (1958).

Wyckoff, S., Wehinger, P. A. and Gehren, T., "Resolution of quasar images," *Astrophys. J.* **247**, 750–761 (1981).

Yoshimura, H. A., "A model of the solar cycle driven by the dynamo action of the global convection in the solar convection zone," *Astrophys. J. Suppl.* **29**, 467–494 (1957a).

Yoshimura, H. A., "Solar-cycle dynamo wave propagation," *Astrophys. J.* **201**, 740–748 (1975b).

Yoshimura, H. A., "Nonlinear astrophysical dynamos: multiple-period dynamo wave oscillations and long-term modulations of the 22-year solar cycle," *Astrophys. J.* **226**, 706–719 (1978).

Yoshimura, H. A., "The solar-cycle period-amplitude relation as evidence of hysteresis of the solar-cycle nonlinear magnetic oscillations and the long-term (55 years) cyclic modulation," *Astrophys. J.* **227**, 1047–1058 (1979).

Zeldovich, Ya. B., "The limiting law of thermoconductivity for an internal problem with small velocities," *JETP* **7**, 1466–1469 (1937).

Zeldovich, Ya. B., "The magnetic field in the two-dimensional motion of a conducting turbulent liquid," *JETP* **31**, 154–156 (1956), [*Sov. Phys. JETP* **4**, 460–462 (1957)].

Zeldovich, Ya. B., "A magnetic model of the Universe," *JETP* **48**, 986–988 (1965), [*Sov. Phys. JETP* **21**, 656–659 (1965)].

Zeldovich, Ya. B., "The separation of uniform matter into parts under the action of gravitation," *Astrophysica*, **6**, 319–335 (1970).

Zeldovich, Ya. B., "The creation of particles and antiparticles in electric and gravitational fields," in *Magic without magic, John Archibald Wheeler* (ed. J. Klauder), San Francisco (1972).

Zeldovich, Ya. B., "On the friction of fluids between rotating cylinders," *Proc. R. Soc. Lond.* A**374**, 299–312 (1981).

Zeldovich, Ya. B., Einasto, Ya. E., and Shandarion, S. F., "Giant voids in the universe," *Nature*, **300**, 407–413 (1982).

Zeldovich, Ya. B. and Khlopov, M. Yu., "On the concentration of relict monopoles in the Universe," *Phys. Lett.* **79B**, 239–241 (1978).

Zeldovich, Ya. B., Molchanov, S. A., Ruzmaikin, A. A. and Sokoloff, D. D., "Kinematic dynamo problem in linear velocity fields," *J. Fluid Mech.*, in press (1983).

Zeldovich, Ya. B. and Novikov, I. D., *Gravitation theory and stellar evolution*, Nauka, Moscow (1971).

Zeldovich, Ya. B. and Novikov, I. D., *The structure and evolution of the Universe*, Nauka, Moscow (1975). English trans., Univ. Press, Chicago (1982).

Zeldovich, Ya. B. and Ruzmaikin, A. A., "The magnetic field in conducting fluid in two-dimensional motion," *JETP* **78**, 980-986 (1980), [*Sov. Phys. JETP* **51**, 493-497 (1980)].

Zeldovich, Ya. B. and Ruzmaikin, A. A., "Dynamo problems in astrophysics," *Sov. Sci. Rev.*, New York (1982).

Index